Y0-BUD-661

The Padé Approximant in Theoretical Physics

This is Volume 71 in
MATHEMATICS IN SCIENCE AND ENGINEERING
A series of monographs and textbooks
Edited by RICHARD BELLMAN, *University of Southern California*

A complete list of the books in this series appears at the end of this volume.

The Padé Approximant in Theoretical Physics

Edited by

GEORGE A. BAKER, JR. *Brookhaven National Laboratory*
Upton, Long Island, New York

and

JOHN L. GAMMEL *Los Alamos Scientific Laboratory*
University of California
Los Alamos, New Mexico

Academic Press *1970* *New York and London*

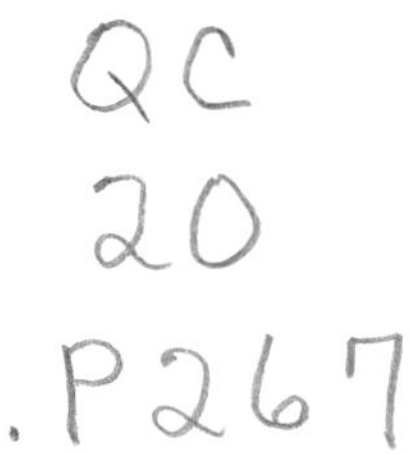

ACADEMIC PRESS, INC.
111 Fifth Avenue, New York, New York 10003

United Kingdom Edition published by
ACADEMIC PRESS, INC. (LONDON) LTD.
Berkeley Square House, London W1X 6BA

LIBRARY OF CONGRESS CATALOG CARD NUMBER: 70-137682

AMS 1970 Subject Classification 41A20

PRINTED IN THE UNITED STATES OF AMERICA

Contents

List of Contributors

Numbers in parentheses indicate the pages on which the authors' contributions begin.

GEORGE A. BAKER, JR. (1), Brookhaven National Laboratory, Upton, Long Island, New York

J. S. R. CHISHOLM (171, 183), University of Kent at Canterbury, Canterbury, England

A. K. COMMON* (183, 241), University of Kent at Canterbury, Canterbury, England

J. L. GAMMEL (303), University of California, Los Alamos Scientific Laboratory, Los Alamos, New Mexico

R. G. GORDON (99), Department of Chemistry, Harvard University, Cambridge, Massachusetts

RICHARD W. HAYMAKER† (257), Department of Physics, University of California, Santa Barbara, California

M. KARPLUS (41), Department of Chemistry, Harvard University, Cambridge, Massachusetts

ROBERT H. KRAICHNAN (129), Dublin, New Hampshire

J. J. KUBIS‡ (303), Cambridge University, Cambridge, England

P. W. LANGHOFF# (41), Department of Chemistry, Harvard University, Cambridge, Massachusetts

D. MASSON (197, 231), University of Toronto, Toronto, Canada

M. T. MENZEL (303), University of California, Los Alamos Scientific Laboratory, Los Alamos, New Mexico

H. M. NIELAND (289), University of Nijmegen, Nijmegen, The Netherlands

J. NUTTALL (219), Texas A & M University, College Station, Texas

LEONARD SCHLESSINGER| (257), Department of Physics, University of Illinois, Urbana, Illinois

J. A. TJON (289), University of Utrecht, Utrecht, The Netherlands

J. C. WHEELER|| (99), Department of Chemistry, Harvard University, Cambridge, Massachusetts

W. R. WORTMAN (333), University of California, Los Alamos Scientific Laboratory, Los Alamos, New Mexico

* Present address: CERN, Geneva, Switzerland.

† Present address: Laboratory of Nuclear Studies, Cornell University, Ithaca, New York.

‡ Present address: Department of Physics, Michigan State University, East Lansing, Michigan.

\# Present address: Department of Chemistry, Indiana University, Bloomington, Indiana.

| Present address: Department of Physics and Astronomy, University of Iowa, Iowa City, Iowa.

|| Present address: Department of Chemistry, University of California, San Diego; La Jolla, California.

Preface

An infinite series of some sort presents itself in many situations in mathematical physics. Since nearly the beginning of mathematics, or more properly, since the development of the Taylor series, the value of such series has been subject to controversy. This book deals with the problem of extracting maximum information from such series using existing mathematical and physical information.

Too many physicists seem to hold the notion that a series must converge to be useful. We are told[1] that after a scientific meeting in which Cauchy had presented his first research on series, Laplace hastened home and remained there in seclusion until he had examined the series in "Méchanique Celeste." Abel wrote: "On the whole, divergent series are a deviltry, and it is a shame to base any demonstration upon them. By using them one can produce any result he wishes, and they are the cause of many calamities and paradoxes." There is no doubt that the negative point of view of Cauchy and Abel prevailed among mathematicians until recent (1890) times, and it still prevails among some physicists.

The mathematical researches of T. J. Stieltjes [Reserches sur les fractions continués, *Ann. Fac. Sci. Toulouse* **8,** J, 1–122 (1894); **9,** A, 1–47 (1894)] established that at least for a class of divergent series (which are most important in theoretical physics as several papers in this book show) rigorous and useful sums can be obtained. The mathematical work with which we are most concerned was done in about 1892 [H. Padé, Sur la représentation approchée d'une fonction par des fractiones rationelles, Thesis, *Ann. Ecole. Nor.* **9,** 1–93 (1892)]. A most useful book presenting these works and recent developments is H. S. Wall's "Continued Fractions" (Van Nostrand, Princeton, New Jersey, 1948). [What seems an almost separate development of the theory of divergent series is presented by G. H. Hardy, "Divergent Series," Oxford Univ. Press, London and New York, 1949.] One of the final sentences in Wall's book reflects the present situation: "It is difficult to appraise the significance of the Padé table in the theory of power series. We feel that an appraisal must await further and deeper investigations" (p. 410). Unfortunately, little attention is given to this subject by mathematicians at the present time. It is ironic that Borel, who started many of the researches in point set theory now so popular, really had on the top of his mind the problem of divergent series

[1] F. Cajori, "A History of Mathematics," p. 337. MacMillan, New York, 1901.

(A. Borel, "Leçons sur les Fonctions Monogènes," Gauthier-Villars, Paris, 1917).

Since about 1960, a small group of theoretical physicists has given attention to the idea that the Padé approximant may be useful in summing series which occur in theoretical physics. A review of work done up until 1964 is given by G. A. Baker, Jr. [*Advan. Theoret. Phys.* **1,** 1 (1965)]. It is shown that the class of series studied by Stieltjes occur frequently. The theory has been most useful and most successful in the problem of critical phenomena. In scattering theory and quantum field theory series of Stieltjes also occur, but the papers on field theory in this volume leave much room for future development.

The present volume is a collection of original papers by some of the authors who have been particularly active in an attempt to sweep away once and for all the notion that only a convergent series well inside its radius of convergence or an asymptotic series of rapidly decreasing terms is useful. There is other work[2] which deserves attention and for which the editors have been unable to obtain full length original articles for inclusion in this volume. When it appears, the proceedings of the Cargèse summer school for 1970 should contain some of this work.

The editors and authors hope that this book will serve to bring the Padé approximant into wider use and establish it as a tool to broaden the powers of the mathematical physicist.

[2] J. L. Basdevant, D. Bessis, and J. Zinn-Justin, Padé approximants in strong interactions. Two body pion and kaon systems, *Nuovo Cimento A* **60,** 185 (1969); D. Bessis and M. Pusterla, Unitary Padé approximants in strong coupling field theory and application to the calculation of the ρ- and f_0-meson Regge trajectories, *Nuovo Cimento A* **54,** 243 (1968).

The Padé Approximant in Theoretical Physics

CHAPTER 1 THE PADÉ APPROXIMANT METHOD AND SOME RELATED GENERALIZATIONS*

George A. Baker, Jr.

Brookhaven National Laboratory

I. INTRODUCTION

In this article we provide, by way of a background a summary of our previous review article (Baker, 1965). We quote, without proof, most of the more important results given therein. In addition we review in considerably more detail advances that have been made subsequently. We have, in the main avoided treating material included in the other articles in this volume.

The Padé approximant method has proved very useful in the provision of quantitative information about the solution of many interesting problems of physics and chemistry. It extracts this information from the perturbation expansion (this is usually easier to obtain than an exact answer) and such qualitative physical information about the solution as is known. The Padé approximant method has been most used, so far, in the field of cooperative phenomena and critical points. Here it is a valuable method of deducing the nature of the critical point singularities. More recently it is being widely used in other fields of research.

Since the subject of this volume is the Padé approximant, it is well to begin with a definition. The Padé approximant is of the form of one polynomial divided by another polynomial. In the $[N, M]$ Padé approximant the numerator has degree M and the denominator degree N. The coefficients are determined by equating like powers of z in the following

* Work performed in part under the auspices of the U.S. Atomic Energy Commission.

equations:

$$f(z)Q(z) - P(z) = Az^{M+N+1} + Bz^{M+N+2} + \dots, \quad Q(0) = 1.0 \tag{1}$$

where $P(z)/Q(z)$ is the $[N, M]$ Padé approximant to $f(z)$. Following Padé (1892) it is customary to arrange the Padé approximants in a table as follows

$$\begin{array}{ccccc} [0,0] & [0,1] & [0,2] & [0,3] & \cdots \\ [1,0] & [1,1] & [1,2] & [1,3] & \cdots \\ [2,0] & [2,1] & [2,2] & [2,3] & \cdots \\ [3,0] & [3,1] & [3,2] & [3,3] & \cdots \\ \cdot\quad\cdot & \cdot\quad\cdot & \cdot\quad\cdot & \cdot\quad\cdot & \cdot \end{array} \tag{2}$$

The first row is composed of the partial sums of the Taylor series. Diagonal sequences (upper left to lower right) have a constant difference in the degree of the numerator and the denominator. The antidiagonal sequences (upper right to lower left) all involve the same number of coefficients.

We view the Padé approximant method as a method of approximate analytic continuation; we show in the next chapter that the "natural" region of convergence of diagonal sequences of Padé approximants is not restricted to the unit circle, as is the case with partial sums of a Taylor series. The full range of convergence of the $[N, N]$ Padé approximants is not known but examples have been investigated (Baker, 1965) which seem to indicate that it greatly exceeds the situations for which it has been proved.

The principle applications of the Padé approximants fall into two classes: (a) the provision of efficient rational approximations to special mathematical functions; and (b) the acquisition of quantitative information about a function for which we have only qualitative information and power-series coefficients. We shall not consider the first class of applications, but confine our attention to the second class.

II. THE THEORY OF THE PADÉ APPROXIMANT METHOD

In this chapter we review briefly some of the more important theorems given in my previous review article (Baker, 1965) plus, in more detail, material which has become known since then. At that time I pointed out that, ideally, we would like to have a theorem similar to that available for Taylor series, i.e., "the sequence of $[N, N]$ Padé approximants converges if and only if . . . ," where the condition is one which is simply checked and depends only on some analytic properties of the function concerned. Some progress in this direction has been made; however such an ideal theorem is not yet available. Strong results are however available for certain fairly extensive special cases.

A. Formal Results

The exact solution for the $[N, M]$ Padé approximant to

$$A(x) = \sum_{j=0}^{\infty} a_j x^j \tag{3}$$

is easily given explicitly in terms of the power-series coefficients a_j. It is

$$[N, M](x) = \frac{\det \begin{vmatrix} a_{M-N+1} & a_{M-N+2} & \cdots & a_{M+1} \\ \vdots & \vdots & \ddots & \vdots \\ a_M & a_{M+1} & \cdots & a_{M+N} \\ \sum_{j=N}^{M} a_{j-N} x^j & \sum_{j=N-1}^{M} a_{j-N+1} x^j & \cdots & \sum_{j=0}^{M} a_j x^j \end{vmatrix}}{\det \begin{vmatrix} a_{M-N+1} & a_{M-N+2} & \cdots & a_{M+1} \\ \vdots & \vdots & \ddots & \vdots \\ a_M & a_{M+1} & \cdots & a_{M+N} \\ x^N & x^{N-1} & \cdots & 1 \end{vmatrix}} \tag{4}$$

where $a_j \equiv 0$ if $j < 0$, and the sums for which the initial point is larger than the final point are to be replaced by 0. This formula can easily be verified by substitution in (1). It is also unique, if we agree to cancel any common factor which may occur so as to present the results as a ratio of polynomials of lowest degree.

For notational convenience let us define

$$\Delta(m, n) = \det \begin{vmatrix} a_m & a_{m+1} & \cdots & a_{m+n} \\ a_{m+1} & a_{m+2} & \cdots & a_{m+n+1} \\ \vdots & \vdots & \ddots & \vdots \\ a_{m+n} & a_{m+n+1} & \cdots & a_{m+2n} \end{vmatrix}, \tag{5}$$

$$[N, N + j] = P_N^{(j)}(x) / Q_N^{(j)}(x) . \tag{6}$$

There are certain simple relations between approximants which are adjacent to each other in the Padé table. For example,

$$\frac{P_{N+1}^{(j)}(x)}{Q_{N+1}^{(j)}(x)} - \frac{P_N^{(j)}(x)}{Q_N^{(j)}(x)} = O(x^{2N+j+1}) \tag{7}$$

because both Padé approximants approximate the same function to order x^{2N+j+2} and x^{2N+j}, respectively. If we now multiply (7) by $Q_{N+1}^{(j)}(x)Q_N^{(j)}(x)$ we have

$$Q_N^{(j)}(x)P_{N+1}^{(j)}(x) - Q_{N+1}^{(j)}(x)P_N^{(j)}(x) = O(x^{2N+j+1}) \,. \tag{8}$$

But the left side of (8) is a polynomial of degree x^{2N+j+1}. Therefore the right hand side must consist of a single power. We can evaluate its coefficient by taking the limit as x goes to infinity. We obtain from (4) the coefficient

$$-\Delta(j+2, N-1)\Delta(j, N+1) + \Delta(j+2, N)\Delta(j, N) \tag{9}$$

This result may be reduced, by identity II. 13 of Baker (1965), to

$$\frac{P_{N+1}^{(j)}(x)}{Q_{N+1}^{(j)}(x)} - \frac{P_N^{(j)}(x)}{Q_N^{(j)}(x)} = \frac{x^{2N+j+1}[\Delta(1+j, N)]^2}{Q_{N+1}^{(j)}(x)Q_N^{(j)}(x)} \,. \tag{10}$$

In a similar manner, various other *adjacent* Padé approximants can be related. The right-hand side is always a single power times a constant over a denominator.

If we consider the cross ratio of four adjacent entries in the Padé table, relations such as (10) can be used to prove simple connections between them. For example, it follows simply that

$$\frac{\{[N, N+j] - [N, N+j+1]\}\{[N+1, N+j] - [N+1, N+j+1]\}}{\{[N, N+j] - [N+1, N+j+1]\}\{[N, N+j+1] - [N+1, N+j]\}} = \text{constant} \tag{11}$$

independent of x. This relation can plainly be solved for one of the four Padé approximants in terms of the other three and could be used, therefore to compute them recursively. An extensive study of recursive schemes for generating Padé approximants has been given by Wynn (1960) and more recently the application of the epsilon algorithm to calculating tables

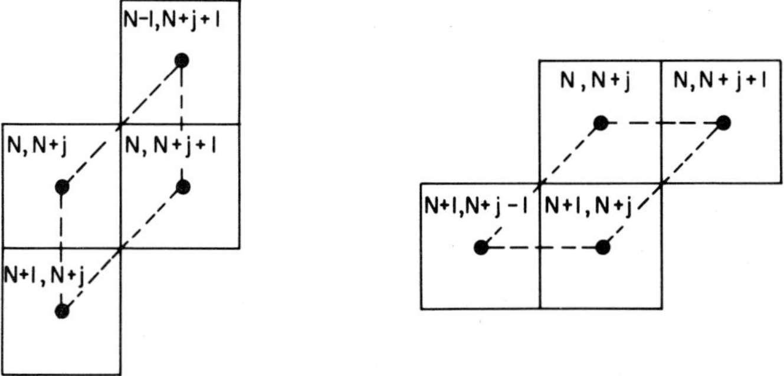

Fig. 1. The two cross ratios used in the derivation of the recursion relations (13).

of approximations to the function value at a point has been considered by Macdonald (1964). We prefer the following recursion scheme for generating Padé approximants. It requires only of the order of N^2 operations and is, apparently, more efficient than any scheme proposed by Wynn (1960). It can be derived by considering the cross ratios for the patterns in Fig. 1. The resulting recursion is given for the sequence

$$\frac{\eta_{2j}(x)}{\theta_{2j}(x)} = [j, n-j], \qquad \frac{\eta_{2j+1}(x)}{\theta_{2j+1}(x)} = [j, n-j-1] \tag{12}$$

by

$$\begin{aligned} \frac{\eta_{2j}(x)}{\theta_{2j}(x)} &= \frac{\bar{\eta}_{2j-1}\eta_{2j-2}(x) - x\bar{\eta}_{2j-2}\eta_{2j-1}(x)}{\bar{\eta}_{2j-1}\theta_{2j-2}(x) - x\bar{\eta}_{2j-2}\theta_{2j-2}(x)}, \\ \frac{\eta_{2j+1}(x)}{\theta_{2j+1}(x)} &= \frac{\bar{\eta}_{2j}\eta_{2j-1}(x) - \bar{\eta}_{2j-1}\eta_{2j}(x)}{\bar{\eta}_{2j}\theta_{2j-1}(x) - \bar{\eta}_{2j-1}\theta_{2j}(x)}, \end{aligned} \tag{13}$$

where $\bar{\eta}_j$ is the coefficient of the highest power of x in $\eta_j(x)$ (i.e., $x^{n-[(j+1)/2]}$). It is to be noted that the starting values are given by

$$\begin{aligned} \eta_0(x) &= \sum_{k=0}^{n} a_k x^k, & \theta_0(x) &= 1.0, \\ \eta_1(x) &= \sum_{k=0}^{n-1} a_k x^k, & \theta_1(x) &= 1.0. \end{aligned} \tag{14}$$

For large values of N this is plainly a more efficient way to calculate the $[N, N]$ than the direct solution of the set of linear equation (1). The difference is, of course, not impressive for the [2, 2].

A more compact expression can be given for (4). Nuttall gives, for the special case $[N, N-1]$ what we call *Nuttall's compact formula* (Nuttall, 1967). It is

$$[N, N-1] = \mathbf{v}^T \mathbf{V}^{-1} \mathbf{v}, \tag{15}$$

where

$$\mathbf{v} = (a_0, a_1, \cdots, a_{N-1}),$$
$$\mathbf{V} = \begin{pmatrix} a_0 - xa_1, & a_1 - xa_2, & \cdots, & a_{N-1} - xa_N \\ \vdots & \vdots & \ddots & \vdots \\ a_{N-1} - xa_N, & a_N - xa_{N+1}, & \cdots, & a_{2N-2} - xa_{2N-1} \end{pmatrix}. \tag{16}$$

We have derived a more general version of the same type. The basic method in the derivation consists of multipling the second column of the matrices in (4) by x and subtracting from the first and then repeating this process with the second and third column and so on. We now define the vector

$$\mathbf{w}_{N,M} = (a_{M-N+1}, a_{M-N+2}, \ldots, a_M) \tag{17}$$

and the matrix

$$\mathbf{W}_{N,M} = \begin{pmatrix} a_{M-N+1} - xa_{M-N+2}, & \cdots, & a_M - xa_{M+1} \\ \vdots & \ddots & \vdots \\ a_M - xa_{M+1}, & \cdots, & a_{M+N-1} - xa_{M+N} \end{pmatrix}. \tag{18}$$

Then

$$[N, M] = \sum_{j=0}^{M-N} a_j x^j + (\mathbf{w}_{N,M}^T \mathbf{W}_{N,M}^{-1} \mathbf{w}_{N,M}) x^{M+1-N}, \tag{19}$$

where $a_j \equiv 0$ if $j < 0$ and the sum in (19) is replaced by 0 if $N > M$. Equation (19) holds for $N > M$ in spite of the negative power of x which appears. A sufficient number of low-order coefficients of x in the matrix element vanish to compensate.

An additional compact expression, involving an off diagonal element of $\mathbf{W}^{-1}$ can be given. It is

$$[N, M] = \sum_{j=0}^{M+n} a_j x^j + x^{M+n+1} (\mathbf{w}_{N,N+M}^T \mathbf{W}_{N,M}^{-1} \mathbf{w}_{N,M+n}) \tag{20}$$

where $0 \leq n \leq N$. These expressions are useful in relating Padé approximants with the Schwinger variational principle (Nuttall, 1967).

B. General Theorems

There are basically two types of sequences of Padé approximants which one can consider. In the first type the degree of the numerator or that of the denominator remains finite while the degree of the other goes to infinity. In the second type, both the degree of the numerator and that of the denominator tend to infinity. The second type seems to be the more powerful, while the first is the simplest for which to obtain rigorous theorems. For the results on the first type we refer to Baker (1965).

The following theorem shows that the "natural" domain of convergence for the sequence of $[N, N]$ Padé approximants is not a circle centered at the origin, but rather a region determined in some way by the location of the nonpolar singularities of the function considered.

THEOREM 1. If $P_N^{(0)}(z)/Q_N^{(0)}(z)$ is the $[N, N]$ Padé approximant to $f(z)$, and $C + Df(0) \neq 0$, then

$$\frac{A + B\left[P_N^{(0)}\left(\dfrac{\alpha w}{1+\beta w}\right) \Big/ Q_N^{(0)}\left(\dfrac{\alpha w}{1+\beta w}\right)\right]}{C + D\left[P_N^{(0)}\left(\dfrac{\alpha w}{1+\beta w}\right) \Big/ Q_N^{(0)}\left(\dfrac{\alpha w}{1+\beta w}\right)\right]} \tag{21}$$

is the $[N, N]$ Padé approximant to

$$\left[A + Bf\left(\frac{\alpha}{1+\beta w}\right)\right]\left[C + Df\left(\frac{\alpha}{1+\beta w}\right)\right]^{-1}. \tag{22}$$

The proof is given in Baker (1965) as is the proof for the following theorem.

THEOREM 2. Let $P_k(z)$ be any infinite sequence of $[N, M]$ Padé approximants to a formal power series where $N + M$ tend to infinity with k. If $|P_k(z)|$ is uniformly bounded in any closed, simply-connected domain D_1 containing the origin as an interior point and $|P_k(z)|^{-1}$ is uniformly bounded in any closed simply-connected domain D_2 containing the origin as an interior point, then the P_k converge to a meromorphic function $f(z)$ in the interior of the union of D_1 and D_2.

This theorem gives a rule of procedure in practice. Select a region of interest in the complex plane in which it is known on physical grounds that the function is meromorphic. This region might, for example, be a neighborhood of a portion of the real axis. From an infinite sequence of the $[N, M]$ Padé approximants select an infinite subsequence which obeys the conditions of Theorem 2 in the region of interest. Then they must, by Theorem 2 converge to the desired answer. The main mathematical problem associated with the convergence problem is to establish the existence of such an *infinite* bounded subsequence when M and N tend to infinity together.

Several theorems have been proved (J. L. Walsh, 1967) on the Padé approximant. His most general result is given by the following theorem.

THEOREM 3. Let Δ be a Jordan region of the extended plane containing the origin, whose boundary is denoted by Γ. Let $w = \varphi(z)$ with $\varphi(0) = 0$ map Δ conformally and one-to-one onto $|w| < 1$ and let Γ_m denote generically the locus $|\varphi(z)| = m$, $0 < m < 1$, in Δ. Let $f(z)$ be analytic at the origin and on Γ, and meromorphic with precisely ν poles in Δ, and suppose the Padé approximants $[n, n]$ are bounded on Γ;

$$|f(z) - [n, n](z)| \leq M, \qquad z \quad \text{or} \quad \Gamma. \tag{23}$$

Suppose $[n, n](z)$ has precisely N_n poles in Δ, $N_n/n \to 0$. Then we have

$$\begin{aligned} &\limsup_{n\to\infty} [\max |f(z) - [n, n](z)|, \qquad z \quad \text{on} \quad T]^{1/n} \\ &\qquad \leq [\max |\varphi(z)|, \qquad z \quad \text{on} \quad T]^2, \end{aligned} \tag{24}$$

where T is an arbitrary closed set in Δ containing no limit point of poles of the $[n, n](z)$.

The proof of this theorem proceeds by the use of Schwartz's lemma. As Walsh points out, we could have selected any subsequence of $[n, n]$ to apply the theorem to rather than the complete sequence. Also we could have selected any sequence $[\mu, \nu]$ provided $\mu + \nu \to \infty$ and $N_\mu/(\mu + \nu) \to 0$ and still proven convergence under the stated conditions.

Chisholm (1966) has proved another theorem on convergence.

THEOREM 4. A function $f(z)$ is regular and nonzero at $z = 0$, and is meromorphic in the region $|z| < R$ and σ is any positive number less than R. Let $\{[m, n]\}$ be an infinite sequence, with $m \to \infty$ and $n \to \infty$ in any way, of Padé approximants to $f(z)$, such that the number of poles and zeros of every approximant in the region $|z| \le \sigma$ are each less than a number $\mu(\sigma)$ independent of m and n. Then the sequence $\{[m, n]\}$ converges uniformly to $f(z)$ in the region Δ defined below.

Let us define $\rho_i = |\zeta_i|$, where ζ_i are the locations of the poles of $f(z)$ inside $|z| < \rho$ and ζ_1 is one of the closest to the origin, and let us define

$$r_1 = \rho(\sqrt{2} - 1)\,, \tag{25}$$

where ρ and ε are any positive numbers such that $\rho < R$, $\sigma < \rho - \varepsilon$. If $r_1 > \rho_1$, then we define r_2 as the unique root of

$$r_2^2(\rho + r_2) = \rho\,\rho_1(\rho - r_2) \tag{26}$$

lying in the range $0 < r_2 < r_1$.

If in addition we define $\sigma_j = |\xi_j|$, where the ξ_j are the poles of $[m, n]$, inside $|z| \le \rho$ then the region Δ is defined by the inequalities

$$\begin{gathered} |z| \le \operatorname{Min}(r_1, r_2) - \varepsilon\,, \\ |z - \zeta_i| \ge \varepsilon\rho_i,\; |z - \xi_j| \ge \varepsilon\sigma_j\,, \end{gathered} \tag{27}$$

where $\varepsilon > 0$.

If we specialize $\mu(\sigma) = 0$, then Δ becomes simply

$$|z| \le (\sqrt{2} - 1)\rho - \varepsilon\,, \tag{28}$$

which is an extension of a result of Baker (1965 Eq. II. 68 *et seq.*) who established only $|z| \le 0.3\sigma$ for Δ.

We add the following theorem. Its application is somewhat intermediate between the two classes of sequences alluded to at the beginning of this section. It is similar to the preceding two theorems in that the locations of the poles are kept closely under control.

THEOREM 5. Let $f(z)$ be meromorphic in the whole complex plane. Then there exists an infinite subsequence of $[N, N + j]$ Padé approximants which converge to $f(z)$ at any point of the complex plane not a pole of $f(z)$, provided, that if a_n are the locations of the poles of $f(z)$ ($|a_n| \le |a_{n+1}|$) that

j goes to infinity sufficiently rapidly so that $\max_{1\le n\le N}\{b_n^{-1}|a_n/a_{N+1}|^j\}\to 0$ as N goes to infinity. We further assume that if b_n are the residues that $\sum |b_n/a_n|$ converges.

Proof. We assume that $f(z)$ is not a rational function, as if it were, a finite Padé approximant would represent it exactly and convergence is then trivial. Likewise, by the definition of meromorphic, $f(z)$ has no limit point of poles in the finite z-plane as this is an essential singularity.

We may write the $[N, N+j]$ in the form

$$[N, N+j] = \sum_{k=0}^{j} \gamma_k z^k + \sum_{l=1}^{N} \frac{\beta_l}{\alpha_l - z}\,. \tag{29}$$

Under the hypotheses on $f(z)$ we may write it as

$$f(z) = \sum_{k=0}^{\infty}\left(\sum_{n=1}^{\infty} b_n\Big/ a_n^{k+1}\right) z^k\,. \tag{30}$$

The equations (1) which determine the α_l and β_l may be reexpressed as

$$\sum_{l=1}^{N} \beta_l\Big/\alpha_l^{k+1} = \sum_{n=1}^{N} b_n\Big/a_n^{k+1} + \sum_{n=N+1}^{\infty} b_n\Big/a_n^{k+1}\,, \qquad k = j+1, j+2, \ldots, j+2N\,. \tag{31}$$

The magnitude of the second term is bounded by

$$\begin{aligned}\left|\sum_{n=N+1}^{\infty} b_n\Big/a_n^{k+1}\right| &\le \sum_{n=N+1}^{\infty} |b_n/a_n^{k+1}| \\ &\le |a_{N+1}|^{-k} \sum_{n=N+1}^{\infty} |b_n/a_n| \\ &\le |a_{N+1}|^{-k} \sum_{n=1}^{\infty} |b_n/a_n| \le M|a_{N+1}|^{-k}\end{aligned} \tag{32}$$

as, by hypothesis, $\sum |b_n/a_n|$ converges. We now choose an infinite subsequence such that $|a_{N+1}| > |a_N|$. We can always do this as there are never an infinite number of poles with the same modulus.

As $|a_n| < |a_{N+1}|$,

$$\sigma(k) = \max_{1\le n\le N} \{b_n^{-1}|a_n/a_{N+1}|^k\} \tag{33}$$

is a decreasing function of k. It is bounded from above by $\sigma(j)$ for k in the range $k = j, \ldots, j+2N$. As $\sigma(j)\to 0$ by hypothesis we can make the second sum in the right-hand side of (31) as small as we like compared to the smallest term in the first sum on the right-hand side. Consequently the α_l and β_l approach the a_n and b_n as closely as we please.

Now consider any point z_0 not a pole of f, we may write

$$\begin{aligned} f(z_0) - [N, N+j](z_0) &= \sum_{n=1}^{N} \left(\frac{b_n}{a_n - z_0} - \frac{\beta_n}{\alpha_n - z_0} \right) \\ &\quad + \sum_{n=N+1}^{\infty} \left(\frac{b_n}{a_n - z_0} \right) - \sum_{k=0}^{j} \gamma_k z_0^k , \end{aligned} \tag{34}$$

where we pick N and j large enough so that the first sum is as small as we please and $|a_n| > |z_0|$ for $n \geq N+1$. But the leading term in z_0 in (34) is z_0^{2N+j}. Thus the last sum, except for arbitrarily small corrections, is approximately a convergent power series and hence converges at $z = z_0$, for N and j large enough, and so is bounded.

Thus by Theorem 2 the $[N, N+j](z_0)$ converges to $f(z_0)$. This theorem is a generalization of that of de Montessus de Balloire (1902) who considered N fixed.

C. Series of Stieltjes

There has been a growing number of applications of the special class of functions called series of Stieltjes to problems in physics and chemistry. These applications will be detailed in other sections of this article and other articles in this book. They have the advantage that a number of rigorous theorems can be proved about them. We will list some of these theorems here and refer to our previous review article for the proofs.

By a series of Stieltjes we mean

$$f(z) = \sum_{j=0}^{\infty} f_j(-z)^j \tag{35}$$

is a series of Stieltjes if and only if, there is a bounded, nondecreasing function $\varphi(u)$, taking on infinitely many values in the interval $0 \leq u < +\infty$ such that

$$f_j = \int_0^{\infty} u^j \, d\varphi(u) . \tag{36}$$

(Note, we do not necessarily assume that the series is convergent.) We introduce the determinants

$$D(m, n) = \det \begin{vmatrix} f_m, & f_{m+1}, & \cdots & f_{m+n} \\ f_{m+1}, & f_{m+2}, & \cdots & f_{m+n+1} \\ \vdots & \vdots & \ddots & \vdots \\ f_{m+n}, & f_{m+n+1}, & \cdots & f_{m+2n} \end{vmatrix} . \tag{37}$$

[Compare signs in (3) and (36) to see difference between D and Δ of (5)].

This definition is equivalent to the conditions

$$D(0, n) > 0 , \qquad D(1, n) > 0 . \tag{38}$$

That the D's are positive is easily seen as $\sum f_{m+p+q}x_p x_q$ is a symmetric quadratic form which is the expected value of a nonnegative-definite quantity and hence has all positive eigenvalues and hence the determinant, as the product of the eigenvalues, is positive.

THEOREM 6. If $\sum f_j(-z)^j$ is a series of Stieltjes, then the poles of the $[N, N+j]$, $j \geq -1$, Padé approximants are on the negative real axis. Furthermore, the poles of successive approximants interlace and all the residues are positive. The roots of the numerator also interlace those of the denominator.

The proof of this theorem is based on the fact that the denominators (18), when reexpressed in terms of the f_j, form a Strum sequence as can be seen from the determinantal formula (Muir, 1960) $A_{11}A_{22} - A_{12}^2 = A \cdot A_{12,12}$ where the subscripted A's represent minors of A.

THEOREM 7. The Padé approximants for series of Stieltjes obey the following inequalities where $f(z)$ is the sum of the series $\sum f_j(-z)^j$, and z is real and nonnegative.

$$(-1)^{1+j}\{[N+1, N+1+j] - [N, N+j]\} \geq 0 , \tag{39a}$$

$$(-1)^{1+j}\{[N, N+j] - [N-1, N+j+1]\} \geq 0 , \tag{39b}$$

$$[N, N] \geq f(z) \geq [N, N-1] \tag{39c}$$

$$[N, N]' \geq f'(z) \geq [N, N-1]' , \tag{39d}$$

where $j \geq -1$. These inequalities have the consequence that the $[N, N]$ and $[N, N-1]$ sequences form the best upper and lower bounds obtainable from the $[N, N+j]$ approximants with a given number of coefficients and that the use of additional coefficients (higher N) improves the bounds. Eq. (39a, b) are valid when differentiated, provided $j \geq 0$ for (39a). We remark that (39c, d) easily imply

$$-\frac{d \ln [N, N]}{dz} \geq \frac{-d \ln f(z)}{dz} \geq \frac{-d \ln [N, N-1]}{dz} . \tag{40}$$

THEOREM 8. Any sequence of $[N, N+j]$ Padé approximants for a series of Stieltjes converges to an analytic function in the cut complex plane $(-\infty \leq z \leq 0)$. If, in addition, $\sum_{p=1}^{\infty} (f_p)^{-1/(2p+1)}$ diverges, then all the sequences tend to a common limit (This condition is roughly equivalent to $|f_p| \leq (2p)!$). If the f_p are a convergent series with a radius of convergence R, then any $[N, N+j]$ sequence convergences in the cut plane $(-\infty \leq z \leq -R)$ to the analytic function defined by the power series.

We remark that for a series of Stieltjes the coefficients of the power-series expansion of the Padé approximants satisfy the inequality

$$|[N, N+j]^{(p)}(z)/(p!)| \leq f_p . \tag{41}$$

Common (1968) has recently raised the question of error bounds on Padé approximants for series of Stieltjes throughout the cut complex plane. Inequalities (39) give very precise bounds for z real and positive, but they provide only lower bounds for z negative and do not provide any information for z complex. Common (1968) derived a number of results in this direction however we will follow the subsequent presentation of Baker (1969). Special cases have been obtained previously by Pfluger and Henrici (1965) and Gargantini and Henrici (1967).

The first step in establishing error bounds is to use the known result that if $f(z)$ is a series of Stieltjes of radius of convergence at least R then $g(z)$, defined by

$$f(z) = f_0/[1 + zg(z)] , \tag{42}$$

is also. That this is so can be seen formally through the use of Hadamard's (1892) theorem on determinants. If $D_g(m, n)$ are the determinants for $g(z)$ analogous to (37) for $f(z)$, then

$$\begin{aligned} D_g(0, p) &= D(1, p)/f_0^{2p+4} > 0 , \\ D_g(1, p) &= D(0, p+1)/f_0^{2p+5} > 0 , \end{aligned} \tag{43}$$

which, via (38) establishes that $g(z)$ is a series of Stieltjes. As $f(z)$ is assumed to have a radius of convergence R, it has the integral representation

$$f(z) = \int_0^{1/R} \frac{d\varphi(u)}{1 + uz} . \tag{44}$$

From this representation it is clear that $f(z)$ is regular in the cut plane $(-\infty \leq z \leq -R)$. Except for possible polar singularities, $g(z)$ must be also, by (42). However, if $g(z)$ had a pole, then $f(z)$ would vanish, but it follows easily from (44) that $f(z) \neq 0$ in the cut z-plane. Thus $g(z)$ also has a radius of convergence of at least R.

As, in (42), $g(z)$ is a member of the same class as $f(z)$, namely, a series of Stieltjes of radius of convergence at least R, we are free to iterate that form. When we do we obtain

$$f(z) = \cfrac{f_0}{1 + \cfrac{za_1}{1 + \cfrac{za_2}{1 + \cdot \, \cdot \, \cdot \, 1 + \cfrac{za_p}{1 + zh_p(z)}}}} , \tag{45}$$

where $h_p(z)$ is again a series of Stieltjes with radius of convergence at least R. Equation (45) may be reexpressed as

$$f(z) = \frac{A_p(z) + zh_p(z)A_{p-1}(z)C_p}{B_p(z) + zh_p(z)B_{p-1}(z)C_p}, \tag{46}$$

where the A's and B's are polynomials in z, and the C_p are constants. That the form is correct can be seen by setting $h_p = 0$ and ∞ in (45) and noting that we obtain the pth and $(p-1)$st expressions, respectively. Wall (1948) shows (Theorems 96.1, 97.1) that the fractions, $A_p(z)/B_p(x)$, are in fact Padé approximants and fill a stairlike sequence in the Padé Table, i.e., [0, 0], [1, 0], [1, 1], [2, 1], [2, 2], ...

Returning to Eq. (42), we note one further restriction that $g(z)$ must satisfy; namely, as $g(z)$ is monotonic $-R \leq z \leq 0$, we must have

$$\lim_{z\to -R} zg(z) \leq 1 \tag{47}$$

in order that $f(z)$ be free of singularities in this range. Therefore

$$g(-R) \leq 1/R\,. \tag{48}$$

As we iterate (42), the $h_p(z)$ obtained will be similarly restricted although the bound is now a function of all the constants in (45) which may be defined from (Wall, 1948) the first p coefficients of the power series for $f(z)$. We choose to redefine $h_p(z)$ so that

$$h_p(-R) \leq 1/R \tag{49}$$

and absorb the change in normalization in the C_p. We may now easily solve for C_p from (46), using the critical equation which makes $z = -R$ a pole of $f(z)$. Thus

$$C_p = B_p(-R)/B_{p-1}(-R)\,. \tag{50}$$

Having established relation (46) subject to (49)–(50), we see that the range of $f(z)$, given the first p coefficients, and given that it is a series of Stieltjes with radius of convergence of at least R, is just the linear fractional transformation of the range of $h_p(z)$.

We now turn our attention to the problem of computing the range of $h_p(z)$, subject to (49). Now we can represent

$$h_p(z) = \int_0^{1/R} \frac{d\phi_p(u)}{1+uz}, \tag{51}$$

which, evaluated at $z = -R$, is

$$h_p(-R) = \int_0^{1/R} \frac{d\phi_p(u)}{1-Ru} \leq 1/R\,. \tag{49'}$$

By (49′),

$$d\omega_p(u) = R\, d\psi_p(u)/(1 - Ru) \tag{52}$$

is also an allowable measure with

$$\int_0^{1/R} d\omega_p(u) \le 1\,. \tag{53}$$

Hence we may rewrite

$$h_p(z) = \frac{1}{R}\int_0^{1/R} \frac{(1 - Ru)\, d\omega_p(u)}{1 + zu}\,, \tag{54}$$

with $d\omega_p$ an arbitrary, nonnegative definite, normalized measure. It follows at once from (54) that, if H_1 and H_2 are possible values of $h_p(z)$, then $\alpha H_1 + (1 - \alpha) H_2$ $(0 \le \alpha \le 1)$ are also, as if ω and ω' are allowed measures in (54) then so is $\alpha\omega + (1 - \alpha)\omega'$. Consequently, the range of $h_p(z)$ is a convex region. The integrand of (54) is a weighted sum of

$$\frac{1}{R}\left(\frac{1 - Ru}{1 + uz}\right), \qquad 0 \le u \le \frac{1}{R}\,, \tag{55}$$

which is a linear fractional transformation of the segment of the u axis. By a well-known property of linear fractional transformations, (Copson, 1948) the result is the arc of a circle. Hence the range of $h_p(z)$ is the convex hull of this arc. The vertical height can be easily calculated to be

$$0 \le -\mathrm{Im}\, h_p(z) \le y\{2R[[(R + x)^2 + y^2]^{1/2} + (R + x)]\}^{-1} \tag{56}$$

where x and y are the real and imaginary parts of z. The range is the complete convex lens-shaped region described above since any point on the circular arc can be obtained for $d\omega_p(u) = \delta(u - u_0)\, du$ where u_0 is appropriately selected. Any point on the real axis portion of the boundary can be obtained as $d\omega(u) = h_p(z)\delta(u)\, du$. All points in the interior can be obtained in an infinite number of ways as linear combinations of boundary points. The range of $f(z)$ is therefore the map of this lens-shaped region under (46). The resulting range, $F_p(z)$ will again be a lens-shaped (Copson, 1948) region. By the method of construction, we have

$$f(z) \in F_p(z) \subset F_{p-1}(z) \subset \cdots \subset F_1(z)\,, \tag{57}$$

where $\subset$ means "contained in".

For every nearest-neighbor pair in the (upper-half) Padé table we can introduce an inclusion region. Using the notation (6), we may write out explicitly, for the even

$$f(z) = \frac{P_m^{(n)}(z)Q_m^{(n-1)}(-R) + zh_{2m,n}(z)P_m^{(n-1)}(z)Q_m^{(n)}(-R)}{Q_m^{(n)}(z)Q_m^{(n-1)}(-R) + zh_{2m,n}(z)Q_m^{(n-1)}(z)Q_m^{(n)}(-R)} \tag{58}$$

and odd

$$f(z) = \frac{P_{m+1}^{(n-1)}(z)Q_m^{(n)}(-R) + zh_{2m+1,n}(z)P_m^{(n)}(z)Q_{m+1}^{(n-1)}(-R)}{Q_{m+1}^{(n-1)}(z)Q_m^{(n)}(-R) + zh_{2m+1,n}(z)Q_m^{(n)}(z)Q_{m+1}^{(n-1)}(-R)} \tag{59}$$

values of p (the first subscript of h), the formulas relating $f(z)$ to $h(z)$. The map of the range of the $h(z)$ found by considering (55) defines the set of ranges $F_{p,n}(z)$. Equation (57) can be generalized for $n = 0, 1, 2, \ldots$,

$$f(z) \in F_{p,n}(z) \subset F_{p-1,n}(z) \cdots \subset F_{1,n}(z), \qquad n = 0, 1, 2, \ldots . \tag{60}$$

These relations generalize the inequalities (39a) and (39c) above to arbitrary values of z in the cut complex plane. We can also prove the inclusion relations

$$F_{p,n}(z) \subset F_{p-1,n+1}(z) , \tag{61}$$

which generalizes (39b) to arbitrary values of z in the cut complex plane. Taken together these imply

$$f(z) \in F_{p,0}(z) \tag{62}$$

is the best-possible bound for $f(z)$ that we may form from the coefficients through z^p, where the $F_{p,0}(z)$ are defined by (58) and (59) using the aforementioned permissible range for the h's.

The best bounds in the cut z-plane $(-\infty < z \leq 0)$ for the problem (36) can be obtained as a special case of these results. This represents the limiting case where $R \to 0$. The formulas (58) and (59) remain valid. By definition $Q_m^{(n)}(0) = \Delta(1 + n, m - 1)$ for all m, n. The range for $h_{p,n}$ follows directly as the limit as $R \to 0$ of (55). It is the complete angular wedge bounded by the real axis and the ray through $R + z^*$. The sequence with $n = 0$ is again the best one. Since a wedge is merely a lens-shaped region with one vertex at infinity, the ranges $F_{p,n}$ for the value of $f(z)$ are again lens-shaped regions.

The problem where

$$f_j = \int_{-\infty}^{+\infty} u^j \, d\varphi(u) \tag{63}$$

was completely treated by Hamburger (1920, 1921) and Gordon (1968). Here the range of values becomes the entire half-plane $\operatorname{Im}(h(z)) \operatorname{Im}(z) < 0$. The map of this is a circle, which is a degenerate lens in which the two sides meet at a straight angle. For real z, it follows from (55) that

$$0 \leq h(z) \leq 1/R . \tag{64}$$

Hence, the inclusion relation (62) becomes, for $z > 0$, the inequalities

$$\frac{P_m^{(0)}(z)Q_m^{(-1)}(-R) + (z/R)P_m^{(-1)}(z)Q_m^{(0)}(-R)}{Q_m^{(0)}(z)Q_m^{(-1)}(-R) + (z/R)Q_m^{(-1)}(z)Q_m^{(0)}(-R)} \leq f(z) \leq [m, m] \tag{65}$$

or

$$[m+1, m] \leq f(z) \leq \frac{P_{m+1}^{(-1)}(z)Q_m^{(0)}(-R) + (z/R)P_m^{(0)}(z)Q_{m+1}^{(-1)}(-R)}{Q_{m+1}^{(-1)}(z)Q_m^{(0)}(-R) + (z/R)Q_m^{(0)}(z)Q_{m+1}^{(-1)}(-R)}, \tag{66}$$

as the last available coefficient is for z^{2m} or z^{2m+1}, respectively. When $-R < z < 0$, the sense of the inequality signs in (65) reverses, and that in (66) remains unchanged.

These bounds $(z > 0)$ reduce to the previous results (39c) in the limit $(z/R) \to \infty$, $(R \to 0)$.

In our previous review article (Baker, 1965) we discussed the 2 or n-point Padé approximant where information about a function at two or more points is incorporated in a rational fraction form. At that time nothing had been proved and experimental investigations had to be relied upon to determine the behavior. By using methods similar to those above it has been shown (Baker, 1969), that error bounds throughout the cut complex plane can be given, when the radius of convergence is known.

We consider now the problem when a set of values at various real points $[f(z_j), j = 1, 2, \ldots, n]$ is given. If in addition any number of successive derivatives at these points is also given, a fairly obvious modification of the following discussion will handle it as well. What emerges from our treatment will be error bounds for an interpolation formula closely similar to Thiele's reciprocal-difference interpolation formula (Milne-Thomson, 1951). We first make the following observation. Let $f(z)$ be given by (44). Then, by the change of variables of integration (w real)

$$v = u/(1 + uw), \tag{67}$$

we have

$$f(z + w) = \int_0^{1/(R+w)} \frac{(1 - vw)\, d\varphi[v/(1 - vw)]}{1 + zv}, \tag{68}$$

which is, by its form, a series of Stieltjes in z, provided $w > -R$, with a radius of convergence of at least $(R + w)$. Consequently, by arguments exactly analogous to those above, if $f(z)$ is a series of Stieltjes and $(z_j, j = 1, 2, \ldots, p)$ are points in the real line $-R < z < +\infty$, then

$$f(z) = \cfrac{a_0}{1 + \cfrac{(z - z_0)a_1}{1 + \cfrac{(z - z_1)a_2}{1 + \cfrac{\ddots}{\quad \cfrac{(z - z_{p-1})a_p}{1 + (z - z_p)g_{p+1}(z)}}}}} \tag{69}$$

defines $g_p(z)$ as a series of Stieltjes, provided the a_p are selected so as to fit $f(z)$ at $z = z_0, z_1, \ldots, z_p$. This is conveniently done through the relations

$$\begin{aligned} g_0(z) &= f(z)\,, \\ g_p(z) &= \frac{g_{p-1}(z_{p-1}) - g_{p-1}(z)}{(z - z_{p-1})g_{p-1}(z)}\,, \qquad p \geq 1\,; \end{aligned} \tag{70}$$

then

$$a_p = g_p(z_p)\,. \tag{71}$$

Equation (69) may be reexpressed as

$$f(z) = \frac{A_p(z) + (z - z_p)g_{p+1}(z)A_{p-1}(z)C_p}{B_p(z) + (z - z_p)g_{p+1}(z)B_{p-1}(z)C_p}\,, \tag{72}$$

where the A's and B's are polynomials in z which result in fitting A_p/B_p through (69)–(71) to the first 0 to p points, and the C_p are constants. That this form is correct may be seen by setting $g_{p+1}(z) = 0$ and ∞ in (69), and noting that we obtain the pth and $(p - 1)$st expressions, respectively. Again, the functions $A_p(z)/B_p(z)$ are rational fractions and the sequence of degrees of the numerator and denominator is the same as in the stair-step sequence [0, 0], [1, 0], [1, 1], [2, 1], [2, 2], ... Let us change the normalization of $g_{p+1}(z)$, absorbing this change by changing C_p correspondingly. We normalize $g_{p+1}(z)$ as

$$(R + z_p)g_{p+1}(-R) \leq 1\,. \tag{73}$$

That $g_{p+1}(-R)$ is bounded follows, as it did before for $h_p(-R)$. If we consider $g_{p+1}(w + z_p) = k_{p+1}(w)$, then we have shown $k_{p+1}(w)$ to be a series of Stieltjes with radius of convergence at least $(R + z_p)$. Hence the bounds on its range given by the convex hull of (55) where $(R + z_p)$ replaces R are valid. We may evaluate C_p similarly, as before, obtaining

$$f(z) = \frac{B_{p-1}(-R)A_p(z) + (z - z_p)g_{p+1}(z)B_p(-R)A_{p-1}(z)}{B_{p-1}(-R)B_p(z) + (z - z_p)g_{p+1}(z)B_{p-1}(z)B_p(-R)}\,. \tag{74}$$

The range of $f(z)$ at any point in the cut $(-\infty < z \leq -R)$ complex plane is a lens-shaped region which is the map under (74) of

$$\text{Range}\,\{g_{p+1}(z)\} = \underset{0\leq u\leq 1(R+z_p)}{\text{convex hull}} \left\{\frac{1}{R + z_p}\left(\frac{1 - (R + z_p)u}{1 + u(z - z_p)}\right)\right\}\,. \tag{75}$$

This lens-shaped region reduces to an interval for z real. This bound on the value of $f(z)$ is the best possible, because it can be shown that we can construct a series of Stieltjes of the required type to take on any value in the range given by (75).

When the set $\{z_j\}$ processes a limit point in the real line $-R < z < +\infty$, then the necessary and sufficient conditions that a set of values $\{f(z_j)\}$ are values of a series of Stieltjes is that

$$f(z_{p+1}) \in F_p(z_{p+1}) \,, \qquad p = 0, 1, \ldots \,, \tag{76}$$

where $F_p(z)$ is the allowable range of $f(z)$ defined by (74) and (75), using the points $z_0, z_1, \ldots, z_p$.

We remark that, in spite of the seemingly unsymmetric method of construction, the function $(A_p(x)/B_p(x))$ does not depend on the order in which the points are selected but is a symmetric function of $\{z_i, \quad i = 1, \ldots, p\}$.

D. Bounded Analytic Functions

A large class of functions which can be reduced to series of Stieltjes is the class of bounded analytic functions. We shall show briefly how this is done by quoting without proof the following theorems given in our previous review article.

THEOREM 9. The most general expression for an analytic function $f(z)$ which has positive real part for $|z| < 1$, and which is real for real values of z is

$$f(z) = \frac{1+z}{1-z}\int_0^{2\pi} \frac{d\varphi(t)}{1 + [4z/(1-z)^2]\sin^2 zt} \,, \tag{77}$$

where φ is monotonic nondecreasing (Riesz, 1911; Herglotz, 1911).

THEOREM 10. Let $f(z)$ be a function possessing all real, right-hand derivatives at $z = 0$. Let $f(z)$ be analytic $|z| < 1$, $\mathrm{Im}(z) \neq 0$, and $\mathrm{Re}(f(z))$ bounded from below by a constant, $-M$, there. Further let $f(z)$ be such that $\lim_{\varepsilon\to 0} x^{-1}\,\mathrm{Im}\,[f(x+i\varepsilon)] \geq 0$ for real $-1 < x < +1$. Then $\hat{F}(u)$ is a series of Stieltjes, where

$$\begin{aligned}
\hat{F}(u) &= (1 + T_1 u)F[u(1 + T_1 u)^{-1}] \,, \\
F(w) &= \{f([(1+w)^{1/2} - 1]/[(1+w)^{1/2} + 1]) + M\}(1+w)^{-1/2} \,, \\
T_1 &= \tfrac{1}{4}(1 - R_1)^2/R_1 \,,
\end{aligned} \tag{78}$$

where R_1 is the minimum of 1 and the least upper bound of real regular points of f. Convergent bounds to $f(z)$ are then given via $\hat{F}(u)$ by Theorems 7 and 8.

In practice it has been found difficult to use the full generality of Theorem 10, because of the difficulty in checking the required conditions, particularly in the "corners" at $x = \pm 1$. The following special case is simpler to use.

Let $f(z)$ be analytic $|z| < 1$, and $\mathrm{Re}(f(z)) \geq 0$, then

$$F(w) = f([(1+w)^{1/2} - 1]/[(1+w)^{1/2} + 1])(1+w)^{-1/2} \tag{79}$$

is a series of Stieltjes.

Gronwall (1932) has given the coefficient formula

$$\begin{gathered} f(z) = \sum_{j=0}^{\infty} f_j(-z)^j, \qquad F(w) = \sum_{j=0}^{\infty} F_j(-w)^j, \\ F_j = 4^{-j} \sum_{k=0}^{j} \binom{2j}{j-0} f_k. \end{gathered} \tag{80}$$

The conditions on $f(z)$ for this special case can be conveniently tested by using a result of Nevanlinna (1919). We note that

$$\mathrm{Re}\,[f(z)] \geq 0, \qquad |z| \leq 1 \tag{81}$$

is equivalent to

$$|\sigma_0(z)| \leq 1, \qquad |z| \leq 1, \tag{82}$$

where

$$\sigma_0(z) = [f(z) - 1]/[f(z) + 1]. \tag{83}$$

Nevalinna proves that it is necessary and sufficient for $|\sigma_0(z)| < 1$ in $|z| < 1$ for $|\sigma_n(0)| \leq 1$ for all n. We define

$$\sigma_n(z) = z^{-1}\left[\frac{\sigma_{n-1}(0) - \sigma_{n-1}(z)}{1 - \sigma_{n-1}^*(0)\sigma_{n-1}(z)}\right], \tag{84}$$

where the asterisk denotes complex conjugate.

From (84) it follows directly that the $\sigma_n(0)$ may be computed from the power-series coefficients of $\sigma_0(z)$ through order n. Using this procedure one can not only check the conditions $|\sigma_n(0)| \leq 1$, for the coefficients available, but also consider the trend of the $\sigma_n(0)$'s in an effort to determine whether a violation will be likely or unlikely if additional coefficients were known. Experimentally, violations of the σ-test occur for the same number of coefficients as violations of the required determinantal inequalities on the F_j, on test cases we have tried.

E. The Padé Conjecture

The theorems given in the preceding sections of this chapter do not by any means exhaust the range of convergence of the Padé approximant methods, as can be seen by numerous examples (See, Baker, 1965). Baker *et al.* (1961) put forward the following conjecture. So far, there are no known counter examples nor any valid proof of it. Even it, which greatly extends the range of convergence, probably does not give the complete range of convergence for the Padé approximant. (We give a weakened version due to Baker, 1965).

PADÉ CONJECTURE. If $P(z)$ is a power series which is regular for $|z| \leq 1$, except for m poles within this circle and except for $z = +1$, at which point the function is assumed continuous when only points $|z| \leq 1$ are considered, then at least a subsequence of the $[N, N]$ Padé approximants are uniformly bounded in the domain formed by removing the interiors of small circles with centers at these poles and uniformly continuous at $z = +1$ for $|z| \leq 1$.

One can prove (Baker, 1965), via Theorem 2, that this conjecture implies uniform convergence in the same closed domain. It will be evident to the reader that repeated application of this conjecture and Theorem 1 expands the region of convergence of at least a subsequence to a domain which is the union of all those circles which include the origin as an interior point and exclude all nonpolar singularities. For details see Baker (1965).

III. GENERALIZED APPROXIMATION PROCEDURES

It is the purpose of this section to widen the class of functions for which convergent upper and lower bounds can be given, by developing an extension of the Padé approximants. We draw these results from Baker (1967). We introduce approximants of the form

$$B_{n,j}(x) = \sum_{m=1}^{n} \alpha_m b(x, \sigma_m) + \sum_{k=0}^{j} \frac{\beta_k}{k!} \left[\left(\frac{\partial}{\partial s} \right)^k b(x, s) \bigg|_{s=0} \right], \tag{85}$$

and consider their use in approximating functions of the type

$$g(z) = \int_0^\infty b(z, s) \, d\varphi(s) \;, \tag{86}$$

where $d\varphi \geq 0$. We prove several general theorems about this type of approximant. We give a sufficient condition that (85) converge, and that it converge to a unique function of z. Among the kernels $b(x, s)$ for which we can prove convergence, we give a simple, necessary, and sufficient condition that (85) form upper and lower bounds to $g(z)$; namely.

$$(-\partial/\partial s)^j b(x, s) \geq 0 \tag{87}$$

for all real, positive x and s, and $j = 0, 1, 2, \ldots$.

Such success as has been had to date in proving convergence theorems concerning Padé approximants has arisen primarily from a representation of the function to be approximated in the form

$$f(z) = \int_0^\infty \frac{d\varphi(u)}{1 + zu}, \tag{88}$$

where $d\varphi \geq 0$ (series of Stieltjes see Section IIC). The Padé approximation then consists of approximating $d\varphi$ by a discrete sum of delta functions whose strengths and locations are determined so that the leading coefficients of the power series expansion of the Padé approximants agree with those of the function being approximated.

We seek to relax form (88) in the hope of expanding the class of functions for which we can establish convergence theorems and bounds on the errors involved in the use of finite approximations. To this end let us consider

$$g(z) = \int_0^\infty b(z, s)\, d\varphi(s), \tag{89}$$

where $d\varphi \geq 0$ and the properties of $b(z, s)$ are not yet specified. Our approximants would again be defined by approximating $d\varphi$ as a sum of delta functions such that the leading power-series coefficients agree with those of $g(z)$.

We shall now introduce a triangular table of approximants.

$$B_{n,j}(x) = \sum_{m=1}^{n} \alpha_m b(x, \sigma_m) + \sum_{k=0}^{j} \frac{\beta_k}{k!} \left[\left(\frac{\partial}{\partial s} \right)^k b(x, s) \Big|_{s=0} \right], \tag{90}$$

where $n = 1, 2, \ldots,$ and $j = -1, 0, 1, \ldots$. These approximants are, for $b(x, s) = 1/(1 + xs)$, the $[n, n + j]$ Padé approximants. When $j = -1$, the second sum in (90) is to be omitted entirely. The defining equations for α, β, σ are obtained as follows: Let us expand

$$b(x, s) = \sum_{m=0}^{\infty} b_m(x)(-s)^m \tag{91}$$

and denote

$$c_n = \int_0^\infty s^n d\varphi(s). \tag{92}$$

Then, equating (89) to (90), we have

$$\sum_{m=0}^{\infty} b_m(x) \sum_{l=1}^{n} \alpha_l(-\sigma_l)^m + \sum_{m=0}^{j} b_m(x)\beta^m(-1)^m = \sum_{m=1}^{\infty} b_m(x)(-1)^m c_m. \tag{93}$$

Let us assume that $b(x, s)$ satisfies a "solvability" condition so that (93) can be made equivalent to a set of equations obtained by equating the coefficients of $b_m(x)$. For example, if $b_m(x) \propto x^m$ as x goes to zero, then we could obtain this equivalence. We take therefore the defining equations to be

$$\begin{aligned} \sum_{l=1}^{n} \alpha_l(\sigma_l)^k + \beta_k &= c_k\,, \qquad 0 \le k \le j\,, \\ \sum_{l=1}^{n} \alpha_l(\sigma_l)^k &= c_k\,, \qquad j < k \le 2n + j\,, \end{aligned} \tag{94}$$

where there are $2n + j + 1$ equations in the same number of parameters and the c_k are determined by the solution of

$$g(z) = \sum_{m=0}^{\infty} b_m(z) c_m (-1)^m\,. \tag{95}$$

If we identify the $(-\sigma_l)^{-1}$ as the location of poles and (α_l/σ_l) as the respective residues, then (94) are exactly the equations for the $[n, n + j]$ Padé approximants to the function

$$C(z) = \sum_{k=0}^{\infty} c_k(-z)^k\,. \tag{96}$$

Thus, for suitably restricted $b(x, s)$, we may reduce the study of this more general approximation procedure to the study of the Padé approximants to a transformed function. The series $C(z)$ will be a series of Stieltjes for $g(z)$ of the form (89).

We are now in a position to give some sufficient conditions on $b(x, s)$ so that the approximants (90) will converge to $g(z)$.

THEOREM 11. Suppose $b(x, s)$ is regular in a uniform neighborhood of the positive, real s axis and $(\ln s)^{1+\mu} \times b(x, s)$ is bounded as $s \to +\infty$, for some $\eta > 0$; then the approximants $B_{n,j}(x)$ converge as n goes to infinity for functions $g(z)$ of the form (89).

Proof. Consider

$$(2\pi i)^{-1} \int_C b(x, s)[n, n + j]\left(\frac{-1}{s}\right) \frac{ds}{s}\,, \tag{97}$$

where the contour C is as shown in Fig. 2. We wish to prove that (97) converges as n tends to infinity. By Theorem 8, $[n, n + j](-1/s)$ converges at every point of C as $n \to \infty$. Furthermore, $s^{-1}[n, n + j](-1/s)$ is bounded at every point of the contour, since $[n, n + j]$ approximates a

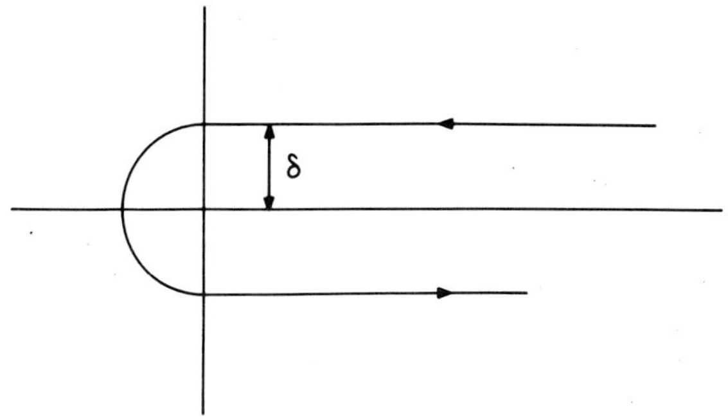

Fig. 2. Integration contour in the s plane for equation (97).

series of Stieltjes. This result follows easily as the α_n, being positive, are bounded by their sum which is c_0. Let us now break the integral (97) into two parts. In one part we consider that portion of C such that $|s| \leq S$ and in the other part $|s| > S$. The second part may be written as

$$(2\pi i)^{-1} \int_{C>S} b(x, s) \left[\sum_{m=1}^{n} \frac{\alpha_m}{s - \sigma_m} + \sum_{k=0}^{j} \beta_k s^{-k-1} \right] ds \,. \tag{98}$$

Now, except for s in the range $|s - \sigma_m| \leq \frac{1}{2}\sigma_m$,

$$1/|s - \sigma_m| < 3/\sigma \,. \tag{99}$$

In the range $|s - \sigma_m| \leq \frac{1}{2}\sigma_m$,

$$1/|s - \sigma_m| = 1/[(\sigma - \sigma_m)^2 + \delta^2]^{1/2} \,, \tag{100}$$

where σ is the real part of s.

Using (99) and (100), the result that the absolute value of an integral is less than or equal to the integral of the absolute value, the fact that all α's and β's are positive real, and (94) for $k = 0$, we may write

$$\begin{aligned} |\text{Eq. } (98)| \leq{}& \frac{2}{2\pi} \int_s^\infty d\sigma \, |b(x, s)| \left[\frac{3c_0}{\sigma} + \sum_{k=1}^{j} \beta_k \sigma^{-k-1} \right] d\sigma \\ &+ \frac{2}{2\pi} \sum_{m=1}^{n}{}' \alpha_m \int_{(1/2)\sigma_m}^{(3/2)\sigma_m} d\sigma \, |b(x, s)| \, [(\sigma - \sigma_m)^2 + \delta^2]^{-1/2} \,, \end{aligned} \tag{101}$$

where the primed summation includes only those terms with $\sigma_m > \frac{1}{2}S$.

Now, assuming the bound on $b(x, s)$ stated in the theorem

$$|b(x, s)| < k/(\ln \sigma)^{1+\eta} \,, \tag{102}$$

we may compute (101) as

$$|\text{Eq. (98)}| \le \frac{6c_0K}{2\pi\eta}(\ln S)^{-\eta} + \frac{2K}{2\pi}(\ln S)^{-1-\eta}\sum_{k=1}^{j} c_j S^{-k} + \sum_{m=1}^{n} \alpha_m \frac{2K\ln[\frac{1}{2}\sigma_m + (\frac{1}{4}\sigma_m^2 + \delta^2)^{1/2}]}{2\pi[\ln(\frac{1}{2}\sigma_m)]^{1+\eta}}. \tag{103}$$

As the function in the last summation is monotonically decreasing in σ_m (as $\sigma_m > \frac{1}{2}S \gg \delta$), we may bound this term by $2Kc_0/(\ln S)^{-\eta}$. Hence

$$|\text{Eq. (98)}| \le \frac{4c_0K}{\pi\eta}(\ln S)^{-\eta} + \frac{K}{\pi}(\ln S)^{-1-\eta}\sum_{k=1}^{j} c_j S^{-k} \tag{104}$$

uniformly in n, for S and j fixed. Thus, given any error, $\varepsilon < 0$, we may pick an S such that the bound of (104) is less than $\frac{1}{2}\varepsilon$. We may then, by Theorem 8 pick an n_0 such that all $[n, n+j]$ Padé approximants with $n > n_0$ differ by less than

$$\varepsilon/[2(S+\pi\delta)\max_{C<S}\{|b(x,s)|\}] \tag{105}$$

at every point of $C < S$, from their limiting value. Consequently the integral (97) converges in the limit as n goes to infinity. However, from the form exemplified by (98) it follows by Cauchy's integral theorem that (97) is exactly $B_{n,j}(x)$. Q.E.D.

Corollary 1. If, in addition,

$$\sum_{p=1}^{\infty} (c_p)^{-1/(2p+1)} \tag{106}$$

diverges, $\lim_{n\to\infty} B_{n,j}(x)$ is independent of j.

Proof. This result follows from Theorem 8 and is the condition that all diagonal sequences of Padé approximants converge to a common limit.

Corollary 2. If the series c_p has a finite radius of convergence R then we may replace the conditions on $b(x, s)$ in Theorem 11 by $b(x, s)$ regular in the neighborhood of $0 \le s \le R^{-1}$. (Note: The condition of Corollary 1 is automatically satisfied here.)

Proof. As all poles of the Padé approximant $[n, n+j](-1/s)$ lie in the range $0 \le s \le R^{-1}$, we may choose, instead of the contour C in (97), C' which encircles that line segment and on which $b(x, s)$ is regular. As the Padé approximants converge (Theorem 8) at every point of C', they converge uniformly and hence the $B_{n,j}(x)$ converge.

One of the advantageous results for Padé approximants to series of Stieltjes is that they form monotonically converging (on the positive real axis) sequences which give upper and lower bounds to the correct answer, and thus allow assessment of the error of the approximants. As the next theorem will show, a remarkably simple, additional property of the function $b(x, s)$ is both necessary and sufficient to ensure the same bounding properties for sequence of $B_{n,j}(x)$ approximants.

THEOREM 12. The approximants $B_{n,j}(x)$ to a function of the form (89) obey the following inequalities where x is real and nonnegative:

$$(-1)^{1+j}\{B_{n+1,j}(x) - B_{n,j}(x)\} \geq 0 , \tag{107a}$$

$$(-1)^{1+j}\{B_{n,j}(x) - B_{n-1,j+2}(x)\} \geq 0 , \tag{107b}$$

$$B_{n,0}(x) \geq g(x) \geq B_{n,-1}(x) , \tag{107c}$$

where $j \geq -1$, if and only if

$$(-\partial/\partial s)^j b(x, s) \geq 0 \tag{108}$$

for all real, nonnegative x and s, and $j = 0, 1, 2, \ldots$. These inequalities have the consequence that the $B_{n,0}(x)$ and $B_{n,-1}(x)$ sequences form the best upper and lower bounds obtainable from the $B_{n,j}(x)$ approximants with a given number of coefficients and that the use of additional coefficients (higher n) improves the bounds.

Proof. The first step in our proof is to establish a representation for the inequalities in terms of partial derivatives of b. To this end let us define

$$\Delta^n f(x_i)/(n)! = \frac{\det\begin{vmatrix} 1 & x_i & x_i^2 & \cdots & x_i^{n-1} & f(x_i) \\ 1 & x_{i+1} & x_{i+1}^2 & \cdots & x_{i+1}^{n-1} & f(x_{i+1}) \\ \vdots & \vdots & \vdots & \ddots & \vdots & \vdots \\ 1 & x_{i+n} & x_{i+n}^2 & \cdots & x_{i+n}^{n-1} & f(x_{i+n}) \end{vmatrix}}{\det\begin{vmatrix} 1 & x_i & x_i^2 & \cdots & x_i^n \\ 1 & x_{i+1} & x_{i+1}^2 & \cdots & x_{i+1}^n \\ \vdots & \vdots & \vdots & \ddots & \vdots \\ 1 & x_{i+n} & x_{i+n}^2 & \cdots & x_{i+n}^n \end{vmatrix}} \tag{109}$$

on the distinct points $(x_i, x_{i+1}, \ldots, x_{i+n})$. It follows easily that

$$\begin{aligned} \Delta^n(x^j) &= 0 , \qquad j < n , \\ &= n! , \qquad j = n , \end{aligned} \tag{110}$$

since, for $j < n$, two columns of the numerator are equal and for $j = n$ the numerator and denominator of (109) cancel. Thus we identify Δ^n as the nth difference operator. (This definition can be shown to be equivalent to the more usual form which is given, for example by Jefferys and Jefferys, 1950.) One can easily establish the usual mean-value-theorem result

$$\Delta^n f(x_i) = f^{(n)}(x_i + \theta(x_{i+n} - x_i)), \qquad 0 < \theta < 1 \tag{111}$$

by considering

$$P(x) = \det \begin{vmatrix} 1 & x & x^2 & \cdots & x^n & f(x) \\ 1 & x_i & x_i^2 & \cdots & x_i^n & f(x_i) \\ \vdots & \vdots & \vdots & \ddots & \vdots & \vdots \\ 1 & x_{i+n} & x_{i+n}^2 & \cdots & x_{i+n}^n & f(x_{i+n}) \end{vmatrix}. \tag{112}$$

This function vanishes for $x = (x_i, x_{i+1}, \ldots, x_{i+n})$ and so by Rolle's theorem (See for example, Franklin, 1940), $P'(x)$ must vanish for

$$x = (x_i^{(1)}, x_{i+1}^{(1)}, \ldots, x_{i+n-1}^{(1)}),$$

where

$$x_i < x_i^{(1)} < x_{i+1} < x_{i+1}^{(1)} < x_{i+2} < \cdots < x_{i+n-1}^{(1)} < x_{i+n}. \tag{113}$$

If we repeat this argument we eventually obtain $P^{(n)}(x_i^{(n)}) = 0$ for $x_i < x_i^{(n)} < x_{i+n}$. If we express this result by directly differentiating (112), we get (111).

Let us now consider inequality (107a) for $j = 0$. It is, by (90),

$$\hat{\beta}_0 b_0(x) + \sum_{l=1}^{n-1} \hat{\alpha}_l n(x, \hat{\sigma}_l) - \beta_0 b_0(x) - \sum_{l=1}^{n} \alpha_l b(x, \sigma_l) = \Omega(b), \tag{114}$$

a functional of b. The points σ_l and $\hat{\sigma}_l$ are all distinct by Theorem 6. They lie in the order $(0, \sigma_n, \hat{\sigma}_{n-1}, \sigma_{n-1}, \ldots, \hat{\sigma}_1, \sigma_1)$. We observe that

$$\Omega(s^j) = 0, \qquad \text{if} \quad j = 0, 1, \ldots, 2n-2 \tag{115}$$

by the fundamental equations (94) since both $B_{n,0}(x)$ and $B_{n-1,0}(x)$ approximate the same series. As (114) is a $2n$-point formula, it is thus a multiple of Δ^{2n-1}. The coefficient can be obtained from Eqs. (10) and (37),

$$\Omega \equiv -\frac{D(1, n-1)\Delta^{(2n-1)}}{D(1, n-2)(2n-1)!}, \tag{116}$$

since by (94) $\Omega(s^{2n-1})$ is $(-1)^{2n-1}$ times the $(2n-1)$st series coefficient in the difference, where

$$D(j, k) = \det \begin{vmatrix} c_j & c_{j+1} & \cdots & c_{j+k} \\ \vdots & \vdots & \ddots & \vdots \\ c_{j+k} & c_{j+k+1} & \cdots & c_{j+2k} \end{vmatrix}. \tag{117}$$

The $D(j, k)$ are all positive [Eq. (38)] as $d\varphi > 0$ in (89). Thus combining our results we have shown that there exists a σ such that

$$(2n-1)!\,[B_{n-1,0}(x) - B_{n,0}(x)] = -\frac{D(1, n-1)}{D(1, n-2)} \frac{\partial^{2n-1}}{\partial s^{2n-1}} b(x, s)\Big|_{s=\sigma}, \tag{118}$$

where $0 < \sigma < \sigma_1$. Hence the derivative condition implies the inequalities. The other inequalities follow in exactly the same way as the one we have just proven.

The converse may be shown as follows: In the same way as before,

$$(2n-2)!\,[B_{n,-1}(x) - B_{n-1,1}(x)] = \frac{D(0, n-1)}{D(0, n-2)} \frac{\partial^{2n-2} b(x, s)}{\partial s^{2n-2}}\Big|_{s=\sigma}, \tag{119}$$

where now $\sigma_m < \sigma < \sigma_1$, as 0 is not one of the set of basic points. However, we may choose these arbitrarily by selecting $d\varphi$ as a sum of delta functions and so, by making (σ_n, σ_1) narrow and sweeping it past any desired point, we may, by Bolzano's theorem, find an example which has σ at any point we desire. Thus, the inequalities imply the derivative conditions for even-order derivatives. Since $b(x, s)$ goes to zero as $s \to \infty$, it follows that the odd derivatives are of fixed negative sign. This result can be seen as follows: Since $\partial^2 b/\partial s^2$ is positive, if $\partial b/\partial s$ ever becomes positive, it cannot decrease in magnitude and therefore $b(x, s)$ diverges at least linearly as s goes to infinity. But $b(x, s)$ in fact goes to zero as s goes to infinity. Thus $\partial b/\partial s$ is nonpositive-definite. It goes to zero as a goes to infinity in order that b remain nonnegative. Similarly, $\partial^2 b/\partial s^2$ must go to zero. We can now repeat this argument for $\partial^2 b/\partial s^2$, $\partial^3 b/\partial s^3$, $\partial^4 b/\partial s^4$ and thus by induction all odd derivatives of b are nonpositive-definite. Consequently the inequalities imply the derivative condition as well as vice versa.

One may inquire what class of functions is characterized by property (108). At least for functions which have certain regularity and boundedness conditions, this class of functions is easily characterized. The following theorem does so and shows that (108) characterizes a large class of functions.

THEOREM 13 (Bernstein's theorem). In the class of functions which are regular in the right-half s plane and go to zero faster than s^{-k}, $k > 1$ for some k as s goes to infinity therein the following statements are equivalent:

$$(-\partial/\partial s)^j f(s) \geq 0, \qquad 0 \leq s < +\infty \tag{120}$$

and

$$f(s) = \int_0^\infty e^{-st}\, d\varphi(t)\ , \tag{121}$$

with $d\varphi \geq 0$.

Proof. If (121) holds for $0 \leq s < +\infty$ we may differentiate under the integral sign $0 < s < +\infty$ and consequently obtain (120), as $d\varphi \geq 0$. (The result for $s = 0$ follows by taking the limit from $s > 0$.)

Suppose now that (120) holds. The conditions in the theorem are sufficient (See, for example Churchill, 1944) to ensure that $f(s)$ can be written as a Laplace transform:

$$f(s) = \int_0^\infty e^{-st} F(t)\, dt\ , \tag{122}$$

where F is continuous and of order $O(1)$ for all $t \geq 0$. Consider

$$(-\partial/\partial s)^j f(s) = \int_0^\infty t^j e^{-st} F(t)\, dt \tag{123}$$

in the limit where s and j go to infinity together in such a way that $j/s = \tau$, a constant. The function $t^j e^{-st}$ can then have an arbitrarily sharp peak at $t = \tau$ for large enough s and j; thus we get approximately

$$(-\partial/\partial s)^j f(s) \approx A \int \delta(t - \tau) F(t) dt \approx AF(\tau) \geq 0 \tag{124}$$

by (120). Hence (120) also implies (121).

We will mention briefly a few examples which fall in the scope of the results of this chapter. The first is obtained by taking

$$b(z, s) = L_\zeta(zs)\ , \tag{125}$$

where $L_\zeta(x)$ is the LeRoy function (Hardy, 1956)

$$L_\zeta(x) = \sum_{n=0}^{\infty} \frac{\Gamma(1 + \zeta n)}{\Gamma(1 + n)} (-x)^n\ , \qquad 0 \leq \zeta < 1\ . \tag{126}$$

Most of its properties can be readily established from its integral representation

$$L_\zeta(x) = \int_0^\infty \exp\left[-(t + xt^\zeta)\right] dt\ . \tag{127}$$

Repeated differentiation of (127) immediately establishes property (108). As one may easily show that

$$| L_\zeta(x) | < \Gamma(1/\zeta + 1) x^{-1/\zeta} \tag{128}$$

from (127) for x real and positive, it follows that the conditions for

Theorems 11 and 12 hold. We see from (126) that $L_\zeta(x)$ is an entire function for $0 \leq \zeta < 1$. The two special cases

$$L_1(x) = (1 + x)^{-1}, \qquad L_0(x) = e^{-x} \tag{129}$$

are what give this family its interest, for it interpolates by means of entire functions between Padé approximants ($\zeta = 1$) and the exponential approximants shown by Theorem 13 to be the most general form of kernel of the class $b(s, z) = b(sz)$ for functions regular, etc., in the right-half plane.

Another family of kernels is given by

$$b(z, s) = [1 + zs/(n + 1)]^{-n}, \qquad 0 < n < \infty . \tag{130}$$

The range $n = 1$ to ∞ again interpolates between the Padé approximants and the exponential approximants of Theorem 13.

IV. APPLICATIONS

We mention here only a few applications of particular interest to us which are not covered in other articles in this book.

A. Cooperative Phenomena

Perhaps more than in any other field, the field of cooperative phenomena has seen the use of the Padé approximant method become a standard workaday procedure. A thorough review of these applications would almost amount to a modern history of the Theory of Cooperative phenomena. As Fisher (1967) has recently written an excellent review of this subject, we will not treat it here.

We will however consider one application which appeared too late to be included in Fisher's review (1967). It was made by Baker (1967).

The starting point for this application is the result of Yang and Lee (1952). They consider the grand partition function for a lattice gas (equivalent to the partition function for the Ising model) and prove a theorem concerning the location of its zeros as a function of fugacity under the assumption of an impermeable hard-core and attractive forces (equivalent to the Ising model with purely ferromagnetic interactions of unspecified range). As they point out, the grand partition function may be written

$$Z_v = \sum_{N=0}^{M} \frac{Q_N}{N!} y^N , \tag{131}$$

where "M is the maximum number of atoms that can be crammed into V" on account of the hard cores. As Z_v is a finite polynomial, it can be factored as

$$Z_v = \prod_{i=1}^{M} (1 - y/y_i) , \tag{132}$$

where the $y_1, \ldots, y_M$ are the roots of the algebraic equation

$$Z_v(y) = 0 . \tag{133}$$

None of the roots can be real and positive since the Q_N are all positive. In the mathematically equivalent Ising model, (For a review of the Ising model see Domb, 1960; Fisher, 1963, 1967.) the fugacity is proportional to the variable

$$\mu = \exp(-2mH/kT) , \tag{134}$$

where m is the magnetic moment, H the magnetic field strength, and T the temperature. Rewriting (132) in terms of μ,

$$Z_v = \prod_{i=1}^{M} (1 - \mu/\mu_i) . \tag{135}$$

The theorem of Yang and Lee (1952) states

$$|\mu_i| = 1 \tag{136}$$

for all ferromagnetic-type interaction (the spins have lower energy when parallel than antiparallel) Ising models. If we now calculate the spontaneous magnetization per spin, we have

$$\begin{aligned} I(\mu)/(Nm) &= 1 - 2\mu(d/d\mu) \ln Z \\ &= \sum_{i=1}^{M} \frac{(1 + \mu/\mu_i)}{(1 - \mu/\mu_i)} . \end{aligned} \tag{137}$$

This result means that if $|\mu| \leq 1$ (note the equality sign),

$$\operatorname{Re}[I(\mu)/Nm] \geq 0 . \tag{138}$$

Consequently, by (79) we have

$$F(w) = (1 + w)^{-1/2} I([(1 + w)^{1/2} - 1]/[(1 + w)^{1/2} + 1]) , \tag{139}$$

which is a series of Stieltjes for all temperatures. This result is perfectly rigorous and without additional assumptions of any kind. Theorems 7 and 8 now assure us that the $[N, N](w)$ and $[N, N-1](w)$ Padé approximants form upper and lower bounds, respectively, to $F(w)$ and hence provide them for $I(\mu)$ over the range $0 \leq \mu < 1$ $(0 \leq w < \infty)$. These bounds must converge $0 \leq \mu < 1$ as $I(\mu)$ is a convergent series with radius 1. The range $1 < \mu \leq \infty$ can be computed from the symmetry of the model between points μ and $1/\mu$. For $T > T_c$, the critical temperature (two- or higher-dimensional models), we know that $I(1) = 0$, and, as the $[N, N-1]$ Padé approximants also equal zero at this point (the $[N, N]$ are infinite here), the consequence is that we have converging bounds to I everywhere in the (H, T) plane, except on $H = 0$, $T < T_c$ (the coexistence curve, for the analogous lattice gas).

Through the use of the results of Section III we can construct bounding approximations to a number of other thermodynamic properties. According to Yang and Lee (1952), the free energy (related to the logarithm of the partition function) is given by

$$\frac{(F + mH)}{kT} = -\int_0^{2\pi} \ln(1 - 2\mu\cos\theta + \mu^2)\, g(\theta)\, d\theta\,, \tag{140}$$

where $g(\theta)$ is nonnegative-definite. Letting $\cos\theta = 1 - 2y$ we may rewrite (140) as

$$\frac{(F + mH)}{kT} = -\int_0^{1} \ln[(1 - \mu)^2 + 4\mu y]\, d\varphi(y)\,. \tag{141}$$

One easily establishes for the allowed range of μ and y that

$$g(\mu, y) = -\ln(1 - \mu)^2 + 4\mu y) \tag{142}$$

possesses property (108). Hence by Corollary 2 and Theorem 12 the approximants (90) bound the free energy,

$$B_{n,-1} \leq (F + mH)/kT \leq B_{n,0}\,, \tag{143}$$

and converge to it [except $B_{n,0}(1) = \infty$] for all temperatures.

Baker (1967) also gives bounds on higher derivatives with respect to magnetic field, and computes numerical examples for the two and three dimensional spin-$\frac{1}{2}$ Ising model to illustrate how well these bounds work.

Baker (1968) uses these results to establish two inequalities. He rewrites (141) in terms of

$$\tau = \tanh(mH/kT) \tag{144}$$

as

$$F/kT = \tfrac{1}{2}\ln[\tfrac{1}{4}(1 - \tau^2)] - \int_0^1 \ln[\tau^2(1 - y) + y]\, d\varphi(y)\,, \tag{145}$$

where use is made of

$$\int_0^1 d\varphi(y) = \frac{1}{2}\,.$$

Now, if we substitute $\omega = y^{-1} - 1$, we obtain

$$\frac{F}{kT} = \frac{1}{2}\ln\left[\frac{1}{4}(1-\tau^2)\right] - \int_0^1 \ln y \, d\varphi(y) - \int_0^\infty \ln(1+\tau^2\omega)\, d\varphi\left(\frac{1}{1+\omega}\right). \tag{146}$$

Hence, differentiating with respect to the magnetic field, we obtain the reduced magnetization per spin as

$$\frac{I}{mN} = \tau + \int_0^\infty \frac{2\tau(1-\tau^2)\omega}{1+\tau^2\omega}\, d\varphi\left(\frac{1}{1+\omega}\right), \tag{147}$$

or the function

$$G(\tau^2) = \frac{(I/mN) - \tau}{\tau(1-\tau^2)} = \int_0^\infty \frac{d\phi(\omega)}{1+\tau^2\omega}, \tag{148}$$

where $d\phi \geq 0$. We remark that for $T > T_c$, the critical temperature, the upper limit of integration is less than infinity, but for $T \leq T_c$ it is infinite. We note that $G(0) = \infty$ for $T \leq T_c$.

The consequence of form (148) is that $G(\tau^2)$ is a series of Stieltjes. This fact means that if we expand

$$G(\tau^2) = G_0(T) - G_1(T)\tau^2 + G_2(T)\tau^4 - \cdots, \tag{149}$$

then

$$D(m, n) = \begin{vmatrix} G_m & G_{m+1} & \cdots & G_{m+n} \\ G_{m+1} & G_{m+2} & & \vdots \\ \vdots & & \ddots & \\ G_{m+n} & & \cdots & G_{m+2n} \end{vmatrix} \geq 0. \tag{150}$$

It is easy to show that the divergence of the G_i at $T = T_c$ is the same as the corresponding (one power of τ or H higher) coefficients in the magnetization. Following the notation of Baker, Gilbert, Eve, and Rushbrooke (1967) ($\gamma_i - \gamma_{i-1} = 2\Delta$ in the notation of Fisher, 1967),

$$G_m(\tau^2, T) \propto (T - T_c)^{-\gamma_m}, \; T - T_c^+ . \tag{151}$$

It follows at once from (150) for $n = 1$ that the critical exponents obey

$$\gamma_{i+1} - 2\gamma_i + \gamma_{i-1} \geq 0 \tag{152}$$

or that the γ_i increase at least linearly with i. These relations are obeyed in every known case within calculational error. See Fisher (1967) for a review. The linear relation

$$\gamma_i = \gamma_0 + 2\Delta i \tag{153}$$

required by the scaling laws (Kadanoff, *et al.*, 1967) is allowed by (152). All available Heisenberg-model data was also tested and no disagreement was found.

We point out that

$$(-1)^n \Delta^n G(\tau^2) \geq 0\,, \tag{154}$$

where Δ is the difference operators with respect to τ^2. This formula should be experimentally testable.

B. Scattering Problems

Quite a bit of attention has been given to scattering problems in the last few years and it will be the subject of several other articles in this volume. We will confine our remarks to a few points where our approach differs slightly from that of the other authors. In Baker (1965) we considered the following problem. Let v be the two-body Hermitian potential energy operator, and $1/b = P/(H_0 - E - i\varepsilon)$ where H_0 is the two-body kinetic energy operator, E the unperturbed energy of the initial state, and P a projection operator which is zero for the initial state and one otherwise. The forward scattering amplitude is then

$$f(0) = -\frac{1}{4\pi}\left\{\psi \left|\left(v\lambda - v\frac{1}{b}v\lambda^2 + v\frac{1}{b}v\frac{1}{b}v\lambda^3 - v\frac{1}{b}v\frac{1}{b}v\frac{1}{b}v\lambda^4 + \cdots\right)\right|\psi\right\}, \tag{155}$$

where λ is regarded as an expansion parameter and ψ is an unperturbed eigenstate of H_0 with eigenvalue E. Below threshold (E negative) we may set $\varepsilon = 0$ as $1/b$ is bounded and nonnegative-definite. Hence in this energy range $1/b$ has a well defined real square root. Let us therefore rewrite (155) as

$$f(0) = -\frac{1}{4\pi}[\langle\psi \,|\, v \,|\, \psi\rangle\lambda - \lambda^2\langle\varphi \,|\, \{1 - \lambda[(1/b^{1/2})v(1/b^{1/2})] + \lambda^2[(1/b^{1/2})v(1/b^{1/2})]^2 - \cdots\} \,|\, \varphi\rangle]\,, \tag{156}$$

where $\varphi = (1/b^{1/2})v\psi$. As the operator $w = (1/b^{1/2})v(1/b^{1}/^{2})$ is Hermitian with real eigenvalues W_k, we may rewrite (156) as

$$f(0) = -\frac{1}{4\pi}\left[\langle\psi \,|\, v \,|\, \psi\rangle\lambda - \lambda^2\left\{\sum_k (1 - W_k\lambda + W_k^2\lambda^2 - W_k^3\lambda^3 + \cdots)\,|\, a_k \,|^2\right\}\right] \tag{157}$$

By the normalization of ψ and the assumed finiteness of the perturbation-series terms, there exists a bounded nondecreasing function $\varphi(u)$, such that, if we denote by $(-1)^p c_p$ the term from the $(p+1)$st power of λ,

$$c_p = \int_{-\infty}^{+\infty} u^p \, d\varphi(u) \tag{158}$$

or

$$f(0) = -\frac{\lambda}{4\pi}\langle \psi \,|\, v \,|\, \psi \rangle + \int_{-\infty}^{+\infty} \frac{\lambda^2 u \, d\varphi(u)}{1 + u\lambda}. \tag{159}$$

If we make new assumptions on v, we can get stronger results. For example, if w (for fixed E) is bounded then the limits of integration in (159) are similarly bounded and then by a linear fractional transformation on λ, $[f(0) + (\lambda/4\pi)\langle \psi \,|\, v \,|\, \rangle]/\lambda^2$ can be manipulated into a series of Stieltjes. Plainly if we are at an energy and coupling constant such that $f(0)$ is an analytic function of E, (159) should give a nonsingular representation of it, and these manipulations should then be possible.

If v is a nonnegative- (or nonpositive-) definite potential operator so that a real $v^{1/2}$ (or $(-v)^{1/2}$) may be defined, then we can rewrite (155) as

$$\begin{aligned} f(0) = -\frac{\lambda}{4\pi}\Big\{ v^{1/2}\psi \Big| \Big(1 - \Big(v^{1/2}\frac{1}{b}\, v^{1/2}\Big)\lambda \\ + \Big(v^{1/2}\frac{1}{b}\, v^{1/2}\Big)^2 \lambda^2 - \cdots \Big) \Big| \, v^{1/2}\psi \Big\} \end{aligned} \tag{160}$$

(or with obvious modifications for v nonpositive-definite instead). As $(v^{1/2} b^{-1} v^{1/2})$ is a Hermitian, nonnegative-definite operator, we may write

$$f(0) = -\lambda \int_0^{\infty} \frac{d\varphi(u)}{1 + \lambda u}, \tag{161}$$

where φ is a bounded, monotonic, nondecreasing function, and hence $-f(0)/\lambda$ is a series of Stieltjes. [If v is nonpositive-definite, λ is replaced by $-\lambda$ in the denominator of (161).] (See also Tani, 1965).

If instead of the forward scattering amplitude, we consider any other amplitude, then (155) becomes

$$f(\theta) = -\frac{1}{4\pi}\Big\{ \gamma \Big| \Big(v\lambda - v\frac{1}{b}\, v\lambda^2 + \cdots \Big) \Big| \psi \Big\} = \langle \gamma \,|\, \mathbf{T} \,|\, \psi \rangle \tag{162}$$

and in (157), $|a_k|^2$ is replaced by $(b_k{}^* a_k)$ where the b_k are the decomposition of γ as the a_k are the decomposition of ψ. Nuttall (1968) has suggested the following method for analyzing the nonforward scattering amplitude and the result is due to him although we do not always follow his methods.

Consider the matrix

$$\mathbf{T} \equiv \begin{pmatrix} \langle\psi \mid \mathbf{T}(\lambda) \mid \psi\rangle & \langle\psi \mid \mathbf{T}(\lambda) \mid \gamma\rangle \\ \langle\gamma \mid \mathbf{T}(\lambda) \mid \psi\rangle & \langle\gamma \mid \mathbf{T}(\lambda) \mid \gamma\rangle \end{pmatrix}. \tag{163}$$

Expand it as a matrix function of λ and form matrix Padé approximants (Gammel and McDonald, 1966) to it through the equations

$$\mathbf{T}(\lambda) - \mathbf{P}_M(\lambda)\mathbf{Q}_N^{-1}(\lambda) = O(\lambda^{N+M+1}), \tag{164}$$

where $\mathbf{P}_M$ and $\mathbf{Q}_N$ are matrix polynomials in λ of degrees M and N. If we consider $(\mathbf{x}^T\mathbf{T}\mathbf{x})$ where $\mathbf{x}^T = (x, y)$, then as

$$(\mathbf{x}^T\mathbf{T}\mathbf{x}) = \langle x\psi + y\gamma \mid \mathbf{T}(\lambda) \mid x\psi + y\gamma\rangle, \tag{165}$$

all the derivations for $\langle\psi \mid \mathbf{T} \mid \psi\rangle$ are also valid for it. Thus $(\mathbf{x}^T\mathbf{T}\mathbf{x})$ is a series of Stieltjes also. But we derive from (164)

$$(\mathbf{x}^T\mathbf{T}(\lambda)\mathbf{x}) - (\mathbf{x}^T\mathbf{P}_M(\lambda)\mathbf{Q}_N^{-1}(\lambda)\mathbf{x}) = O(\lambda^{N+M+1}). \tag{166}$$

Hence, by the uniqueness theorem for Padé approximants (see Chisholm, 1963, for a fuller discussion) $(\mathbf{x}^T\mathbf{P}_M(\lambda)\mathbf{Q}_N^{-1}(\lambda)\mathbf{x})$ is the $[N, M]$ Padé approximant to the series $(\mathbf{x}^T\mathbf{T}(\lambda)\mathbf{x})$, which is a series of Stieltjes. Consequently the inequalities (39) apply for any $\mathbf{x}$. Translating back into matrix language, $\mathbf{A} \geq 0$ means $\mathbf{A}$ is a nonnegative-definite matrix. In particular,

$$[\mathbf{N}, \mathbf{N}](\lambda) - \mathbf{T}(\lambda) \geq 0, \tag{167a}$$

$$\mathbf{T}(\lambda) - [\mathbf{N}, \mathbf{N} - \mathbf{1}](\lambda) \geq 0. \tag{167b}$$

Taking the determinant of (167a), we must have

$$\begin{aligned}([\mathbf{N}, \mathbf{N}]_{11} - \mathbf{T}_{11})([\mathbf{N}, \mathbf{N}]_{22} - \mathbf{T}_{22})& \\ \geq |[\mathbf{N}, \mathbf{N}]_{12} - \mathbf{T}_{12}|^2 \geq 0&\end{aligned} \tag{168}$$

and the same from (167b) with $[\mathbf{N}, \mathbf{N} - \mathbf{1}]$ substituted for $[\mathbf{N}, \mathbf{N}]$. Equation (168) provides direct bounds on the value of $\mathbf{T}_{12}$ in terms of the direct lower bounds which we can provide through ordinary $[N, N-1]$ Padé approximants to $\mathbf{T}_{11}$ and $\mathbf{T}_{22}$. Hence we are in a position to treat non-forward as well as forward scattering amplitudes in a converging and bounded manner. Generalized value inclusion regions for complex values of λ and other ranges of u in (161) besides $(0, \infty)$ can be developed in a manner similar to (42)–(76).

Baker and Chisholm (1966) have gone beyond potential scattering theory and considered the perturbation expansion for the S-matrix for the Peres-model field theory (Peres, 1963). The two fields in the model are harmonic oscillators of the same frequency, described by operators x and

y. The interaction is taken as

$$g\,\delta(t)x^2y \tag{169}$$

analogous to an instantaneous interaction of the form $\bar{\psi}\psi\varphi$ between a fermion field ψ and a boson field φ. The operators x and y can cause single quantum jumps (creations or annihilations) between the oscillator states. Baker and Chisholm (1966) proved that every element of the S-matrix is a finite linear combination of series of Stieltjes with real coefficients. Consequently it must be that

$$S_{\alpha\beta}(g^2) = \int_0^\infty \frac{d\varphi_{\alpha\beta}(a)}{1 + g^2u}, \tag{170}$$

where $d\varphi_{\alpha\beta}$ is no longer ≥ 0, but is at least of bounded variation,

$$\int_0^\infty |\,d\varphi_{\alpha\beta}(u)\,| < M_{\alpha\beta} < \infty\,. \tag{171}$$

In the simple Peres model, it was possible to split $d\varphi_{\alpha\beta}$ up into a positive and a negative part and then use the series of Stieltjes inequalities on each series separately. It is not unimaginable that in a more realistic field theory, the S matrix would retain something of the same structure, but we would no longer be able to separate the positive measure and negative measure portions *a priori*. Thus there is some interest in the question, given a function of the form (170), can we construct two functions $A(x)$ and $B(x)$ such that

$$S_{\alpha\beta}(x) = A(x) - B(x)\,, \tag{172}$$

where $A(x)$ and $B(x)$ are each series of Stieltjes. If the radius of convergence is finite [upper limit in (170) is finite], then this problem has been completely solved by Riesz and Hausdorf. We follow the summary given by Wall (1948). He shows that a $d\varphi_{\alpha\beta}(u)$ exists such that

$$\begin{aligned} S_{\alpha\beta}(x) &= \sum_{p=0}^{\infty} c_p(-x)^p\,, \\ c_p &= \int_0^1 u^p\, d\varphi_{\alpha\beta}(u)\,, \\ &\int_0^1 |\,d\varphi_{\alpha\beta}(u)\,| < M_{\alpha\beta} \end{aligned} \tag{173}$$

if and only if the constants c_p satisfy the inequalities

$$M_n = \sum_{p=0}^{n} \binom{n}{p} |\,\Delta^{n-p}c_p\,| \leq M, \qquad n = 0, 1, 2, \ldots\,, \tag{174}$$

where $\binom{n}{p}$ is the usual binomial coefficient, and M is independent of n. The functions $A(x)$ and $B(x)$ may be constructed as

$$A(x) = \sum_{p=0}^{\infty} a_p(-x)^p\,, \qquad B(x) = \sum_{p=0}^{\infty} b_p(-x)^p\,, \tag{175}$$

$$a_p = \lim_{n\to\infty} a_p^{(n)}\,, \qquad b_p = \lim_{n\to\infty} b_p^{(n)}\,, \tag{176}$$

where

$$a_p^{(n)} = \tfrac{1}{2}(\sigma_p^{(n)} + c_p)\,, \qquad b_p^{(n)} = \tfrac{1}{2}(\sigma_p^{(n)} - c_p) \tag{177}$$

and the limits have been proved to exist. We differ from Wall in the definition of $\sigma_p^{(n)}$ but each set has the same limit as n goes to infinity.

$$\sigma_p^{(n)} = \sum_{j=0}^{n-p} \binom{n-p}{j} |\Delta^{n-p-j} c_{p+j}| \qquad (p \leq n)\,. \tag{178}$$

Our $\sigma_p^{(n)}$ depends only on $c_p, \ldots, c_n$. Each approximant set of coefficients $\{a_p^{(n)}, b_p^{(n)};\ p = 0, 1, \ldots, n\}$ is totally monotone, the condition that it correspond to a $d\varphi \geq 0$.

If we wish to consider the case where the radius of convergence tends to zero it is only necessary to take the limit of the above results as $R \to 0$, however care must be taken to take account of n of the order of $1/R$. Let us define

$$\begin{aligned} D_\nu(c_p) &= \int_0^\infty e^{-\nu u} u^p \, d\varphi(u) \\ &= c_p - \frac{\nu}{1!} c_{p+1} + \frac{\nu^2}{2!} c_{p+2} - \frac{\nu^3}{3!} c_{p+3} + \cdots\,, \end{aligned} \tag{179}$$

where $\varphi(u)$ is the measure function, assumed to be of bounded variation. (Note that $D_\nu(c_p) \geq 0$ for all ν and p is equivalent to $d\varphi(u) \geq 0$ for all u.) In addition define

$$\sigma_p^{(\nu)} = \sum_{m=0}^{\infty} \frac{\nu^m}{m!} |D_\nu(c_{p+m})|\,. \tag{180}$$

(Note that, if $c_0, c_1, \ldots,$ is a series of Stieltjes, all the $D_\nu > 0$ and $\sigma_p^{(\nu)} \equiv c_p$.) Now let

$$\sigma_p = \lim_{\nu\to\infty} [\sigma_p^{(\nu)}]\,. \tag{181}$$

This limit exists, if the c_j are of the required form, and $|d\varphi|$ vanishes sufficiently fast at infinity. We can show that

$$\int_u^\infty |d\varphi| < \Gamma e^{-\varepsilon u}\,, \qquad \varepsilon > 0\,, \tag{182}$$

is a sufficient condition for the existence of (181). (The c_p may diverge like $p!$ in this case.) The functions $A(x)$ and $B(x)$ are defined by (175) and

$$a_p = \tfrac{1}{2}(\sigma_p + c_p)\,, \qquad b_p = \tfrac{1}{2}(\sigma_p - c_p) \tag{183}$$

Consequently it is possible to construct A and B from the series coefficients alone and thus as each function is summable by Padé methods, a sequence of c_p's, which does not diverge too rapidly, defines a unique and in principle determined function. [We wish to acknowledge several helpful discussions, with Dr. J. L. Gammel, of the material discussed in Eq. (172)–(183).]

REFERENCES

Baker, G. A., Jr. (1965). "Advances in Theoretical Physics" (K. A. Brueckner, ed.), Vol. 1, pp. 1–58. Academic Press, New York.

Baker, G. A., Jr. (1967). *Phys. Rev.* **161,** 434.

Baker, G. A., Jr. (1968). *Phys. Rev. Letters* **20,** 990.

Baker, G. A., Jr. (1969). *J. Math. Phys.* **10,** 814.

Baker, G. A., Jr. and Chisholm, J. S. R. (1966). *J. Math. Phys.* **7,** 1900.

Baker, G. A. Jr., Gammel, J. L., and Wills, J. G. (1961). *J. Math. Anal. Appl.* **2,** 405.

Baker, G. .A. Jr., Gilbert, H. E., Eve, J., and Rushbrooke, G. S. (1967). *Phys. Rev.* **164,** 800.

Berstein, F. (1928). *Acta Math.* **51,** 56.

Chisholm, J. S. R. (1963). *J. Math. Phys.* **4,** 1506.

Chisholm, J. S. R. (1966). *J. Math. Phys.* **7,** 39.

Churchill, R. V. (1944). "Modern Operational Mathematics in Engineering." McGraw-Hill, New York.

Common, A. K. (1968). *J. Math. Phys.* **9,** 32.

Copson, E. T. (1948). "An Introduction to the Theory of Functions of a Complex Variable." Oxford Univ. Press, London.

de Montessus de Balloire, R. (1902). *Bull. Soc. Math. France* **30,** 28.

Domb, C. (1960). *Phil. Mag. Suppl.* **9,** 149.

Fisher, M. E. (1963). *J. Math. Phys.* **4,** 278.

Fisher, M. E. (1967). *Repts. Prog. Phys.* **30** (part II), 615.

Franklin, P. (1940). "A Treatise on Advanced Calculus." Wiley, New York.

Gammel, J. L. and McDonald, F. A. (1966). *Phys. Rev.* **142,** 1245.

Gargantini, J. and Henrici, P. (1967). *Math. Comput.* **21,** 18.

Gordon, R. G. (1968). *J. Math. Phys.* **9,** 1087.

Gronwall, T. H. (1932). *Ann. Math.* **33,** 101.

Hadamard, J. (1892). *J. de Math.* (4) **8,** 101.

Hamburger, H. (1920–1921). *Math. Annal.* **81,** 235; **82,** 120, 168, as reviewed by Shahat, J. A., and Tamarkin, J. D. (1943). "*The Problem of Moments*," Mathematical Surveys, No. I. American Mathematical Society, New York.

Hardy, G. H. (1956). "Divergent Series." Oxford Univ. Press, London and New York.

Herglotz, G. (1911). *Leipzig Ber.* **63,** 501.

Jeffreys, H. and Jeffreys, B. S. (1950). "Methods of Mathematical Physics." Cambridge University Press, London.

Kadanoff, L. P., Götze, W., Hamblen, D., Hecht, R., Lewis, E. A. S., Palciauskas, V. V., Razl, M., Swift, J., Aspnes, D., and Kane, J. (1967). *Rev. Mod. Phys.* **39,** 395.
Macdonald, J. R. (1964). *J. Appl. Phys.* **35,** 3034.
Milne-Thompson, L. M. (1951). "The Calculus of Finite Differences." Macmillan, New York.
Muir, T. (1960). "A Treatise on the theory of Determinants" (revised by W. H. Metzer), p. 372. Dover, New York.
Nevanlinna, R. (1919). *Ann. Acad. Sci. Fennicae, Ser. A* **13,** 1.
Nuttall, J. (1967). *Phys. Rev.* **157,** 1312.
Nuttall, J., (1968). Private communication.
Padé, H. (1892). *Ann. Ecole Super. Normale* **9,** Suppl., 1.
Peres, A. (1963). *J. Math. Phys.* **4,** 332.
Pfluger, A., and Henrici, P. (1966). Contemporary problems in the theory of analytic functions (in Russian). *Int. Conf. Theory of Analytic Functions, Erevan, 1965,* M. A. Lavrent'ev, (ed). Izdat. "Nauka," Moscow, 1966.
Riesz, F. (1911). *Ann. Ecole. Super. Normale* **28,** 34.
Tani, S. (1965). *Phys. Rev.* **139,** B1011.
Wall, H. S. (1948). "Analytic Theory of Continued Fractions." Van Nostrand, Princeton, New Jersey.
Walsh, J. L. (1967). *SIAM J. Numer. Anal.* **4,** 211.
Wynn, P. (1960). *Math. Comput.* **14,** 147.
Yang, C. N., and Lee, T. D. (1952). *Phys. Rev.* **87,** 404, 410.

CHAPTER 2 APPLICATION OF PADÉ APPROXIMANTS TO DISPERSION FORCE AND OPTICAL POLARIZABILITY COMPUTATIONS*

P. W. Langhoff[†] *and M. Karplus*

Department of Chemistry, Harvard University

I. INTRODUCTION

The investigation of long-range interactions between atomic and molecular systems has long been a fruitful and active field of research (Hirschfelder and Meath, 1967). Although a variety of techniques have been employed for the quantitative estimation of the van der Waals (dispersion force) contribution (Pauly and Toennies, 1965; Bernstein and Muckerman, 1967; Dalgarno and Davison, 1966; Dalgarno, 1967) to such interactions, theoretical methods for obtaining error bounds for the estimates have been introduced only recently. One approach utilizes Padé approximants for obtaining the error bounds (Langhoff and Karplus, 1967). It is our purpose here to describe the Padé approximant technique, with emphasis on the leading (dipole-dipole) dispersion force term (Langhoff and Karplus, 1970b), and to demonstrate its applicability to the computation of the closely related optical polarizabilities as well (Langhoff and Karplus, 1970a).

The starting point of the present treatment is the replacement of the standard perturbation theory expression for the dispersion force (Eisenschitz and London, 1930) between a pair of nondegenerate, nonoverlapping atoms by the Casimir–Polder (1948) formula involving an integral over imaginary frequencies of the appropriate dynamic polarizabilities for the interacting species; dipole polarizabilities for the dipole-dipole term, dipole and quadru-

* Supported in part by a grant from the National Science Foundation.

[†] Present address: Department of Chemistry, Indiana University, Bloomington, Indiana 47401.

pole polarizabilities for the dipole-quadrupole term, and so on. The dynamic (dipole) polarizability of an atom or molecule is given by the familiar Kramers–Heisenberg dispersion formula (Slater, 1960), which can be expressed as an integral over the (dipole) oscillator-strength distribution of the system. The properties of the integrand serve to identify the polarizability as a series of Stieltjes (Baker, 1965), for which bounds are provided by Padé approximants determined from the moments of the oscillator-strength distribution. Introduction of the Padé approximants into the Casimir–Polder integral formula gives the desired bounded estimates for the dispersion force coefficients.

In Sect. II, we present the necessary formalism for the dispersion force, demonstrate that the polarizability is a series of Stieltjes, and show how the Padé approximant to this series leads to upper and lower bounds on the polarizability at imaginary frequencies. Attention is given not only to the construction of bounds to the dipole-dipole dispersion force, but also to the general multipole-multipole interaction, the three-body dipole interaction, and the leading relativistic correction to the two-body interaction. In Sect. III, we concern ourselves with the computational aspects of the Padé approximant technique, beginning with the determination of the necessary moments of the oscillator-strength distribution. For one- and two-electron atoms, we obtain accurate moment values by first-order time-dependent perturbation theory. For more complex atoms (inert gases, alkali atoms) and for diatomic molecules, measured optical dispersion or absorption data are used to evaluate the four lowest-order moments, a fitting procedure in which the Stieltjes determinantal constraints (Baker, 1965) play an important role. From the computed moments for the one- and two-electron atoms, the Padé approximants are obtained by the standard algorithm, and their roots and residues (effective transition frequencies and oscillator strengths, respectively) are determined for comparison with the experimental values. Some of the numerical difficulties that can arise in applications of the Padé procedure are considered. Section IV is concerned with the results obtained in utilizing the Padé approximants to find bounds to polarizabilities and dispersion force coefficients. For the polarizability at real frequencies, lower bounds are obtained from the customary Padé approximants. The effectiveness of the Padé approximant in summing the Cauchy series for the polarizability within its radius of convergence is illustrated for atomic hydrogen and helium, as well as for more complex systems; comparisons are made with both theoretical and experimental dispersion curves (Langhoff and Karplus, 1969). By a variable transformation (Reisz–Herglotz–Gronwall transformation) (Baker, 1965) and the newer method of Common (Langhoff and Karplus, 1969; Common, 1968; Baker, 1969), Padé approximants are used to find upper bounds to the polarizability along the real axis. Along the imaginary frequency axis, the rapidly convergent upper and lower bounds on the polarizability for atomic hydrogen and helium are described. The convergence of the associated dispersion force coefficients is also demonstrated for these two systems, as well as for more complex atoms. Comparisons are made where possible with dispersion force coefficients inferred from molecular beam scattering measurements and with previous theoretical estimates. Numerical results are also given for three-body dipole interaction coefficients and for the leading relativistic correction to the two-body dispersion interaction. In Sect. V, a number of alternative methods for obtaining bounded estimates of dispersion force coefficients are discussed. Emphasis is given to the relation between these procedures and the Padé approximant technique. Some concluding remarks are presented in Sect. VI.

II. FORMALISM

We present here the formalism relating the dispersion force to the frequency-dependent polarizabilities and use the fact that the latter form series of Stieltjes to obtain bounds on the former by means of Padé approximants.

A. The Eisenschitz–London (1930) and Casimir–Polder (1948) Dispersion Force Expressions

The second-order perturbation energy corresponding to the dipole-dipole interaction between two nonoverlapping, spherically symmetric (closed-shell) atoms a and b is (atomic units are used throughout)

$$E^{(2)}(a, b) = -C_{ab}/R_{ab}^6, \tag{1}$$

where R_{ab} is the interatomic distance and C_{ab} is the dispersion force coefficient given by

$$C_{ab} = \frac{3}{2} \sum_{n\neq 0} \sum_{m\neq 0} \frac{f_{n0}^{(a)} f_{m0}^{(b)}}{\omega_{n0}^{(a)} \omega_{m0}^{(b)} (\omega_{n0}^{(a)} + \omega_{m0}^{(b)})}. \tag{2}$$

The dipole oscillator strength $f_{n0}^{(a)}$ appearing in Eq. (2) is defined by

$$f_{n0}^{(a)} = \tfrac{2}{3}\omega_{n0}^{(a)} |\langle \phi_0^{(a)} | \mathbf{r} | \phi_n^{(a)} \rangle|^2 \tag{3a}$$

for the dipole transition from the ground state $\phi_0^{(a)}$ to the excited state $\phi_n^{(a)}$, with $\omega_{n0}^{(a)}$ the associated transition frequency and $\mathbf{r}$ the total dipole moment operator. The summation symbols in Eq. (2) signify either summation over the discrete portion of the spectrum or integration over the continuum. It is evident that the formal expression of Eq. (2) implies a double summation or integration over the entire oscillator-strength distributions of the interacting species.

An alternative expression for C_{ab} can be obtained by introduction of the dynamic dipole polarizability, $\alpha(z)$, of an atomic system. From the Kramers–Heisenberg formula (Slater, 1960) we have

$$\alpha(z) = P\int_0^\infty \frac{(df/d\varepsilon)}{\varepsilon^2 - z^2}\,d\varepsilon\,, \tag{4}$$

where the complete oscillator-strength distribution, $df/d\varepsilon$, is

$$df/d\varepsilon = \sum_{n\neq 0}^{\infty} f_{n0}\delta(\omega_{n0} - \varepsilon) + dg/d\varepsilon\,, \tag{5}$$

with $dg/d\varepsilon$ corresponding to the continuum portion of the spectrum; that is, for the continuum Eq. (3a) is replaced by

$$dg/d\varepsilon = \tfrac{2}{3}\varepsilon\,|\langle\phi_0^{(a)}|\,\mathbf{r}\,|\phi(\varepsilon)^{(a)}\rangle|^2\,, \tag{3b}$$

where $\phi(\varepsilon)^{(a)}$ is the continuum function at energy $(\varepsilon + \varepsilon_0^{(a)})$. In Eq. (4), the principle value of the integral is taken when the variable z falls within an absorption band.

From Eqs. (3) and (4) and the identity

$$\frac{1}{r+s} = \frac{2}{\pi}\int_0^\infty \frac{rs\,d\omega}{(r^2+\omega^2)(s^2+\omega^2)}\,, \qquad r, s > 0\,, \tag{6}$$

which can be proved by complex integration, we have

$$C_{ab} = \frac{3}{\pi}\int_0^\infty \alpha_a(iy)\alpha_b(iy)\,dy\,, \tag{7}$$

the so-called Casimir–Polder formula relating the dispersion energy coefficient to an integral of polarizabilities over imaginary frequencies.

One advantage of Eq. (7) over Eq. (2) is that the former focuses attention on the "one-center" polarizabilities of Eq. (4), whereas the latter is a "two-center" problem. Furthermore, the polarizability at imaginary frequencies

$$\alpha(iy) = \int_0^\infty \frac{(df/d\varepsilon)}{\varepsilon^2 + y^2}\,d\varepsilon \tag{8a}$$

has certain very useful properties in the interval $0 \leq \mathrm{Re}(y) < \infty$, as described below.

B. The Polarizability as a Series of Stieltjes

The essential property of the polarizability expression, Eq. (8a), is that it is a series of Stieltjes, as can be demonstrated by a number of procedures

(Langhoff and Karplus, 1970a; Baker, 1965). We introduce the variable $u = 1/\varepsilon^2$ $(0 \le \varepsilon, u < \infty)$, in terms of which the polarizability can be written as

$$\alpha(iy) = \int_0^\infty \frac{d\phi(u)}{1 + y^2 u}, \tag{8b}$$

where

$$\phi(u) = \int_0^u (u^{-1/2}/2)(df/d\varepsilon)\, du . \tag{9}$$

From the definition of the oscillator strength f_{n0} [Eq. (3)] and the fact that we are concerned with the ground state $(\omega_{n0} > 0)$, it is evident that $df/d\varepsilon$ given by Eq. (5) is positive for all values of ε in the interval $0 \le \varepsilon < \infty$. This insures that the distribution function $\phi(u)$ is a nondecreasing function of u. Making a power-series expansion of Eq. (8b) (in the variable y^2) about the origin, we have

$$\alpha(iy) = \sum_{k=0}^{\infty} \mu_k(-y^2)^k , \tag{10}$$

where the moments, μ_k, are given by

$$\mu_k = \int_0^\infty u^k\, d\phi(u) , \qquad k = 0, 1, 2, \ldots . \tag{11}$$

Thus, the polarizability $\alpha(iy)$ is expressed in the standard form and is demonstrated to be a series of Stieltjes (Baker, 1965). In terms of the variable ε and the oscillator-strength distribution, $df/d\varepsilon$, the moments μ_k can be written as

$$\begin{aligned} \mu_k &= \int_0^\infty \varepsilon^{-(2k+2)}(df/d\varepsilon)\, d\varepsilon , \qquad k = 0, 1, 2, \ldots \\ &= \sum_{n=1}^{\infty} \frac{f_{n0}}{\omega_{n0}^{2k+2}} + \int_{\omega_\infty}^{\infty} \frac{(dg/d\varepsilon)\, d\varepsilon}{\omega(\varepsilon)^{2k+2}}, \qquad \omega_\infty = \text{threshold.} \end{aligned} \tag{12}$$

The moments μ_k are seen to be the Cauchy coefficients of the commonly used expansion for the frequency-dependent polarizability or the closely related refractive index (Korff and Breit, 1932). We shall exhibit, in a subsequent section (Sect. III), procedures whereby the moments appearing in Eqs. (11) and (12) can be computed theoretically or inferred from experimental dispersion or absorption data. For the present, we assume that the necessary moments are available.

Given the series of Stieltjes, Eq. (10), which converges within the interval $0 \le |y^2| < \omega_{10}^2$, where ω_{10} is the resonance transition frequency, we introduce the $[n, m]_\alpha$ Padé approximant (Baker, 1965)

$$[n, m]_\alpha = P_m(y)/Q_n(y) , \tag{13}$$

which furnishes the analytic continuation of Eq. (10) outside the radius of convergence and effectively sums the series within the radius of convergence. The polynomials $P_m(y)$ and $Q_n(y)$ are defined by the relations

$$P_m(y) = \sum_{i=0}^{m} a_i(-y^2)^i , \tag{14a}$$

$$Q_n(y) = 1 + \sum_{i=1}^{n} b_i(-y^2)^i , \tag{14b}$$

where the $(n + m + 1)$ coefficients a_i, b_i can be determined by requiring that the power-series expansion of Eq. (13) (in the variable y^2) equal Eq. (10) term-by-term to order $n + m$; equivalently, we set

$$P_m - Q_n\left[\sum_{k=0}^{n+m} \mu_k(-y^2)^k\right] = 0 \tag{15}$$

term-by-term to order $n + m$. Of particular utility for the present discussion are the $[n, n]_\alpha$ and $[n, n - 1]_\alpha$ approximants (i.e., $[n, n + k]_\alpha$, $k = 0$, -1 in Baker's (1965) notation), which satisfy the inequalities

$$[n, n]_\alpha \geq \alpha(iy) \geq [n, n - 1]_\alpha , \qquad 0 \leq \mathrm{Re}(y^2) < \infty . \tag{16}$$

Moreover, $[n, n]_\alpha$ and $[n, n - 1]_\alpha$, respectively, furnish the best upper and lower bounds on the polarizability which can be obtained from the given moment values. On the negative real y^2 axis, the inequalities

$$[n, n]_\alpha , [n, n - 1]_\alpha \leq \alpha(\omega) , \qquad 0 \geq \mathrm{Re}(y^2) > -\omega_{10}^2 , \tag{17}$$

are satisfied, so that lower bounds on the polarizability for real frequency $(\omega^2 = -y^2)$ are available. In a subsequent section, we consider the construction of accompanying upper bounds on the polarizability at real frequency $(0 \leq \omega < \omega_{10})$.

For the $[n, n]_\alpha$ and $[n, n - 1]_\alpha$ approximants, Eq. (15) results in the linear inhomogeneous equations

$$\sum_{j=1}^{n} A_{ij}b_j = -\mu_{n+k+i} , \qquad i = 1, 2, \ldots, n , \quad k = 0, -1 , \tag{18}$$

with

$$A_{ij} = \mu_{n+k+i-j} , \qquad k = 0, -1 , \tag{19}$$

from which the coefficients b_j can be obtained. The a_i coefficients are given by $(b_0 = 1)$

$$a_i = \sum_{j=0}^{i} \mu_{i-j}b_j , \qquad i = 0, 1, \ldots, n + k , \quad k = 0, -1 , \tag{20}$$

which follow from Eq. (15). It is of importance to note that the existence of nontrivial solutions to Eqs. (18) is insured by the Stieltjes conditions (Baker, 1965)

$$D(n, m) = \begin{vmatrix} \mu_n & \mu_{n+1} & \cdots & \mu_{n+m} \\ \mu_{n+1} & \mu_{n+2} & \cdots & \mu_{n+m+1} \\ \vdots & \vdots & & \vdots \\ \mu_{n+m} & \mu_{n+m+1} & \cdots & \mu_{n+2m} \end{vmatrix} > 0, \qquad m = 0, 1, \ldots; \tag{21}$$

i.e., Eq. (21) guarantees that the matrix (A_{ij}) appearing in Eqs. (18) is nonsingular.

C. Bounds on Dipole-Dipole Dispersion Force Coefficients

From the Casimir–Polder formula [Eq. (7)] and the inequalities [Eq. (16)], the Padé approximant bounds to dipole-dipole dispersion force coefficients take the form

$$\frac{3}{\pi}\int_0^\infty [n, n]_\alpha^{(a)} [n, n]_\alpha^{(b)}\, dy \geq C_{ab}, \qquad n = 1, 2, \ldots, \tag{22a}$$

$$\frac{3}{\pi}\int_0^\infty [n, n-1]_\alpha^{(a)} [n, n-1]_\alpha^{(b)}\, dy \leq C_{ab}, \qquad n = 1, 2, \ldots. \tag{22b}$$

Equation (22b) can be cast into a particularly simple and suggestive form by a partial fraction reduction for $[n, n-1]_\alpha$, which leads to the expression

$$[n, n-1]_\alpha = \sum_{i=1}^{n} \frac{f_i}{\omega_i^2 + y} \leq \alpha(iy), \tag{23}$$

where the f_i and ω_i are "effective" oscillator strengths and transition frequencies, respectively; that is, the correspondence between Eq. (23) and the Kramers–Heisenberg dispersion formula [Eqs. (4) and (5)] shows that the $[n, n-1]_\alpha$ Padé approximant introduces a fictitious finite spectrum to approximate the exact spectrum of the atom. The f_i and ω_i appearing in Eq. (23) are upper bounds to the first n oscillator strengths f_{n0} and transition frequencies ω_{n0} of the real system (see Baker, 1965, pp. 10–16). Substituting Eq. (23) into Eq. (22b), we have

$$\frac{3}{2}\sum_{i=1}^{na}\sum_{j=1}^{nb} \frac{f_i^{(a)} f_j^{(b)}}{\omega_i^{(a)}\omega_j^{(b)}(\omega_i^{(a)} + \omega_j^{(b)})} \leq C_{ab}, \tag{24a}$$

an expression corresponding in form to the exact result given in Eq. (2). Although the equality holds strictly only in the limit, $n \to \infty$, for which the exact polarizability on the imaginary axis and that of the Padé approximant coincide, it will be shown subsequently (Sect. IV) that rapid numerical convergence is obtained from Eq. (24a). In fact, accurate dispersion force coefficients result from a fictitious spectrum that is a rather poor approximation to the actual spectrum.

For the upper bounds of Eq. (22a), a reduction similar to Eqs. (23) and (24a) cannot be achieved, since the $[n, n]_\alpha$ approximants do not satisfy the Thomas–Reiche–Kuhn sum rule constraint (Langhoff and Karplus, 1967)

$$\alpha(iy \to \infty) = \int_0^\infty \frac{(df/d\varepsilon)\, d\varepsilon}{y^2} = \frac{N}{y^2} \to 0\,, \tag{25}$$

where N is the number of electrons in the system. The approximant $[n, n]_\alpha$ instead approaches the finite value (a_n/b_n) in the limit $iy \to \infty$, and the upper bounds obtained by direct use of Eq. (22a) diverge to ∞. One way of modifying Eq. (22a) to obtain a practical bound is to introduce the asymptotic form for $\alpha(iy)$ given by Eq. (25), which itself forms an upper bound to $\alpha(iy)$ in the interval $0 \le \mathrm{Re}\,(y^2)$; that is, we join the $[n, n]_\alpha$ approximant to the function N/y^2 at their crossing point and obtain a piecewise function that is a finite upper bound to $\alpha(iy)$ over the required frequency range and obeys the global constraint of Eq. (25), as well (Langhoff and Karplus, 1967). The integral of Eq. (22a), constructed from functions with such a "piecewise" form yields useful results but is somewhat awkward, and an alternate procedure is desirable. This can be obtained by considering the function (Langhoff and Karplus, 1970b)

$$f(iy) = N - y^2\alpha(iy) = \int_0^\infty \frac{\varepsilon^2(df/d\varepsilon)}{\varepsilon^2 + y^2}\, d\varepsilon\,, \tag{26}$$

which can be shown to be a series of Stieltjes, since $\varepsilon^2(df/d\varepsilon)$ is a positive function in the range $0 \le \varepsilon < \infty$. Consequently,

$$[n, n-1]_f \le f(iy) = N - y^2\alpha(iy) \tag{27}$$

or

$$(N - [n, n-1]_f)/y^2 \ge \alpha(iy)\,. \tag{28}$$

Explicit consideration of the upper bound in Eq. (28) shows that it can be expressed in a form similar to that of Eq. (23); that is,

$$(N - [n, n-1]_f)/y^2 = \sum_{i=1}^{n} \frac{\bar{f}_i}{\bar{\omega}_i^{\,2} + y^2} \ge \alpha(iy)\,, \tag{29}$$

where the $\bar{f}_i$ and $\bar{\omega}_i$ appearing in Eq. (29) are, of course, distinct from those of Eq. (23). From Eq. (29), it is evident that the alternative bounding function has the proper behavior as $y \to \infty$ since $\sum_{i=1}^{n} \bar{f}_i = N$. Making use of

Eqs. (7) and (29), we have

$$\frac{3}{2}\sum_{i=1}^{n_a}\sum_{j=1}^{n_b}\frac{\bar{f}_i^{(a)}\bar{f}_j^{(b)}}{\bar{\omega}_i^{(a)}\bar{\omega}_j^{(b)}(\bar{\omega}_i^{(a)}+\bar{\omega}_j^{(b)})}\geq C_{ab}\,. \tag{24b}$$

Numerical results obtained with the equations derived in this section are presented in Sect. IV, after a consideration of certain computational aspects of the evaluation of the moment coefficients and the Padé algorithm (Sect. III).

D. Bounds on Other Dispersion Force Coefficients

In the foregoing, we have restricted the discussion to the leading (dipole-dipole) dispersion interaction term between nondegenerate systems. The bounding procedures we have developed, however, are applicable to the higher-order multipole interaction terms, to three-, four-, and more-body nonadditive dispersion force terms, and to additional terms that arise from relativistic effects.

The complete electrostatic interaction Hamiltonian for atoms a and b can be written as

$$H^{(1)}(a,b)=\frac{Z_aZ_b}{R_{ab}}-\sum_{i=1}^{N_a}\frac{Z_b}{r_{bi}}-\sum_{j=1}^{N_b}\frac{Z_a}{r_{aj}}+\sum_{i=1}^{N_a}\sum_{j=1}^{N_b}\frac{1}{r_{ij}}, \tag{30}$$

where r_{bi} and r_{aj} are the distances of electrons i and j from the atom centers b and a, respectively, the atoms a, b have nuclear charges Z_a, Z_b, and they have electrons $i = 1, N_a$, $j = 1, N_b$. Expansion of Eq. (30) for large R_{ab} in a Taylor series around a and b yields the multipole series (Dalgarno and Davison, 1966; Dalgarno, 1967)

$$H^{(1)}(a,b)=\sum_{l_a=0}^{\infty}\sum_{l_b=0}^{\infty}V(l_a,l_b)/R_{ab}^{l_a+l_b+1}, \tag{31}$$

where

$$V(l_a,l_b)=\sum_{m=-\min(l_a,l_b)}^{\min(l_a,l_b)}a(l_a,l_b,m)\left[\sum_{i=1}^{N_a}r_i^{l_a}Y_{l_a}^{m}(i)\right]\left[\sum_{j=1}^{N_b}r_j^{l_b}Y_{l_b}^{-m}(j)\right], \tag{32}$$

and

$$a(l_a,l_b,m)=\frac{(-1)^{l_b}4\pi(l_a+l_b)!}{[(2l_a+1)(2l_b+1)(l_a-m)!\,(l_a+m)!\,(l_b-m)!\,(l_b+m)!]^{1/2}}. \tag{33}$$

Right-hand coordinate systems are used on both atoms with the z axes coinciding with the internuclear line. For neutral atoms, the leading term of the multipole series [Eq. (31)] is the dipole-dipole term ($l_a = l_b = 1$)

that we have considered in Sects. II. A–C. The complete second-order energy for the operator $H^{(1)}(a, b)$ can be obtained by perturbation theory with the assumption that the complete set of functions associated with the excited states of atoms a and b, respectively, are nonoverlapping. Restricting attention to spherically symmetric systems for simplicity, we find

$$E^{(2)}(a, b) = -\sum_{l_a=1}^{\infty} \sum_{l_b=1}^{\infty} C^{(2)}(l_a, l_b) / R_{ab}^{2(l_a+l_b+1)}, \tag{34}$$

where

$$C^{(2)}(l_a, l_b) = b(l_a, l_b) \sum_{k\neq 0} \sum_{l\neq 0} \frac{f_{k0}(l_a) f_{l0}(l_b)}{\omega_{k0}^{(a)} \omega_{l0}^{(b)} (\omega_{k0}^{(a)} + \omega_{l0}^{(b)})}, \tag{35}$$

and

$$b(l_a, l_b) = \frac{(2l_a + 2l_b)!}{4(2l_a)!\,(2l_b)!}. \tag{36}$$

In Eq. (35), the $2l_a$th pole oscillator strength has the form

$$f_{k0}(l_a) = \frac{8\pi}{2l_a + 1} \omega_{k0}^{(a)} \left| \langle \phi_0^{(a)} | \sum_{i=1}^{N_a} r_i^{l_a} Y_{l_a}^0(i) \, | \phi_k^{(a)} \rangle \right|^2, \tag{37}$$

with $\omega_{k0}^{(a)}$ the associated transition frequency; a corresponding expression holds for $f_{l0}(l_b)$. Introducing the multipole polarizability

$$\alpha_l(z) = \sum_k \frac{f_{k0}(l)}{\omega_{k0}^2 - z^2}, \tag{38}$$

we see that

$$C^{(2)}(l_a, l_b) = b(l_a, l_b) \frac{2}{\pi} \int_0^{\infty} \alpha_{l_a}(iy) \alpha_{l_b}(iy)\, dy, \tag{39}$$

which is a generalization of the Casimir–Polder integral formula of Eq. (7).

Since Eq. (38) can be recognized to be a series of Stieltjes, the bounding procedures developed for the C_{ab} dipole-dipole coefficient applies to the general multipole-multipole interaction term as well. As in the dipole case, Cauchy moment coefficients for the multipole polarizability of Eq. (38) are needed to carry out the Padé procedure. Because these are difficult to obtain experimentally, the most practical method of determining their values is by generalization of the quantum-mechanical treatments described in Sects. III. A–C. We do not discuss them further in this review (Dalgarno and Davison, 1966; Dalgarno, 1967; McQuarrie *et al.*, 1969).

The nonseparable interaction potential for three nonoverlapping atomic systems was first considered in detail by Axilrod and Teller (1943) (Muto, 1943). They obtained the leading (third-order) dipole term for three spherical atoms in the form

$$E^{(3)}(a, b, c) = (1 + 3\cos\theta_{ab}\cos\theta_{ac}\cos\theta_{bc})\frac{C_{abc}}{R_{ab}^3 R_{ac}^3 R_{bc}^3}, \tag{40}$$

where

$$C_{abc} = \frac{3}{4}\sum_k\sum_l\sum_m \frac{f_{k0}^{(a)} f_{l0}^{(b)} f_{m0}^{(c)}}{\omega_{k0}^{(a)}\omega_{l0}^{(b)}\omega_{m0}^{(c)}}\left\{\frac{1}{(\omega_{k0}^{(a)} + \omega_{l0}^{(b)})}\frac{1}{(\omega_{k0}^{(a)} + \omega_{m0}^{(c)})} + \frac{1}{(\omega_{k0}^{(a)} + \omega_{l0}^{(b)})}\frac{1}{(\omega_{l0}^{(b)} + \omega_{m0}^{(c)})} + \frac{1}{(\omega_{k0}^{(a)} + \omega_{m0}^{(c)})}\frac{1}{(\omega_{l0}^{(b)} + \omega_{m0}^{(c)})}\right\}. \tag{41}$$

In analogy with the Casimir–Polder procedure, we can express Eq. (41) as an integral,

$$C_{abc} = \frac{3}{\pi}\int_0^\infty \alpha_a(iy)\alpha_b(iy)\alpha_c(iy)\,dy\,, \tag{42}$$

involving the frequency-dependent dipole polarizabilities of the three atoms. Equation (42) is in a form that allows construction of bounds to C_{abc} by use of the Padé approximants to the polarizabilities in a manner analogous to the two-body case. There is very little experimental information available concerning the three-body force except possibly in connection with third virial coefficients (Kihara, 1958; Graben and Present, 1962; Sherwood and Pransnitz, 1964) and in the condensed phase (Jansen, 1964; Meyer, 1969), for which the dispersion approximation is of questionable validity. An illustrative application of the bounding procedure is given in Sect. IV.

In the preceding, we have focused attention on the Coulomb Hamiltonian and obtained the familiar multipole interaction series. For some situations, this Hamiltonian is inadequate and relativistic effects must be included. The $(1/R)$ expansion for the interactions of two nondegenerate atoms with the relativistic correction obtained from the Breit–Pauli Hamiltonian has been developed by Meath and Hirschfelder (1966) (Hirschfelder and Meath, 1967) and others (Casimir and Polder, 1948). The leading relativistic term has the form

$$E_R^{(2)}(a, b) = D_{ab}\alpha^2/R_{ab}^4\,, \tag{43}$$

where α is the fine structure constant. The coefficient D_{ab} is given by

$$D_{ab} = \frac{1}{2}\sum_k\sum_l \frac{f_{k0}^{(a)} f_{l0}^{(b)}}{(\omega_{k0}^{(a)} + \omega_{l0}^{(b)})}, \tag{44}$$

where $f_{k0}^{(a)}, f_{l0}^{(b)}$ are the dipole oscillator strengths and $\omega_{k0}^{(a)}, \omega_{l0}^{(b)}$ are the associated transition frequencies. Introducing the function

$$\beta(z) = \sum_k \frac{\omega_{k0} f_{k0}}{(\omega_{k0}^2 - z^2)}\,, \tag{45}$$

we can write Eq. (44) in the form

$$D_{ab} = \frac{1}{\pi} \int_0^\infty \beta_a(iy)\beta_b(iy)\,dy\,, \tag{46}$$

analogous to the Casimir–Polder expression. We recognize $\beta(iy)$ of Eq. (45) as a series of Stieltjes, for which bounds can be constructed from the Padé approximants, given the appropriate moments (Chang and Karplus, 1970). From the power-series expansion

$$\beta(iy) = \sum_{k=0}^{\infty} \beta_k(-y^2)^k\,, \tag{47}$$

with

$$\beta_k = \sum_{n=1}^{\infty} \frac{f_{n0}}{\omega_{n0}^{2k+1}} + \int_{\omega_\infty}^{\infty} \frac{(dg/d\varepsilon)\,d\varepsilon}{\omega(\varepsilon)^{2k+1}}\,, \tag{48}$$

we see that the required moment coefficients, β_k, are distinct from the μ_k of Eq. (12). Values for the β_k cannot be determined directly from experiment. However, useful approximate values can be obtained from the general sum-rule expression

$$S(k) = \sum_i f_i \omega_i{}^k \tag{49}$$

and the formula

$$S(k) = S(0)\left[a + \frac{b}{(2.5-k)} + \frac{c}{(2.5-k)^2} + \frac{d}{(2.5-k)^3}\right]^k, \tag{50}$$

which has been suggested by Dalgarno and Kingston (1960) and by Bell (1965) as a reliable approximation to the sum-rule values. The coefficients a, b, c, and d are chosen to reproduce the moment values

$$\mu_k = S(-2k-2)\,, \qquad k = 0, 1, 2, \ldots, \tag{51}$$

and the β_k values are given by

$$\beta_k = S(-2k-1)\,. \tag{52}$$

While Eqs. (50) and (52) furnish useful approximate moments, some care must be exercised to insure that the resulting values satisfy the Stieltjes constraints [Eq. (21)] and the requirement that the series of Eq. (47) have a known radius of convergence. Moment coefficients which do not satisfy these requirements can be unsatisfactory (see Sects. III D and IV D).

As in the dipole case, the $[1, 0]_\beta$ and $[2, 1]_\beta$ approximants provide the first two lower bounds to $\beta(iy)$ and D_{ab}, whereas the function $\gamma(iy)$,

$$\gamma(iy) = \beta_{-1} - y^2\beta(iy) = \sum_{k=0}^{\infty} \beta_{k-1}(-y^2)^k , \tag{53}$$

provides upper bounds, with

$$\beta(iy) \leq (\beta_{-1} - [n, n-1]_\gamma)/y^2 . \tag{54}$$

In Sect. IV, we consider the inert gas interaction as examples. Experimental comparisons are not available, although relativistic corrections may be important in the appropriate beam scattering experiments or in other collision processes.

III. COMPUTATION PROCEDURES

Some techniques for evaluating the moments of oscillator-strength distributions are described in this section. We consider perturbation theory calculations applicable to simple atoms and molecules and discuss the analysis of measured refractivity or absorption data for arbitrary systems. Once the necessary moments are known, the appropriate Padé approximants can be obtained and their poles and residues can be determined.

A. Time-Dependent Perturbation Theory

The Kramers–Heisenberg dispersion formula [Eq. (4)] is customarily derived from the Schrödinger equation for the interaction between an atom and an electromagnetic wave by use of first-order, time-dependent perturbation theory. The first-order equation (Karplus and Kolker, 1963)

$$\left[h^{(0)}(\mathbf{r}) - i\frac{\partial}{\partial t}\right]\Psi^{(1)}(\mathbf{r}, t) + H^{(1)}(\mathbf{r}, t)\Psi^{(0)}(\mathbf{r}, t) = 0 \tag{55}$$

is solved by introducing the Dirac expansion

$$\Psi^{(1)}(\mathbf{r}, t) = \sum_{n\neq 0}^{\infty} a_n^{(1)}(t)\phi_n^{(0)}(\mathbf{r})\, e^{-i\varepsilon_n t} + \int_{\omega_\infty}^{\infty} a^{(1)}(\varepsilon, t)\phi^{(0)}(\mathbf{r}, \varepsilon)e^{-i(\varepsilon+\varepsilon_0)t}\, d\varepsilon . \tag{56}$$

For the perturbation Hamiltonian, $H^{(1)}(\mathbf{r}, t)$, the dipole approximation is used; e.g.,

$$H^{(1)}(\mathbf{r}, t) = \sum_{i=1}^{N} r_i \cos\theta_i \cos\omega t = h^{(1)}(\mathbf{r}) \cos\omega t \tag{57}$$

for an incident monochromatic plane wave of frequency ω polarized in the z direction. Evaluation of the $a_n^{(1)}(t)$ and $a^{(1)}(\varepsilon, t)$ from Eq. (55) and computation of the induced dipole moment to first-order gives the Kramers–Heisenberg formula [Eq. (4)], with the oscillator strength distribution defined in Eqs. (3) and (5). Once the oscillator–strength distribution has been determined, the moments can be calculated by Eq. (12). However, a difficulty with this formally correct approach is that, for most cases (except atomic hydrogen), the complete set of discrete and continuum state eigenfunctions is not known. If they were available, a direct calculation of the various properties considered here would be possible, so that the Padé approximant technique would not be required.

A computationally more convenient expression for the moments μ_k can be derived by direct expansion of the first-order perturbation function $\Psi^{(1)}(\mathbf{r}, t)$ [Eqs. (55) and (56)] (Podolsky, 1928; Dalgarno and Kingston, 1960; Musulin and Epstein, 1964; Langhoff and Karplus, 1970a). Separating out the time dependence of Eq. (55) by the ansatz (Karplus and Kolker, 1963)

$$\Psi^{(1)}(\mathbf{r}, t) = (\phi_+^{(1)}(\mathbf{r})e^{i\omega t} + \phi_-^{(1)}(\mathbf{r})e^{-i\omega t}), \tag{58}$$

we obtain the time-independent equations

$$(h^{(0)}(\mathbf{r}) - \varepsilon^{(0)} \pm \omega)\phi_\pm^{(1)}(\mathbf{r}) + h^{(1)}(\mathbf{r})\phi^{(0)}(\mathbf{r}) = 0\,. \tag{59}$$

If the functions $\phi_\pm^{(1)}(\mathbf{r})$, which depend implicitly on ω, are expressed as a power series in ω

$$\phi_\pm^{(1)} = \sum_{n=0}^{\infty} \chi_n^{(1)}(\pm)\omega^n\,, \tag{60}$$

and substituted into Eq. (59), there results the set of coupled linear inhomogeneous equations,

$$(h^{(0)} - \varepsilon^{(0)})\chi_0^{(1)}(\pm) + h^{(1)}\phi^{(0)} = 0\,, \tag{61a}$$

$$(h^{(0)} - \varepsilon^{(0)})\chi_n^{(1)}(\pm) \pm \chi_{n-1}^{(1)}(\pm) = 0\,, \qquad n = 1, 2, \ldots. \tag{61b}$$

Equations (61b) can be solved sequentially, starting with the solution to Eq. (61a).

The functions $\phi_\pm^{(1)}(\mathbf{r})$ can be used to write the frequency-dependent polarizability [Eq. (4)] in the form

$$\alpha(\omega) = \langle\phi_+^{(1)} \,|\, h^{(1)} \,|\, \phi_0\rangle + \langle\phi_-^{(1)} \,|\, h^{(1)} \,|\, \phi_0\rangle\,. \tag{62}$$

Substitution of the expansion [Eq. (60)] into Eq. (62) and comparison with Eq. (10) shows that the moments μ_k are given by

$$\mu_k = 2\langle\chi_{2k}^{(1)}(+) \,|\, h^{(1)} \,|\, \phi^{(0)}\rangle\,, \tag{63}$$

where we have made use of the relationship

$$\chi_n^{(1)}(-) = (-1)^n \chi_n^{(1)}(+) , \tag{64}$$

evident from Eqs. (61). The utility of Eqs. (61) and (63) will be illustrated by specific applications (see below).

B. Moment Computations for Atomic Hydrogen

The hydrogen atom furnishes a useful example and test of the procedure described in Sect. III, A. For the ground state, we have

$$h^{(0)} = -\tfrac{1}{2}\nabla^2 - r^{-1} , \qquad \phi^{(0)} = (\pi)^{-1/2} e^{-r} , \qquad \varepsilon^{(0)} = -\tfrac{1}{2} , \tag{65}$$

and, from Eq. (57),

$$h^{(1)} = r \cos\theta . \tag{66}$$

Substitution of Eqs. (65) and (66) into Eq. (61a) yields the exact solution

$$\chi_0^{(1)}(\pm) = \phi^{(0)} P_1(\cos\theta)\,(a_0^{(1)} r + a_0^{(2)} r^2) , \tag{67}$$

with $a_0^{(1)} = 1$ and $a_0^{(2)} = \frac{1}{2}$. From this result, the solutions to Eq. (61b) for all n are obtained in the form

$$\chi_n^{(1)}(+) = \phi^{(0)} P_1(\cos\theta) \sum_{i=1}^{n+2} a_n^{(i)} r^i , \tag{68a}$$

$$\chi_n^{(1)}(-) = (-1)^n \chi_n^{(1)}(+) , \tag{68b}$$

where the $a_n^{(i)}$ obey the recursion formulas ($n \geq 1$)

$$i a_n^{(i)} = [i(i+3)/2]\, a_n^{(i+1)} - a_{n-1}^{(i-1)} , \qquad i = 2, 3, \ldots, n+1 , \tag{69a}$$

$$a_n^{(n+2)} = (-1)^n/(n+2)! , \tag{69b}$$

$$a_n^{(1)} = 2a_n^{(2)} . \tag{69c}$$

Introducing these expressions for $\chi_n^{(1)}(+)$ into Eq. (63), we obtain the oscillator-strength moments for atomic hydrogen:

$$\mu_k = \frac{8}{3} \sum_{i=1}^{2k+2} a_{2k}^{(i)} \frac{(i+3)!}{2^{(i+4)}} . \tag{70}$$

We have calculated a large number of moments for atomic hydrogen by use of Eqs. (69) and (70); the first 15 moments are listed in the second column of Table I. As a useful numerical check, we have also computed μ_k values directly for atomic hydrogen by substituting into Eq. (12),

$$\mu_k = \sum_{n=1}^{\infty} f_{n0}/\omega_{n0}^{2k+2} + \int_{\omega_\infty}^{\infty} \frac{(dg/d\varepsilon)}{\omega(\varepsilon)^{2k+2}}\, d\varepsilon , \tag{12}$$

the known expressions (Harriman, 1956; Bethe and Salpeter, 1957) for the oscillator strengths and transition frequencies,

$$f_{n0} = \tfrac{16}{3}(n+1)^{-3}\omega_{n0}^{-4}[n/(n+1)]^{(2n+2)}, \tag{71a}$$

$$\omega_{n0} = \tfrac{1}{2}[1 - 1/(n+1)^2], \tag{71b}$$

$$\frac{dg}{d\varepsilon} = \frac{16}{3}\,\frac{\exp\{-4(2\varepsilon)^{-1/2}\operatorname{arc\,cot}[(2\varepsilon)^{-1/2}]\}}{\omega^4[1-\exp(-2\pi(2\varepsilon)^{-1/2})]}, \tag{71c}$$

$$\omega(\varepsilon) = \varepsilon + \tfrac{1}{2}. \tag{71d}$$

Table I exhibits the contributions to μ_k from the $2p$ state alone, from all of the discrete states, and from the continuum states. Evidently, the contribution to μ_k from the continuum decreases rapidly with increasing k, and the contribution from the $2p$ state alone dominates the high-order μ_k values. This is not unexpected, since the higher derivatives of an analytic function are determined largely by the residue of the singularity nearest the point of evaluation. Also shown in Table I are the ratios μ_k/μ_{k+1},

TABLE I

Cauchy Moments for Atomic Hydrogen[a]

k	$(\mu_k)^b$	Contribution to μ_k from[c]			$(\mu_k/\mu_{k+1})^d$
		$2p$ state	Discrete spectrum	Continuous spectrum	
0	0.450000×10^{1}	0.295962×10^{1}	0.366326×10^{1}	0.083674×10^{1}	0.169279
1	0.265833×10^{2}	0.210462×10^{2}	0.243948×10^{2}	0.021885×10^{2}	0.154385
2	0.172188×10^{3}	0.149662×10^{3}	0.165694×10^{3}	0.006494×10^{3}	0.148171
3	0.116209×10^{4}	0.106426×10^{4}	0.114145×10^{4}	0.002964×10^{4}	0.145077
4	0.801017×10^{4}	0.756808×10^{4}	0.794171×10^{4}	0.006846×10^{4}	0.143375
5	0.558685×10^{5}	0.538175×10^{5}	0.556346×10^{5}	0.002339×10^{5}	0.142378
6	0.392395×10^{6}	0.382702×10^{6}	0.391578×10^{6}	0.000817×10^{6}	0.141768
7	0.276787×10^{7}	0.272144×10^{7}	0.276502×10^{7}	0.000285×10^{7}	0.141383
8	0.195771×10^{8}	0.193524×10^{8}	0.195667×10^{8}	0.000104×10^{8}	0.141134
9	0.138713×10^{9}	0.137617×10^{9}	0.138675×10^{9}	0.000038×10^{9}	0.140970
10	0.983991×10^{9}	0.978613×10^{9}	0.983852×10^{9}	0.000139×10^{9}	0.140861
11	0.698555×10^{10}	0.695902×10^{10}	0.698504×10^{10}	0.000051×10^{10}	0.140787
12	0.496177×10^{11}	0.494864×10^{11}	0.496158×10^{11}	0.000019×10^{11}	0.140737
13	0.352556×10^{12}	0.351903×10^{12}	0.352349×10^{12}	0.000007×10^{12}	0.140703
14	0.250567×10^{13}	0.250242×10^{13}	0.250565×10^{13}	0.000002×10^{13}	0.140679
15	0.178112×10^{14}	0.177950×10^{14}	0.178111×10^{14}	0.000001×10^{14}	0.140663

[a] Coefficients appearing in the expansion, $\alpha(\omega) = \sum_{k=0}^{\infty} \mu_k(\omega^2)^k$, in atomic units.
[b] Values obtained from Eq. (70) in the text.
[c] Values obtained from Eqs. (12) and (71) in the text.
[d] The radius of convergence of the series is $R = \lim_{k\to\infty} \mu_k/\mu_{k+1} = \omega_{10}^2 = 0.140625$.

which are converging to ω_{10}^2, the radius of convergence of the series. We shall make use of the moments appearing in Table I to evaluate Padé approximants to the polarizability series in Sects. III. E and IV.

C. Moment Computations for Atomic Helium in the Hartree–Fock Approximation

For many-electron systems, the exact zero-order and first-order perturbed functions are generally not known. An approximate procedure that yields results of reasonable accuracy is provided by the coupled Hartree–Fock approximation (Dalgarno and Victor, 1966). For atomic helium, the first-order perturbation equation analogous to Eq. (59) is (Dalgarno, 1966)

$$(h^{(0)} - \varepsilon^{(0)} \pm \omega)\phi^{(1)}(\pm) + [h^{(1)} + v^{(1)}(\pm)]\phi^{(0)} = 0\,, \tag{72}$$

where

$$h^{(0)} = -\tfrac{1}{2}\nabla^2 - (2/r) + \langle\phi^{(0)}\,|\,(r_{12})^{-1}(2 - P_{12})\,|\,\phi^{(0)}\rangle\,, \tag{73a}$$

$$\begin{aligned} v^{(1)}(\pm) = {}& \langle\phi^{(1)}(\mp)\,|\,(r_{12})^{-1}(2 - P_{12})\,|\,\phi^{(0)}\rangle \\ &+ \langle\phi^{(0)}\,|\,(r_{12})^{-1}(2 - P_{12})\,|\,\phi^{(1)}(\pm)\rangle\,, \end{aligned} \tag{73b}$$

$$h^{(1)} = r\cos\theta\,. \tag{73c}$$

Here $\varepsilon^{(0)}$ and $\phi^{(0)}$ are the Hartree–Fock one-electron energy and orbital, respectively (Clementi, 1963), and P_{12} is the permutation operator. Expansion of the first-order perturbed orbital as in Eq. (60),

$$\phi^{(1)}(\pm) = \sum_{n=0}^{\infty} \chi_n^{(1)}(\pm)\omega^n\,, \tag{74}$$

and of the operator $v^{(1)}(\pm)$,

$$v^{(1)}(\pm) = \sum_{n=0}^{\infty} v_n^{(1)}(\pm)\omega^n\,, \tag{75}$$

with

$$\begin{aligned} v_n^{(1)}(\pm) = {}& \langle\chi_n^{(1)}(\mp)\,|\,(r_{12})^{-1}(2 - P_{12})\,|\,\phi^{(0)}\rangle \\ &+ \langle\phi^{(0)}\,|\,(r_{12})^{-1}(2 - P_{12})\,|\,\chi_n^{(1)}(\pm)\rangle\,, \end{aligned} \tag{76}$$

yields the equations

$$(h^{(0)} - \varepsilon^{(0)})\chi_0^{(1)}(\pm) + [h^{(1)} + v_0^{(1)}(\pm)]\phi^{(0)} = 0\,, \tag{77a}$$

$$(h^{(0)} - \varepsilon^{(0)})\chi_n^{(1)}(\pm) + v_n^{(1)}(\pm)\phi^{(0)} \pm \chi_{n-1}^{(1)}(\pm) = 0\,, \quad n = 1, 2, \ldots. \tag{77b}$$

The moment coefficients for helium have the form given in Eq. (63), with the $\chi_{2k}^{(1)}(+)$ obtained by solution of Eqs. (77).

Although Eqs. (77) have not been solved exactly, variational techniques can provide accurate approximate solutions (Dalgarno, 1962). Introducing the functionals

$$J_0(\chi_0^{(1)}) = \langle \chi_0^{(1)} \,|\, h^{(0)} - \varepsilon^{(0)} \,|\, \chi_0^{(1)} \rangle + \langle \chi_0^{(1)} \,|\, v_0^{(1)} \,|\, \phi^{(0)} \rangle + 2\langle \chi_0^{(1)} \,|\, h^{(1)} \,|\, \phi^{(0)} \rangle , \tag{78a}$$

$$J_n(\chi_n^{(1)}) = \langle \chi_n^{(1)} \,|\, h^{(0)} - \varepsilon^{(0)} \,|\, \chi_n^{(1)} \rangle + \langle \chi_n^{(1)} \,|\, v_n^{(1)} \,|\, \phi^{(0)} \rangle + 2\langle \chi_n^{(1)} \,|\, \chi_{n-1}^{(1)} \rangle , \qquad n = 1, 2, \ldots , \tag{78b}$$

where the $\chi_n^{(1)}$ are variational trial functions, we see that Eqs. (77) for $\chi_n^{(1)}(+)$ are the Euler equations associated with stationary values of the J_n. We choose the discrete basis set,

$$f_i(\mathbf{r}) = P_1(\cos\theta) r^i e^{-\alpha_i r} , \tag{79}$$

and expand all the $\chi_n^{(1)}$,

$$\chi_n^{(1)} = \sum_{i=1}^{N} a_n^{(i)} f_i . \tag{80}$$

Substitution of Eq. (80) into Eqs. (78) yields the equations

$$J_0 = \bar{a}_0 \cdot A \cdot \bar{a}_0 + 2\bar{b} \cdot \bar{a}_0 , \tag{81a}$$

$$J_n = \bar{a}_n \cdot B \cdot \bar{a}_n + 2\bar{a}_n \cdot S \cdot \bar{a}_{n-1} , \qquad n = 1, 3, \ldots \text{ odd} , \tag{81b}$$

$$J_n = \bar{a}_n \cdot A \cdot \bar{a}_n + 2\bar{a}_n \cdot S \cdot \bar{a}_{n-1} , \qquad n = 2, 4, \ldots \text{ even} , \tag{81c}$$

where

$$(A)_{ij} = \langle f_i \,|\, -\tfrac{1}{2}\nabla^2 - (2/r) - \varepsilon^{(0)} \,|\, f_j \rangle + \langle f_i \phi^{(0)} \,|\, (r_{12})^{-1}(1 + 2P_{12}) \,|\, f_j \phi^{(0)} \rangle , \tag{82a}$$

$$(B)_{ij} = \langle f_i \,|\, -\tfrac{1}{2}\nabla^2 - (2/r) - \varepsilon^{(0)} \,|\, f_j \rangle + \langle f_i \phi^{(0)} \,|\, (r_{12})^{-1} \,|\, f_j \phi^{(0)} \rangle , \tag{82b}$$

$$(S)_{ij} = \langle f_i \,|\, f_j \rangle , \tag{82c}$$

$$(\bar{a}_n)_i = a_n^{(i)} , \tag{82d}$$

$$(\bar{b})_i = \langle f_i \,|\, h^{(1)} \,|\, \phi^{(0)} \rangle . \tag{82e}$$

If we use fixed exponential parameters α_i in Eq. (79), the linear coefficients $\bar{a}_n$ can be determined from the variation conditions

$$\delta J_0 = \delta\bar{a}_0 \cdot (A \cdot \bar{a}_n + \bar{b}) = 0 , \tag{83a}$$

$$\delta J_n = \delta\bar{a}_n \cdot (B \cdot \bar{a}_n + S \cdot \bar{a}_{n-1}) = 0 , \qquad n = 1, 2, \ldots \text{ odd} , \tag{83b}$$

$$\delta J_n = \delta\bar{a}_n \cdot (A \cdot \bar{a}_n + S \cdot \bar{a}_{n-1}) = 0 , \qquad n = 2, 4, \ldots \text{ even} , \tag{83c}$$

which lead to the linear inhomogeneous equations,

$$A\cdot\bar{a}_0 + \bar{b} = 0\,, \tag{84a}$$

$$B\cdot\bar{a}_n + S\cdot\bar{a}_{n-1} = 0\,, \qquad n = 1, 3, \ldots \text{ odd}\,, \tag{84b}$$

$$A\cdot\bar{a}_n + S\cdot\bar{a}_{n-1} = 0\,, \qquad n = 2, 4, \ldots \text{ even}\,. \tag{84c}$$

These equations have solution

$$\bar{a}_0 = -A^{-1}\cdot\bar{b}\,, \tag{85a}$$

$$\bar{a}_n = -B^{-1}\cdot S\cdot\bar{a}_{n-1}\,, \qquad n = 1, 3, \ldots \text{ odd}\,, \tag{85b}$$

$$\bar{a}_n = -A^{-1}\cdot S\cdot\bar{a}_{n-1}\,, \qquad n = 2, 4, \ldots \text{ even}\,, \tag{85c}$$

in terms of which the moments [Eq. (63)] take on the simple form

$$\mu_k = 4\bar{a}_{2k}\cdot\bar{b}\,, \qquad k = 0, 1, 2, \ldots\,, \tag{86}$$

where we have included the contributions from both electrons in helium. The variational procedure outlined in Eqs. (78)–(86) reduces the problem of the determination of moment coefficients to one of simple matrix inversion and multiplication, similar to that encountered in static polarizability calculations (Dalgarno, 1962; Langhoff *et al.*, 1966).

Success of the foregoing procedure for obtaining μ_k values depends critically upon the rate of convergence of Eqs. (80) and (86). In Table II, we present results for the moments μ_k $(k = 0, 1, \ldots)$ of atomic helium. The nonlinear parameters used in Eq. (79) were determined by observing the rate of convergence for different choices, although a full nonlinear optimization was not performed. The values $\alpha_i = 27/16$ (i odd), $\alpha_i = 1$ (i even) insured relatively uniform convergence for both low- and high-order moment coefficients. We see from Table II that all of the first 10 moments for atomic helium have converged. Furthermore, the numerical values are in reasonable agreement with previous results obtained by an alternative procedure (Dalgarno and Victor, 1967). It is of interest that a number of the moments in Table II exhibit oscillations, although convergence results in all cases. From the second variation of the functionals, Eqs. (81), we see that a minimum value (as opposed to a stationary value) will be obtained at the solution point $\delta J_n = 0$ if the matrices A and B of Eqs. (82) are positive. This is the case for the computations performed here, and, therefore, although some of the Cauchy moments experience oscillations prior to convergence, the numerical values of the associated functionals, J_n, converge monotonically to a minimum in every case. The Cauchy moments in Table II are used to construct Padé approximants in Sects. III. E and IV.

TABLE II

Convergence of Variationally Determined Moment Coefficients for Atomic Helium in the Hartree–Fock Approximation[a]

Number of terms (N)[b]	μ_0	μ_1	μ_2	μ_3	μ_4
2	1.30606	1.41773	1.79294	2.33712	3.06287
3	1.32014	1.38502	1.74645	2.38696	3.36890
4	1.32165	1.38624	1.73009	2.34780	3.34257
5	1.32186	1.38597	1.72978	2.34953	3.34614
6	1.32186	1.38597	1.72975	2.34955	3.34648
7	1.32186	1.38598	1.72976	2.34955	3.34645
8	1.32186	1.38598	1.72976	2.34955	3.34645
Previous result[c]	1.32222	1.38627	1.72935	2.34811	3.34457

Number of terms (N)[b]	μ_5	μ_6	μ_7	μ_8	μ_9
2	4.01775	5.27118	6.91583	9.07367	11.9048
3	4.81218	6.90378	9.91997	14.2619	20.5083
4	4.90349	7.33102	11.0911	16.9025	25.8736
5	4.90679	7.33478	11.1078	16.9726	26.0930
6	4.90709	7.33344	11.1028	16.9653	26.0981
7	4.90709	7.33367	11.1037	16.9673	26.1017
8	4.90709	7.33367	11.1037	16.9673	26.1017
Previous result[c]	4.90743	7.34164	11.1263	17.0009	26.1511

[a] All values in atomic units. Zeroth-order Hartree–Fock function taken from E. Clementi, *J. Chem. Phys.* **38,** 996 (1963).

[b] Number of terms in the variational function of Eq. (80) in the text.

[c] Obtained from series expansion of the polarizability of A. Dalgarno and G. A. Victor, *Proc. Phys. Soc.* (*London*) **90,** 605 (1967).

D. Determination of Cauchy Moments from Optical Dispersion and Absorption Data

Although the computational technique described in the foregoing section can be applied to arbitrarily large atoms and molecules for which ground-state Hartree–Fock or more accurate wave functions are available, it is often convenient to follow an empirical procedure in obtaining the Cauchy moments for larger systems. The dynamical polarizability, $\alpha(\omega)$, is related to the measurable refractive index, $n(\omega)$, of a dilute gas by the familiar Lorentz–Lorenz expression (Korff and Breit, 1932)

$$n(\omega) - 1 = 2\pi N_0 \alpha(\omega) \,, \tag{87a}$$

where N_0 is the number of gaseous atoms per cubic centimeter. Measurements of $[n(\omega) - 1]$ of Eq. (87a), commonly termed the refractivity, are usually performed in the normal dispersion region, $0 \leq \omega < \omega_{10}$. From such data, the first few Cauchy moments can be inferred by writing

$$n(\omega) - 1 = 2\pi N_0 \sum_{k=0}^{n} \mu_k (\omega^2)^k , \tag{87b}$$

and using a least-squares fitting procedure to determine numerical values for the μ_k ($k = 0, 1, 2, \ldots, n$). The number of moment coefficients which can be obtained in this way depends critically upon the frequency range and accuracy of the measured refractivity.

Although values for the first two Cauchy moments can usually be obtained by such a simple linear least-squares fit to refractivity data, the higher-order moments cannot be extracted directly because of lack of either sufficient curvature or high enough accuracy in the measured data; that is, the moments can be ill-defined because of the difficulty of separating the contribution of each order to the measured data. However, the Stieltjes conditions (Baker, 1965) of Eqs. (21) offer constraints that aid in the determination of the higher-order moments by restricting the space of allowable values. Additional constraints are furnished by the conditions

$$\mu_k / \mu_{k+1} \geq \omega_{10}^2 , \tag{88}$$

obtained from the fact that the radius of convergence of the polarizability series is given by the square of the first dipole-allowed transition frequency (Schwartz, 1966). If Eqs. (88) are not satisfied, the moments estimated from refractivity data can be unsatisfactory (Dalgarno and Kingston, 1960).

In the application of a constrained least-squares fitting procedure to refractivity data for the inert gases and the diatomic molecules hydrogen, nitrogen, and oxygen, we have found it possible to evaluate the first four moments, corresponding to the expression

$$\alpha(\omega) = \mu_0 + \mu_1\omega^2 + \mu_2\omega^4 + \mu_3\omega^6 . \tag{89}$$

The explicit form of the constraints for this case is

$$D(n, 0) > 0, \qquad n = -1, 0, 1, 2, 3 , \tag{90a}$$

$$D(-1, 1) = \mu_{-1}\mu_1 - \mu_0^2 > 0 , \tag{90b}$$

$$D(0, 1) = \mu_0\mu_2 - \mu_1^2 > 0 , \tag{90c}$$

$$D(1, 1) = \mu_1\mu_3 - \mu_2^2 > 0 , \tag{90d}$$

$$\begin{aligned} D(-1, 2) = {} & \mu_{-1}(\mu_1\mu_3 - \mu_2^2) - \mu_0(\mu_0\mu_3 - \mu_1\mu_3) \\ & + \mu_1(\mu_0\mu_2 - \mu_1^2) > 0 , \end{aligned} \tag{90e}$$

$$\mu_n / \mu_{n+1} \geq \omega_{10}^2 , \qquad n = 0, 1, 2, 3 ; \tag{90f}$$

the f-sum rule ($\mu_{-1} = N$) has been included in these constraining conditions. Introduction of the new variables (Langhoff and Karplus, 1969)

$$\theta_1 = 1/\mu_0 , \tag{91a}$$

$$\theta_n = \mu_{n-2}/\mu_{n-1} , \qquad n = 2, 3, 4 , \tag{91b}$$

linearizes some of the constraints of Eqs. (90), which take the form

$$\theta_n > 0 , \qquad n = 1, 2, 3, 4 , \tag{92a}$$

$$N\theta_1 - \theta_2 > 0 , \tag{92b}$$

$$\theta_2 - \theta_3 > 0 , \tag{92c}$$

$$\theta_3 - \theta_4 > 0 , \tag{92d}$$

$$N\theta_1\theta_2(\theta_3 - \theta_4) - \theta_2\theta_3(\theta_2 - \theta_4) + \theta_3\theta_4(\theta_2 - \theta_3) > 0 , \tag{92e}$$

$$\theta_n - \omega_{10}^2 \geq 0 , \qquad n = 2, 3, 4 , \tag{92f}$$

whereas Eq. (89) becomes

$$\alpha(\omega) = \left(\frac{1}{\theta_1}\right)\left\{1 + \frac{\omega^2}{\theta_2}\left[1 + \frac{\omega^2}{\theta_3}\left(1 + \frac{\omega^2}{\theta_4}\right)\right]\right\} . \tag{93}$$

We can now minimize the least-squares deviation of Eq. (93) from the measured data and simultaneously insure that the linear constraints [Eqs. (92b), (92c), (92d), and (92f)] are satisfied using an available computational algorithm (Langhoff and Karplus, 1969). The nonlinear constraint of Eqs. (90e) and (92e) is used as a check on the resulting θ_n values, small adjustment being made if necessary.

The Cauchy moments resulting from application of the foregoing procedure to the inert gases and to molecular hydrogen, nitrogen, and oxygen are shown in Table III. In every case, the constraints were found to play a significant role in the extraction of the higher-order moments. The resulting fits to measured refractivity data are superior to those obtained with moments previously given in the literature (Dalgarno and Kingston, 1960; Barker and Leonard, 1964).

If complete absorption data were available, the procedure considered here would be somewhat superfluous; that is, the polarizabilities and dispersion energy coefficients could be obtained directly from the oscillator-strength distributions, which are simply related to measured absorption coefficients. It is usually the case, however, that the measured discrete line or band oscillator strengths are not accurate enough to predict reliable polarizabilities (Samson, 1966; Fano and Cooper, 1968) by means of the Kramers–Heisenberg dispersion formula. In fact, adjustments are made to the inferred oscillator-strength distributions so that known moments or sum rules are reproduced (Dalgarno and Davison, 1966; Dalgarno, 1967).

TABLE III

Cauchy Moments for Atomic and Molecular Systems[a]

System[b]	μ_0	μ_1	μ_2	μ_3
He	1.3838	1.550	2.066	2.95
Ne	2.6680	2.863	5.682	13.4
Ar	11.091	28.16	96.78	387
Kr	16.740	55.53	238.7	1200
Xe	27.340	116.2	928.3	8520
H_2	5.4390	20.20	81.61	350
N_2	11.743	30.17	99.21	374
O_2	10.600	36.97	132.0	480
Li	1.637×10^2	3.527×10^4	7.641×10^6	1.656×10^9
Na	1.674×10^2	2.734×10^4	4.568×10^6	7.683×10^8
K	2.880×10^2	7.916×10^4	2.245×10^7	6.371×10^9
Rb	3.159×10^2	9.269×10^4	2.821×10^7	8.588×10^9
Cs	3.565×10^2	1.284×10^5	4.655×10^7	1.693×10^{10}

[a] All values in atomic units.

[b] Values for the inert and diatomic gases obtained from measured refractivity data. See P. W. Langhoff and M. Karplus, *J. Opt. Soc. Am.* **59,** 863 (1969). Values for the alkali atoms obtained from approximate oscillator-strength distributions. See Langhoff and Karplus (1970b) and Dalgarno and Davison (1967) for discussion of the optical spectrum. Contributions from the x-ray region estimated from K. Siegbahn (ed.), "Beta- and Gamma-Ray Spectroscopy", Wiley (Interscience), New York, 1955. The resulting static polarizabilities for the alkali atoms are in general agreement with the experimental results of G. E. Chamberlain and J. C. Zorn, *Phys. Rev.* **129,** 677 (1963).

The resulting oscillator-strength distribution can then be used directly in Eq. (2) to obtain the dipole-dipole dispersion energy coefficient. Although the values so obtained may be reliable, there are no quantitative criteria for judging their accuracy.

For the alkali atoms, we attempt to use such experimentally inferred oscillator-strength distributions (Korff and Breit, 1932; Dalgarno and Davison, 1967) to obtain values for the Cauchy moments. Each distribution is adjusted slightly to reproduce the measured value of μ_0, the static polarizability (Chamberlain and Zorn, 1963; Salop *et al.*, 1961). The resulting moments (Langhoff and Karplus, 1969, 1970b) are shown in Table III. It is important to recognize that the complete oscillator-strength distributions we have employed in constructing the Cauchy moments for the alkali atoms can also be used in Eqs. (2) and (4) to obtain dispersion force coefficients and dynamic polarizabilities directly. However, the Padé approximants obtained from the moments of Table III result in, at most, simple two-term effective oscillator-strength distributions which, nevertheless, provide dispersion force coefficients and polarizabilities which are identical with those obtained from the complete spectrum. Consequently, in the case of the alkali atoms, we can view the Padé procedure as effectively contracting a complete spectrum to a small number of terms which yield accurate polarizabilities and dispersion force coefficients.

While the theoretical moments of Tables I and II are accurate to the number of significant figures shown, error limits must be assigned for the semiempirical moments of Table III. Based on the accuracy of the absorption and dispersion data we have employed and on our procedure for the determination of the moments, we estimate the maximum error limits for the inert and diatomic gases to be: $\mu_0(\pm 1\%)$, $\mu_1(\pm 5\%)$, $\mu_2(\pm 10\%)$, $\mu_3(\pm 15\%)$, and for the alkali atoms to be: $\mu_0(\pm 5\%)$, $\mu_1(\pm 10\%)$, $\mu_2(\pm 10\%)$, $\mu_3(\pm 10\%)$.

E. The Poles and Residues of Padé Approximants

The Padé approximants $[n, n]_\alpha$ and $[n, n-1]_\alpha$ to the frequency-dependent polarizability $\alpha(iy)$ have poles and residues that are upper bounds to the first n transition frequencies and oscillator strengths, respectively, of

TABLE IV

Atomic Hydrogen Transition Frequencies and Oscillator Strengths Obtained by Padé Continuation of the Cauchy Series[a]

	Transition frequencies from $[n, n-1]_\alpha/[n, n]_\alpha$				
n	ω_1	ω_2	ω_3	ω_4	ω_5
1	0.4114/0.3929	—	—	—	—
2	0.3797/0.3772	0.6166/0.5510	—	—	—
3	0.3756/0.3752	0.4949/0.4773	0.8323/0.7132	—	—
4	0.3751/0.3750	0.4622/0.4561	0.6090/0.5113	1.056 /0.8817	—
5	0.3750/0.3750	0.4509/0.4486	0.5366/0.5199	0.7294/0.6711	1.285 /1.054
6	0.3750/0.3750	0.4468/0.4459	0.5044/0.4960	0.6168/0.5892	0.8538/0.7754
7	0.3750/0.3750	0.4452/0.4449	0.4882/0.4837	0.5629/0.5475	0.7006/0.6621
Exact[b]	0.3750	0.4444	0.4688	0.4800	0.4861
	Oscillator strengths from $[n, n-1]_\alpha/[n, n]_\alpha$				
n	f_1	f_2	f_3	f_4	f_5
1	0.7618/0.6336	—	—	—	—
2	0.4940/0.4583	0.4083/0.3625	—	—	—
3	0.4307/0.4229	0.2648/0.2223	0.2535/0.2514	—	—
4	0.4183/0.4170	0.1723/0.1477	0.2114/0.1907	0.1674/0.1783	—
5	0.4164/0.4163	0.1232/0.1105	0.1595/0.1411	0.1638/0.1557	0.1169/0.1309
6	0.4162/0.4162	0.0988/0.0926	0.1193/0.1052	0.1386/0.1279	0.1277/0.1259
7	0.4162/0.4162	0.0867/0.0842	0.0914/0.0807	0.1115/0.1022	0.1188/0.1127
Exact[b]	0.4162	0.07911	0.02899	0.01394	0.007799

[a] All values are in atomic units.

[b] See Eqs. (71a) and (71b) in the text.

the system. To examine whether these bounds are useful, we have employed the coefficients in Table I for atomic hydrogen to determine $[n, n]_\alpha$ and $[n, n-1]_\alpha$ for $n = 1$ to 7. The procedure requires inversion of the A matrix of Eq. (19), which is guaranteed nonsingular by Eq. (21). Approximations to the transition frequencies of atomic hydrogen are given by the roots of the polynomial Q_n; approximations to the oscillator strengths are given by the associated residues. The transition frequencies and oscillator strengths so obtained are shown in Table IV (Langhoff and Karplus, 1970a). The $[n, n-1]_\alpha$ approximants are determined by the values in Table IV, as in Eq. (23); this is not true for the $[n, n]_\alpha$ approximants, which do not allow a corresponding partial fraction reduction.

It is apparent from Table IV that rapid convergence to the exact value has been obtained for the resonance frequency, ω_{10}, but that the higher transition frequencies are not given accurately by even the $[7, 6]_\alpha$ approximant. A similar result is found for the oscillator strengths, where convergence has been achieved for f_{10} but not for the f values associated with higher transitions.

It might appear that the results of Table IV could be improved by introduction of the $[n, n]_\alpha$ and $[n, n-1]_\alpha$ approximants with larger n. Severe numerical difficulties ensue, however, in any such procedure. In particular, with increasing dimension the A matrix of Eq. (19) rapidly becomes near-singular, complicating the process of its inversion (Schwartz, 1966). This follows from the fact that the ratio $\mu_k/\mu_{k+1} \to \omega_{10}^2$ becomes constant to many significant figures for large k and, therefore, that adjoining rows and/or columns of the A matrix become related by a constant factor. Thus, although Eq. (21) insures that the determinant of the A matrix is nonzero, its value becomes numerically small for large n. The approach to near-singularity of the A matrix can also be viewed as a direct consequence of the fact that the higher-order moments are dominated by the contribution of the $2p$ state (see Table I). Of course, the necessity of using very accurate coefficients in the continuation of a power series far from the origin is a general requirement of analytic continuation procedures and not a particular limitation of Padé approximants.

Although numerical difficulties arise in attempts to invert an A matrix of high order, once the b_i coefficients of Eq. (14b) are determined (accurate or not), there are no problems in obtaining the roots of the polynomial Q_n to high accuracy. However, in order to maintain numerical significance for all results quoted here, double precision arithmetic has been utilized throughout.

The variationally determined moments of Table II for atomic helium in the Hartree–Fock approximation have also been used to construct Padé approximants. Proceeding as for atomic hydrogen, we have obtained the poles and residues exhibited in Table V (Langhoff and Karplus, 1970a).

Transition frequencies in reasonable agreement with previous Hartree–Fock predictions are obtained (Dalgarno and Victor, 1967; Sengupta and Mukherji, 1967), although the higher-order transition frequencies are in poor agreement with the correct results (Moore, 1949). Furthermore, except for the case of f_{10}, the resulting oscillator strengths bear little resemblance to accurately calculated results and available experimental values (Schiff and Pekeris, 1964; Dalgarno and Stewart, 1960; Wiese *et al.*, 1968). The onset of numerical instability is evident in Table V from the irregular behavior of the $n = 5$ approximants. This instability is a direct consequence of the fact that the moment coefficients of Table II for atomic helium are accurate only to six significant figures; additional significance

TABLE V

Transition Frequencies and Oscillator Strengths for Atomic Helium in the Coupled Hartree–Fock Approximation Obtained by Padé Continuation of the Cauchy Series[a]

	Transition frequencies from $[n, n-1]_\alpha/[n, n]_\alpha$			
n	ω_1	ω_2	ω_3	ω_4
1	0.9766/0.8951	—	—	—
2	0.8353/0.8191	1.408 /1.212	—	—
3	0.8069/0.8026	1.052 /0.9951	1.850 /1.540	—
4	0.7996/0.7985	0.9467/0.9262	1.285 /1.193	2.308 /1.891
5	0.7981/0.7979	0.9150/0.9104	1.140 /1.138	1.644 /1.750
Experimental[b]	0.7797	0.8484	0.8725	0.8836
Previous calculation[c]	0.8020	0.9942	1.477	2.853
Previous calculation[d]	0.7965	0.8637	0.9083	—
	Oscillator strengths from $[n, n-1]_\alpha/[n, n]_\alpha$			
n	f_1	f_2	f_3	f_4
1	1.261 /0.8898	—	—	—
2	0.5403/0.4371	1.085 /0.8959	—	—
3	0.3489/0.3136	0.6025/0.4848	0.8257/0.7633	—
4	0.2853/0.2742	0.3631/0.3062	0.5841/0.5012	0.6229/0.6161
5	0.2643/0.2656	0.2740/0.2632	0.4378/0.4460	0.5322/0.6483
Previous calculation[c]	0.3140	0.4832	0.6402	0.4550
Accurate[e]	0.2762	0.0734	0.0304	0.0153

[a] All values in atomic units.

[b] C. E. Moore, Atomic Energy Levels, Vol. I. Nat. Bur. of Stds. Cir. 467 (1949).

[c] A. Dalgarno and G. A. Victor, *Proc. Phys. Soc.* (*London*) **90,** 605 (1967).

[d] S. Sengupta and A. Mukherji, *J. Chem. Phys.* **47,** 260 (1967).

[e] B. Schiff and C. L. Pekeris, *Phys. Rev.* **134,** A640 (1964); A. Dalgarno and A. L. Stewart, *Proc. Phys. Soc.* (*London*) **76,** 49 (1960); W. L. Wiese, M. W. Smith, and B. M. Glennon, Atomic Transition Probabilities, Natl. Bur. Std. Ref. Data System **4,** Vol. 1, p. 11. U.S. Government Printing Office, Washington, D. C., 1968.

would be required to eliminate the instability and construct higher-order approximants.

From the foregoing, it is clear that for both atomic hydrogen and helium the Padé approximants furnish a fictitious oscillator-strength distribution, which bears little resemblance to the true distribution except for the f_{10} value and the first few ω_{n0} values. Nevertheless, we shall find in Sect. IV that the Padé approximants converge rapidly in the regions of primary physical interest, i.e., for real frequencies in the normal dispersion range and for nonnegative imaginary frequencies.

IV. APPLICATIONS

The formalism developed in Sect. II, with the Cauchy moments computed in Sect. III, is applied in this section to the refractive index in the normal dispersion region, where the Padé approximants bound and sum the power series expansion, and to the determination of bounds on dipole and other contributions to the dispersion energy.

A. Summation of the Cauchy Dispersion Equation

The power-series expansion of the refractivity, or Cauchy dispersion equation (Sect. III. D),

$$n(\omega) - 1 = 2\pi N_0 \sum_{k=0}^{n} \mu_k (\omega^2)^k \,, \tag{87b}$$

reproduces the low-frequency data used to determine the moments μ_k but, in general, deviates significantly at higher frequencies. Since the $[n, n]_\alpha$ and $[n, n-1]_\alpha$ Padé approximants furnish lower bounds to the polarizability for real frequencies [Eq. (17)], they can be used with the Lorentz–Lorenz relation [Eq. (87a)] to obtain a bound on the refractive index; that is,

$$2\pi N_0[n, n]_\alpha \,, 2\pi N_0[n, n-1]_\alpha \leq n(\omega) - 1 \,, \qquad 0 \leq \omega < \omega_{10} \,. \tag{94}$$

The effectiveness of the Padé approximants in summing the Cauchy equation is most clearly demonstrated by specific examples. In Fig. 1 are shown some results for atomic hydrogen. The first four theoretical moments given in Table I are used to obtain the four-term Cauchy equation and the indicated Padé approximants. Also included in the figure is the exact refractivity for atomic hydrogen (Karplus and Kolker, 1963). We see that, although the four-term Cauchy equation is a good approximation only for $\omega \leq 0.2$ *a.u.*, the $[2, 1]_\alpha$ approximant coincides with the exact curve over a much larger frequency range. Furthermore, the lower-order approximants exhibit bounded convergence, in agreement with Eq. (94).

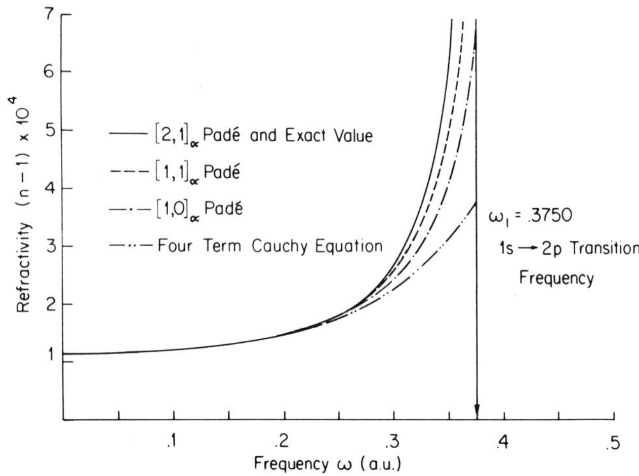

Fig. 1. Atomic hydrogen refractivity from the $[n, n]_\alpha$ and $[n, n-1]_\alpha$ Padé approximants. Exact value taken from M. Karplus and H. J. Kolker, *J. Chem. Phys.* **39,** 1493 (1963).

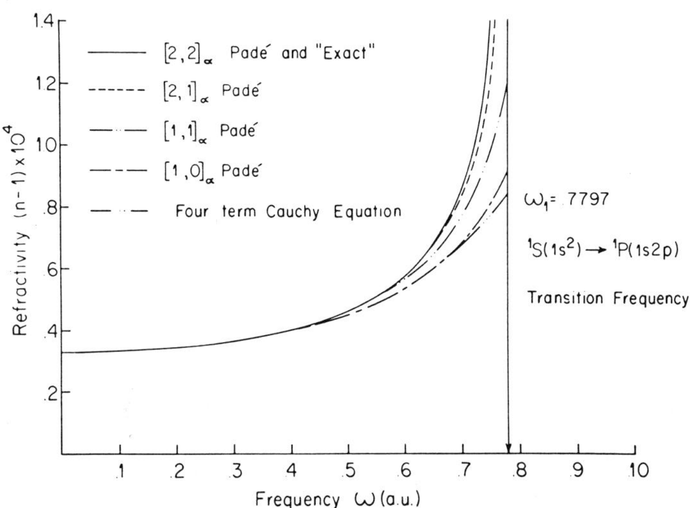

Fig. 2. Atomic helium refractivity in the Hartree–Fock approximation from the $[n, n]_\alpha$ and $[n, n-1]_\alpha$ Padé approximants. "Exact" or accurately calculated Hartree–Fock value taken from A. Dalgarno and G. A. Victor, *Proc. Roy. Soc.* (*London*) **A291,** 291 (1966).

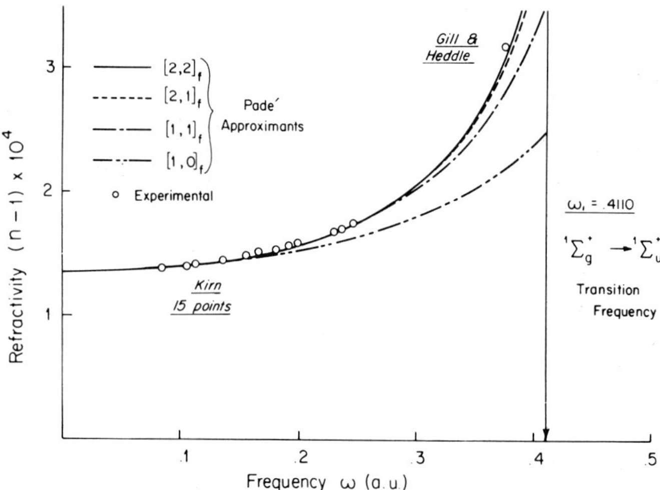

Fig. 3. Molecular hydrogen refractivity from the $[n, n]_f$ and $[n, n-1]_f$ Padé approximants. Experimental values from M. Kirn, *Ann. Physik*. **64,** 566 (1921), and P. Gill and D. W. O. Heddle, *J. Opt. Soc. Am.* **53,** 847 (1963).

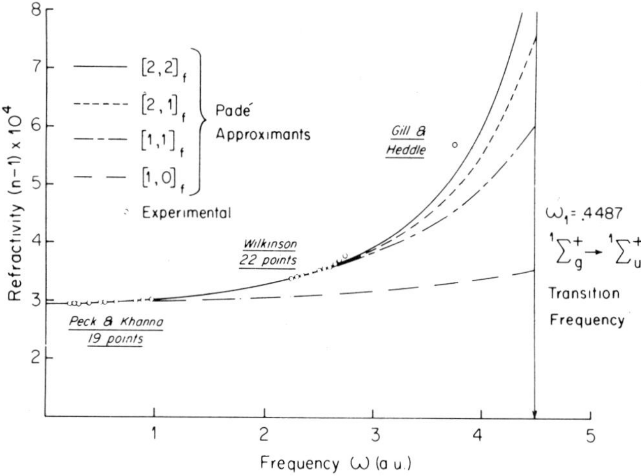

Fig. 4. Molecular nitrogen refractivity from the $[n, n]_f$ and $[n, n-1]_f$ Padé approximants. Experimental values from E. R. Peck and B. N. Khanna, *J. Opt. Soc. Am.* **56,** 1059 (1966), P. G. Wilkinson, *ibid*. **50,** 1002 (1960), and P. Gill and D. W. O. Heddle, *ibid*. **53,** 847 (1963).

Similar results are obtained for atomic helium with the moments of Table II, as shown in Fig. 2. The four-term Cauchy equation and the approximants obtained from it are compared with an accurate theoretical refractivity curve. Evidently, the $[2, 2]_\alpha$ approximant coincides with the accurate result over the frequency range considered, while the $[1, 0]_\alpha$, $[1, 1]_\alpha$, and $[2, 1]_\alpha$ approximants exhibit convergence from below. On the other hand, the four-term Cauchy equation is accurate only for $\omega < 0.4$ *a.u.*

Correspondingly improved convergence is obtained by using moments evaluated from dispersion data (Langhoff and Karplus, 1969). In Fig. 3, results for molecular hydrogen are shown. Refractivity measurements in the visible and near-ultraviolet (Koch, 1912; Kirn, 1921) have been used to determine moments that furnish Padé approximants that agree with the one available refractivity measurement in the far-vacuum ultraviolet (Heddle *et al.*, 1963; Gill and Heddle, 1963). In Fig. 3, we have made use of the $[n, n]_f$ and $[n, n-1]_f$ approximants defined by Eq. (27), rather than the $[n, n]_\alpha$ and $[n, n-1]_\alpha$ approximants; note, however, that $[1, 1]_f = [1, 0]_\alpha$ and $[2, 2]_f = [2, 1]_\alpha$. Similar results for molecular nitrogen are shown in Fig. 4. Here dispersion data in the visible (Peck and Khanna, 1966) have been used for the moments. Evidently, the approximants are in agreement with the vacuum ultraviolet measurements (Wilkinson, 1960), although convergence to the single far-vacuum ultraviolet measurement (Heddle *et al.*, 1963; Gill and Heddle, 1963) has not been obtained. Presumably, higher-order approximants will result in convergence in this shorter-wavelength region.

B. The Riesz–Herglotz–Gronwall Transformation

A useful sequence of upper bounds to the refractivity at real frequency can be constructed by a variable transformation. Introducing $u = \omega^2/\omega_{10}^2$, with ω_{10} the resonance frequency, and the transformation

$$u = [(1 + z)^{1/2} - 1]/[(1 + z)^{1/2} + 1]\,, \tag{95}$$

we find that the refractivity can be written (Baker, 1965) as

$$n(\omega) - 1 = 2\pi N_0 g(z)(1 + z)^{1/2}\,, \tag{96}$$

where

$$g(z) = \sum_{k=0}^{\infty} g_k(-z)^k \tag{97}$$

is a series of Stieltjes with

$$g_k = 4^{-k} \sum_{i=0}^{k} (-1)^i \binom{2k}{k-i} \omega_{10}^{2i}\mu_i\,; \tag{98}$$

the symbol $\binom{2k}{k-i}$ represents a binomial coefficient. Since Eq. (97) is a

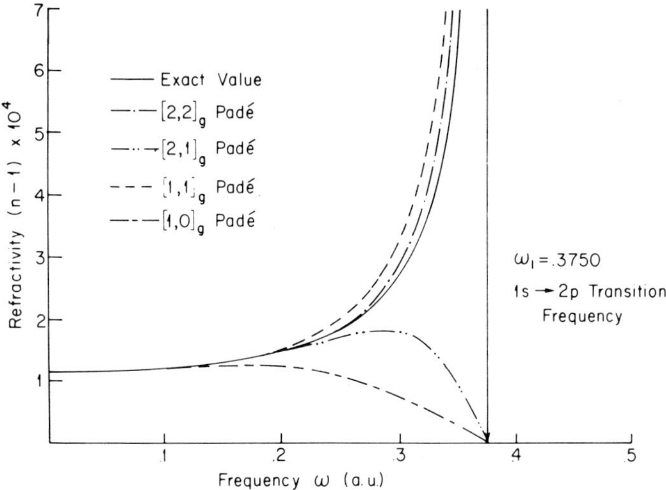

Fig. 5. Atomic hydrogen refractivity from the $[n, n]_g$ and $[n, n-1]_g$ Padé approximants. Exact value taken from M. Karplus and H. J. Kolker, *J. Chem. Phys.* **39**, 1493 (1963).

series of Stieltjes, we have

$$[n, n]_g \geq g(z) \geq [n, n-1]_g\,, \qquad 0 \leq \mathrm{Re}(z)\,, \tag{99}$$

and

$$2\pi N_0(1+z)^{1/2}[n, n]_g \geq n(\omega) - 1 \geq 2\pi N_0(1+z)^{1/2}[n, n-1]_g\,, \tag{100}$$

for real frequency $(0 \leq \omega \leq \omega_{10})$. Equation (100) furnishes an upper bound to the refractivity in the region of interest, as well as an alternative lower bound.

For atomic hydrogen with the moments of Table I, the four lowest-order bounds obtained from Eq. (100) are shown in Fig. 5 and compared with the exact refractivity. The $[1, 0]_g$ and $[2, 1]_g$ lower bounds are evidently of little practical value in this case (they go to zero at $\omega = \omega_{10}$), whereas the $[1, 1]_g$ and $[2, 2]_g$ upper bounds are numerically closer to the correct result, although they have not converged. Use of the higher-order approximants in Eq. (100) results in smooth convergence of the upper bounds given by the $[n, n]_g$ approximants to the exact refractivity. Although the results obtained with the $[n, n]_g$ approximants are of interest, alternative expressions that converge more rapidly would be desirable (see below).

C. The Common Bounds

An alternate set of upper bounds to the refractivity may be obtained following a recent development of Padé theory by Common (1968; Baker,

1969). Introducing $z = -\omega^2$ and the function

$$k(z) = \sum_{i=0}^{\infty} k_i(-z)^i , \tag{101}$$

where

$$k_i = (i+1)^{-1}(\mu_0/\omega_{10}^{2i+2} - \mu_{i+1}) , \tag{102}$$

we can show that the polarizability is given by

$$\alpha(z) = \frac{\mu_0\omega_{10}^2}{\omega_{10}^2 + z} + z\left[k(z) + z\frac{dk}{dz}\right]. \tag{103}$$

Furthermore, $k(z)$ is a series of Stieltjes, and introduction of the approximants $[n, n]_k$ and $[n, n-1]_k$, which give lower bounds to $k(z)$ in the interval $0 \geq z > -\omega_{10}^2$, furnishes upper bounds to $\alpha(z)$ in this interval by Eq. (103). For the refractivity, we obtain

$$n(\omega) - 1 \leq 2\pi N_0\left(\frac{\mu_0\omega_{10}^2}{\omega_{10}^2 - \omega^2} - \omega^2\left\{[n, n]_k - \omega^2[n, n]_k^{(1)}\right\}\right), \quad 0 \leq \omega < \omega_{10} , \tag{104}$$

$$n(\omega) - 1 \leq 2\pi N_0\left(\frac{\mu_0\omega_{10}^2}{\omega_{10}^2 - \omega^2} - \omega^2\left\{[n, n-1]_k - \omega^2[n, n-1]_k^{(1)}\right\}\right), \quad 0 \leq \omega < \omega_{10} , \tag{105}$$

where the superscript (1) indicates the first derivative with respect to $z = -\omega^2$.

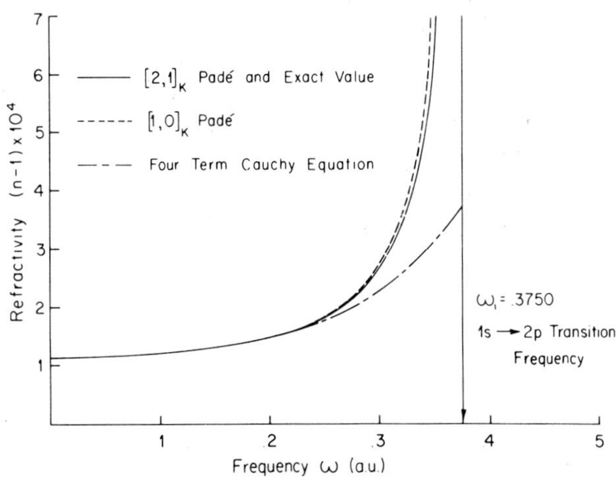

Fig. 6. Atomic hydrogen refractivity from the $[n, n-1]_k$ Padé approximants. Exact value taken from M. Karplus and H. J. Kolker, *J. Chem. Phys.* **39,** 1493 (1963).

The general theory of Padé approximants for series of Stieltjes (Baker, 1965) assures us that the bounded approximants appearing in Eqs. (104) and (105) converge in the limit $(n \to \infty)$ to the correct function. The extent to which the approximants furnish practical, rapid convergence must be determined by specific application. In Fig. 6, we show the bounds obtained from Eqs. (104) and (105) in the case of atomic hydrogen. Rapid convergence to the exact refractivity results from the $[1, 0]_k$, $[1, 1]_k$ (not shown), and $[2, 1]_k$ approximants. Even the $[1, 0]_k$ approximant is quite close to the exact result, to which the $[2, 1]_k$ approximant has converged; the $[1, 1]_k$ approximant falls between the $[1, 0]_k$ result and the exact value.

A second application of the bounds afforded by Eqs. (104) and (105)

TABLE VI

Comparison of Measured and Predicted Molecular Nitrogen Refractivity $[(n - 1) \times 10^4]$ in the Vacuum Ultraviolet[a]

		Present results[c]				
λ(Å)	Measured[b]	$[1, 1]_f$	$[2, 1]_f$	$[2, 2]_f$	$[1, 1]_k$	$[1, 0]_k$
2042.48	3.424	3.368	3.387	3.389	3.390	3.395
2019.72	3.417	3.380	3.399	3.402	3.403	3.408
2007.69	3.411	3.386	3.406	3.408	3.410	3.415
1988.27	3.435	3.396	3.417	3.420	3.421	3.427
1970.88	3.435	3.405	3.427	3.430	3.432	3.438
1962.01	3.444	3.410	3.433	3.436	3.437	3.444
1955.12	3.448	3.414	3.437	3.440	3.442	3.449
1944.73	3.445	3.420	3.444	3.447	3.448	3.456
1938.89	3.445	3.423	3.447	3.450	3.452	3.460
1938.01	3.448	3.424	3.448	3.451	3.453	3.460
1930.04	3.451	3.429	3.453	3.456	3.458	3.466
1874.25	3.489	3.464	3.492	3.496	3.498	3.508
1845.87	3.516	3.483	3.513	3.518	3.521	3.532
1824.30	3.538	3.498	3.531	3.535	3.539	3.551
1785.05	3.568	3.528	3.564	3.570	3.574	3.588
1758.28	3.605	3.550	3.589	3.595	3.600	3.616
1750.04	3.616	3.557	3.597	3.603	3.609	3.625
1746.06	3.631	3.561	3.601	3.607	3.613	3.629
1742.20	3.633	3.564	3.605	3.611	3.617	3.634
1716.78	3.670	3.587	3.631	3.638	3.645	3.663
1691.09	3.730	3.611	3.659	3.667	3.675	3.696
1649.19	3.834	3.654	3.708	3.718	3.729	3.754

[a] All refractivity values quoted are at NTP (0°C, 1 atm).

[b] P. G. Wilkinson, *J. Opt. Soc. Am.* **50,** 1002 (1960).

[c] For previous prediction of molecular nitrogen refractivity in the ultraviolet using absorption data and the Kramers–Heisenberg dispersion formula see A. Dalgarno, T. Degges, and D. A. Williams, *Proc. Phys. Soc.* (*London*) **92,** 291 (1967).

is given in Table VI, where we compare a number of refractivity values in the ultraviolet for molecular nitrogen (Wilkinson, 1960) with the predictions of the Padé approximant (see also Fig. 4). Both the upper and lower bounding approximants exhibit convergence and are in good agreement with the measured values, a striking demonstration of the ability of the Padé approximants to sum the Cauchy equation accurately in the ultraviolet. The small discrepancy between experimental and predicted values may, in fact, be due to the presence of oxygen contamination in the measured samples (Heddle *et al.*, 1963; Gill and Heddle, 1963). It is important to realize that the refractivity data used to obtain the moments for molecular nitrogen (Table III) correspond to the region 20,587–4,679 Å (Peck and Khanna, 1966), whereas the Padé results are in general agreement with experiment at the considerably shorter wavelengths listed in Table VI.

A final illustration of the Common bounds is given in Table VII, where four measured refractivities for Ar are compared with the Padé predictions. The Padé results for the first two wavelength intervals have converged to values in agreement with the measurements, whereas the predictions for the last two intervals have not fully converged. Nevertheless, the approximants bound the experimental values, and the averages of the $[2, 1]_f$ and $[1, 1]_k$ results are close to the experimental values.

TABLE VII

Comparison of Measured and Predicted Argon Refractivity $[(n-1) \times 10^4]$ in the Ultraviolet[a]

			Present results				
λ(Å)[b]	Measured[c]	Previous calculation[d]	$[1, 1]_f$	$[2, 1]_f$	$[2, 2]_f$	$[1, 1]_k$	$[1, 0]_k$
1850±80	3.25 ±0.03	3.32 ±0.06	3.28 ±0.05	3.32 ±0.06	3.32 ±0.06	3.32 ±0.06	3.34 ±0.06
1805±80	3.28 ±0.03	3.36 ±0.07	3.31 ±0.06	3.35 ±0.07	3.36 ±0.07	3.36 ±0.06	3.38 ±0.06
1210±30	5.59 ±0.10	5.39 ±0.40	4.34 ±0.12	4.47 ±0.20	4.96 ±0.26	6.59 ±1.1	7.56 ±1.5
1216	5.65, 5.75[e]	5.30, 4.98[f]	4.31	4.71	4.91	6.39	7.29

[a] All refractivity values quoted are at NTP (0°C, 1 atm).

[b] The uncertainties on the wavelength values correspond to the range used in the measurements.

[c] D. W. O. Heddle, R. E. Jennings, and A. S. L. Parsons, *J. Opt. Soc. Am.* **53,** 840 (1963); P. Gill and D. W. O. Heddle, *ibid.* **53,** 847 (1963).

[d] A. E. Kingston, *J. Opt. Soc. Am.* **54,** 1145 (1964).

[e] F. Marmo, as listed in Liggett and Levinger, Ref. *f*.

[f] G. Liggett and J. S. Levinger, *J. Opt. Soc. Am.* **58,** 109. (1968).

D. Dipole Dispersion Force Results

We have shown in Sect. II. C that convenient analytical forms, Eqs. (23) and (29), result from the Padé approximants $[n, n-1]_\alpha$ and $[n, n-1]_f$ to the polarizability and $f(iy)$ function, respectively. Prior to describing the numerical values obtained for the dispersion force with these approximants, it is of interest to write out expressions for the lowest-order upper and lower bounds. From Eq. (23) with $n = 1$, we have

$$[1, 0]_\alpha = \frac{\mu_0^2/\mu_1}{\mu_0/\mu_1 + y^2} \leq \alpha(iy) , \tag{106}$$

which yields

$$\tfrac{3}{2}\,\mu_0^{(a)}\mu_0^{(b)} \frac{\omega_a\omega_b}{(\omega_a + \omega_b)} \leq C_{ab} \tag{107}$$

when substituted into Eq. (24a); here $\omega_a{}^2 = \mu_0^{(a)}/\mu_1^{(a)}$ and $\omega_b{}^2 = \mu_0^{(b)}/\mu_1^{(b)}$. Equation (107) has the form of London's formula (London, 1930), which makes use of only the static polarizabilities, $\mu_0^{(a)}$ and $\mu_0^{(b)}$, and some effective excitation energies ω_a and ω_b. If the latter are chosen from dispersion data by fitting the first two Cauchy moments, in a manner similar to that originally suggested (London, 1930; Margenau, 1939), the London formula is a lower bound to C_{ab}; if ω_a and ω_b are chosen otherwise (e.g., ionization potentials), Eq. (107) is not necessarily a bound, but may still be a very good approximation.

From Eq. (29) for $n = 1$, we obtain

$$\frac{(N - [1, 0]_f)}{y^2} = \frac{N}{N/\mu_0 + y^2} \geq \alpha(iy) , \tag{108}$$

which gives rise to the upper bound [Eq. (24b)]

$$\tfrac{3}{2}\,\mu_0^{(a)}\mu_0^{(b)} \frac{\omega_a\omega_b}{(\omega_a + \omega_b)} \geq C_{ab} , \tag{109}$$

where now $\omega_a{}^2 = N^{(a)}/\mu_0^{(a)}$ and $\omega_b{}^2 = N^{(b)}/\mu_0^{(b)}$. Equation (109) is the familiar Slater–Kirkwood (1931) formula, which is an upper bound to the dipole dispersion coefficient if the exact number of electrons $N^{(a)}$ and $N^{(b)}$ is used in each case; if an effective number of dispersion electrons is introduced instead, the bounding property is lost, although reasonable values can result.

It is satisfying that the Padé procedure incorporates the two foregoing well-known London and Slater–Kirkwood formulas for the dipole dispersion energy in a natural manner. Use of higher-order Padé approximants in Eqs. (24a) and (24b) leads to improved values for the lower and upper bounds to C_{ab}.

TABLE VIII

Dipole Dispersion Force Coefficients C_{ab} (*a.u.*) for Atomic Hydrogen and Helium Obtained from Padé Approximants $[n, n-1]_f/[n, n-1]_\alpha$[a]

n	H–H	H–He	He–He
1	7.16/6.25	3.04/2.58	1.61/1.28
2	6.54/6.47	2.75/2.70	1.40/1.36
3	6.51/6.49	2.72/2.72	1.38/1.37
4	6.50/6.50	2.72/2.72	1.37/1.37
5	6.50/6.50	2.72/2.72	1.37/1.37
Previous results	6.499[b]	—	1.376[c]

[a] Values obtained from Eqs. (24a) and (24b) as discussed in the text.

[b] L. Pauling and J. Y. Beach, *Phys. Rev.* **47,** 686 (1935); J. O. Hirschfelder and P. O. Löwdin, *Mol. Phys.* **2,** 229 (1959); Y. M. Chan and A. Dalgarno, *Mol. Phys.* **9,** 349 (1965); W. D. Davison, *Proc. Phys. Soc.* (*London*) **87,** 133 (1966).

[c] A. Dalgarno and G. A. Victor, *Proc. Phys. Soc.* (*London*) **90,** 605 (1967).

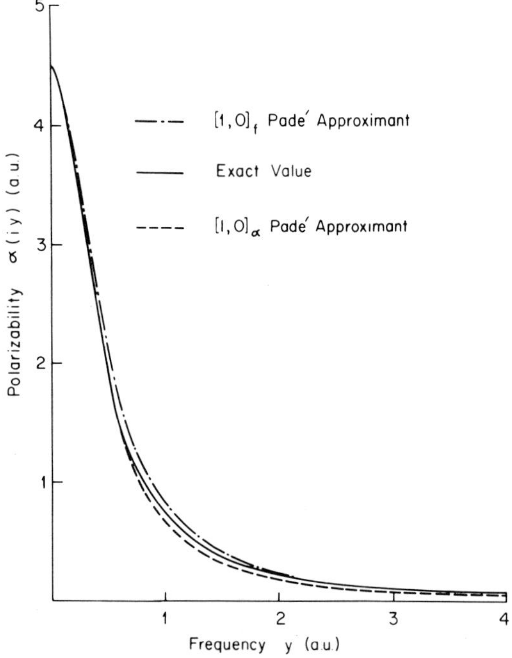

Fig. 7. Atomic hydrogen polarizability on the imaginary axis from the $[1, 0]_\alpha$ and $[1, 0]_f$ Padé approximants. Exact value taken from M. Karplus and H. J. Kolker, *J. Chem. Phys.* **39,** 1493 (1963).

As an illustration, we show in Table VIII dispersion force coefficients for atomic hydrogen and helium obtained from the Cauchy moments in Tables I and II. Smooth bounded convergence is obtained to values in agreement with the available accurate calculations. Most important, the $n = 2$ results of Table VIII are almost convergent. This suggests that, for larger systems, for which only the first four moment values are available, we can anticipate satisfactory convergence for the dispersion coefficients (see below). Even the lowest-order approximants furnish useful bounds to the dispersion force values. This is a consequence of the fact that the approximants [Eqs. (106) and (108)] furnish accurate lower and upper bounds, respectively, to the polarizability. In Fig. 7, we plot the lowest-order bounds for atomic hydrogen, in comparison with the exact polarizability. The higher-order approximants (not shown) exhibit smooth convergence to the correct curve. Similar results for atomic helium in the Hartree–Fock approximation are shown in Fig. 8. We note that, as for atomic hydrogen, the $[1, 0]_\alpha$ result for helium is somewhat more accurate at low frequency, whereas the $[1, 0]_f$ approximant is closer to the correct result at high frequency.

The empirical Cauchy moments given in Table III are used to obtain bounds for the C_{ab} coefficients for a variety of gases. The resulting values

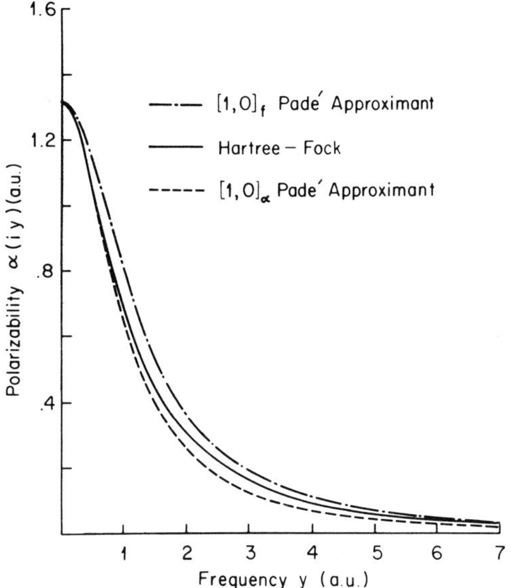

Fig. 8. Atomic helium Hartree–Fock polarizability on the imaginary axis from the $[1, 0]_\alpha$ and $[1, 0]_f$ Padé approximants. The exact Hartree–Fock value is that obtained from the convergent higher-order $[n, n-1]_\alpha$ and $[n, n-1]_f$ approximants.

are listed in Tables IX–XIV, together with previously published theoretical and semiempirical estimates (Dalgarno and Davison, 1966; Dalgarno, 1967) and experimental results (Pauly and Toennies, 1965; Bernstein and Muckerman, 1967). The inert gas coefficients listed in Table IX are typical of all the results obtained. Of course, the bounds presented are limited by the experimental accuracy of the refractivity data from which the moments were extracted. We see that in most cases tight bounds are obtained by the Padé procedure, and that these bounds generally agree with experiment. Only for the He–Kr and He–Xe interactions do there appear significant

TABLE IX

Dipole Dispersion Force Coefficients C_{ab} (*a.u.*) for the Inert Gases

Interacting pair	Upper bounds[a] $[1,0]_f$	Upper bounds[a] $[2,1]_f$	Lower bounds[a] $[2,1]_\alpha$	Lower bounds[a] $[1,0]_\alpha$	Semi-empirical values[b]	Experimental[c]
He–He	1.73	1.48	1.45	1.36	1.47	2[d]
He–Ne	4.11	3.28	3.05	2.64	3.0	3.9[d], 3.3[e]
He–Ar	14.2	10.5	9.46	8.68	9.6	13[d], 9.9[e], 8.5[f]
He–Kr	23.0	15.1	13.0	12.1	13	21[d]
He–Xe	31.8	26.7	22.4	18.2	19	36[d]
Ne–Ne	10.3	7.59	6.48	5.15	6.3	6.6[g]
Ne–Ar	34.1	23.2	19.5	16.9	20	21[f]
Ne–Kr	55.9	33.5	26.5	23.4	27	—
Ne–Xe	89.1	60.3	46.1	35.3	38	—
Ar–Ar	118	75.0	63.6	57.9	65	61[f], 70[g]
Ar–Kr	190	109	88.6	81.6	91	86[f]
Ar–Xe	304	191	151	124	130	130[f]
Kr–Kr	308	157	124	115	130	117[f], 140[g]
Kr–Xe	493	275	210	177	190	—
Xe–Xe	788	489	356	272	270	320[g]

[a] Values obtained from Eqs. (24a) and (24b) following the procedures discussed in the text.

[b] Semiempirical estimates recommended by A. Dalgarno, *Advan. Chem. Phys.* **12,** 143 (1967), obtained following the procedures of A. Dalgarno, I. H. Morrison, and R. M. Pengelly, *Intern. J. Quant. Chem.* **1,** 161 (1967).

[c] Experimental uncertainties estimated to be approximately 15–20% of the values quoted; see R. B. Bernstein and J. T. Muckerman, *Advan. Chem. Phys.* **12,** 389, 414 (1967).

[d] Values obtained from molecular beam scattering data reviewed by R. B. Bernstein and J. T. Muckerman, *Advan. Chem. Phys.* **12,** 389 (1967).

[e] R. Durin, R. Helbing, and H. Pauly, *Z. Physik* **188,** 468 (1965); R. W. Landorff and C. R. Mueller, *J. Chem. Phys.* **45,** 240 (1966).

[f] E. W. Rothe and R. H. Neynaber, *J. Chem. Phys.* **42,** 3206 (1965); *ibid.* **43,** 4177 (1965).

[g] R. J. Munn, *J. Chem. Phys.* **42,** 3032 (1965); E. A. Mason, R. J. Munn, and F. J. Smith, *Discussions Faraday Soc.* **40,** 27 (1965).

disagreements between our results and experiment. Since the experimental results are so close to the $[1, 0]_f$ upper bound, which, in general, is an overestimate, we suggest that the measured values should be redetermined. Precise error limits for the experimental values are difficult to estimate; uncertainties on the order of ± 15–20% are suggested (Pauly and Toennies, 1965; Bernstein and Muckerman, 1967). The uncertainties induced in our Padé approximant dispersion force coefficients by the error limits in the moments of Table III are less than 10–15% in all cases.

The semiempirical estimates (Dalgarno *et al.*, 1967) given in Table IX are obtained by a procedure that is discussed in a subsequent section (Sect.

TABLE X

Dipole Dispersion Force Coefficients C_{ab} (*a.u.*) for the Alkali Atoms

Interacting pair	Upper Bounds[a] $[1, 0]_f$	$[2, 1]_f$	Lower bounds[a] $[2, 1]_\alpha$	$[1, 0]_\alpha$	Semi-empirical values[b]	Experimental[c]
H-H	7.16	6.54	6.47	6.25	6.50	—
H-Li	116	66.4	65.3	64.6	67	—
H-Na	188	81.1	77.8	74.3	73	—
H-K	323	118	111	102	100	—
H-Rb	423	131	124	109	110	—
H-Cs	516	118	113	112	140	—
Li-Li	2,721	1,380	1,375	1,369	1,390	—
Li-Na	3,641	1,526	1,515	1,497	1,470	—
Li-K	6,269	2,331	2,310	2,262	2,290	—
Li-Rb	7,525	2,529	2,507	2,439	2,510	—
Li-Cs	8,813	2,628	2,609	2,601	3,160	—
Na-Na	5,388	1,708	1,683	1,645	1,580	920
Na-K	9,278	2,591	2,541	2,463	2,440	1,130
Na-Rb	11,630	2,818	2,762	2,652	2,670	1,440
Na-Cs	13,890	2,881	2,844	2,819	3,350	1,510
K-K	15,980	3,979	3,888	3,752	3,820	1,590
K-Rb	20,020	4,336	4,227	4,049	4,190	1,630
K-Cs	23,920	4,463	4,399	4,331	5,300	1,920
Rb-Rb	25,610	4,736	4,604	4,369	4,600	1,670
Rb-Cs	30,890	4,854	4,776	4,678	5,820	2,150
Cs-Cs	37,440	5,089	5,035	5,023	7,380	3,460

[a] Values obtained from Eqs. (24a) and (24b) following the procedures discussed in the text.

[b] Semiempirical estimates recommended by A. Dalgarno, *Advan. Chem. Phys.* **12,** 143 (1967), obtained following the procedures of A. Dalgarno and W. D. Davison, *Mol. Phys.* **13,** 479 (1967).

[c] Values obtained from molecular beam scattering data by V. Buck and H. Pauly, *Z. Phys.* **185,** 155, (1965). Experimental uncertainties estimated to be approximately 15–20% of the values quoted; see R. B. Bernstein and J. T. Muckerman, *Advan. Chem. Phys.* **12,** 389, 414 (1967).

TABLE XI

Dipole Dispersion Force Coefficients C_{ab} (*a.u.*) for Diatomic Gases

Interacting pair	Upper bounds[a] $[1, 0]_f$	Upper bounds[a] $[2, 1]_f$	Lower bounds[a] $[2, 1]_\alpha$	Lower bounds[a] $[1, 0]_\alpha$	Semiempirical values[b]
H_2–H_2	13.5	12.2	12.0	11.6	13
H_2–N_2	37.4	30.2	28.7	27.2	30
H_2–O_2	35.1	23.8	23.3	22.8	—
N_2–N_2	113	78.1	69.5	64.5	73
N_2–O_2	108	59.3	55.8	53.8	—
O_2–O_2	104	46.7	45.4	45.1	—

[a] Values obtained from Eqs. (24a) and (24b) following the procedures discussed in the text.

[b] Semiempirical estimates recommended by A. Dalgarno, *Advan. Chem. Phys.* **12,** 143 (1967), obtained following the procedures of A. Dalgarno, I. H. Morrison, and R. M. Pengelly, *Intern. J. Quant. Chem.* **1,** 161 (1967).

V.A). We see that they are in general agreement with the Padé predictions, although they are usually nearer the lower bounds. Only the semiempirical result for the Xe interactions is significantly lower than the best Padé result.

In Table X are shown the bounds to dispersion force coefficients for interacting pairs of alkali atoms (including hydrogen). Very tight bounds are obtained in excellent agreement with the semiempirical estimates (Dalgarno and Davison, 1967) in all cases except for the interactions involving Cs. The latter disagreement can be traced to the use, in the semiempirical procedure, of an oscillator-strength distribution that leads to a static polarizability, μ_0, for Cs which is larger than that of Table III, the experimental value (Chamberlain and Zorn, 1963; Salop, *et al.*, 1961), or previous semiempirical estimates (Dalgarno and Kingston, 1959). We note that the experimental results in Table X are approximately half as large as the Padé or semiempirical results. This serious disagreement between theory and experiment has been ascribed to the contribution of overlap interactions to the effective potential involved in the scattering experiments (Dalgarno and Davison, 1967; Smith, 1965).

Table XI presents the dispersion force coefficients for diatomic hydrogen, nitrogen, and oxygen, the listed values corresponding to interactions averaged over all orientations. No experimental results are available for comparison with the predicted bounds, although the semiempirical results are in good agreement with the Padé values.

Our values for the "mixed" interacting pairs are shown in Tables XII–XIV. There is general agreement between the Padé results and the semiempirical and experimental values, although in a number of cases the predicted bounds are apparently violated. Specifically, some of the inert gas-alkali atom experimental results are substantially above the [2, 1] upper bounds.

TABLE XII

Dipole Dispersion Force Coefficients C_{ab} (*a.u.*) for Inert Gas-Alkali Atom Interactions

Interacting pair	Upper bounds[a] $[1, 0]_f$	Upper bounds[a] $[2, 1]_f$	Lower bounds[a] $[2, 1]_\alpha$	Lower bounds[a] $[1, 0]_\alpha$	Semi-empirical values[b]	Experimental[c]
He–H	3.16	3.85	2.80	2.68	2.83	—
He–Li	41.3	22.7	21.9	21.6	22	—
He–Na	73.4	29.4	26.8	25.1	25	—
He–K	127	43.6	38.0	33.9	34	—
He–Rb	175	49.8	43.8	36.1	37	—
He–Cs	219	40.9	37.1	36.9	45	—
Ne–H	6.83	5.86	5.64	5.20	5.6	—
Ne–Li	82.9	44.5	42.5	41.7	42	43[d]
Ne–Na	152	58.9	52.3	48.5	48	—
Ne–K	261	88.2	74.1	65.4	66	74[e]
Ne–Rb	368	103	86.0	69.6	72	—
Ne–Cs	466	82.1	72.0	71.3	87	—
Ar–H	25.8	20.7	19.7	18.5	20	—
Ar–Li	333	176	170	167	180	210[d]
Ar–Na	594	225	206	194	190	—
Ar–K	1,024	332	292	264	270	305[e], 320[d]
Ar–Rb	1,418	378	332	281	290	—
Ar–Cs	1,781	318	291	288	350	340[d]
Kr–H	40.3	29.9	28.0	26.6	29	—
Kr–Li	509	262	253	249	260	280[d]
Kr–Na	917	333	305	288	280	—
Kr–K	1,581	492	433	393	400	520[f], 570[g]
Kr–Rb	2,201	558	489	419	430	—
Kr–Cs	2,773	475	434	430	520	510[d]
Xe–H	65.1	50.9	46.6	41.1	42	—
Xe–Li	829	430	412	401	410	480[d]
Xe–Na	1,488	553	498	463	450	—
Xe–K	2,565	822	708	634	630	710[g], 790[d]
Xe–Rb	3,565	940	803	675	690	—
Xe–Cs	4,488	785	707	695	830	770[g], 790[d]

[a] Values obtained from Eqs. (24a) and (24b) following procedures discussed in the text.

[b] Semiempirical estimates recommended by A. Dalgarno, *Advan. Chem. Phys.* **12,** 143 (1967), obtained following the procedures of A. Dalgarno and W. D. Davison, *Mol. Phys.* **13,** 479 (1967).

[c] Experimental uncertainties estimated to be approximately 15–20% of the values quoted; see R. B. Bernstein and J. T. Muckerman, *Advan. Chem. Phys.* **12,** 389, 414 (1967).

[d] Values obtained from molecular beam scattering data reviewed by R. B. Bernstein and J. T. Muckerman, *Advan. Chem. Phys.* **12,** 389 (1967).

[e] P. R. Brooks, *Bull. Am. Phys. Soc.* **10,** 382 (1965).

[f] E. W. Rothe and R. B. Bernstein, *J. Chem. Phys.* **31,** 1619 (1959). See also E. W. Rothe and R. H. Neynaber, *ibid.* **43,** 4177 (1965).

[g] E. W. Rothe and R. H. Neynaber, *J. Chem. Phys.* **42,** 3306 (1965).

TABLE XIII

Dipole Dispersion Force Coefficients C_{ab} (*a.u.*) for Inert Gas-Diatomic Molecule Interactions

Interacting pair	Upper bounds[a] $[1,0]_f$	Upper bounds[a] $[2,1]_f$	Lower bounds[a] $[2,1]_\alpha$	Lower bounds[a] $[1,0]_\alpha$	Semi-empirical values[b]	Experi-mental[c]
He–H_2	4.55	4.05	3.98	3.79	4.1	—
He–N_2	13.9	10.7	9.85	9.16	10	—
He–O_2	13.4	7.96	7.71	7.52	—	—
Ne–H_2	10.1	8.47	8.08	7.37	8.2	11.7
Ne–N_2	32.8	23.3	20.2	17.8	21	—
Ne–O_2	31.9	16.8	15.6	14.6	—	—
Ar–H_2	37.2	29.2	27.4	25.8	28	29
Ar–N_2	115	76.4	66.5	61.1	69	—
Ar–O_2	110	57.4	53.2	51.0	—	—
Kr–H_2	58.6	42.0	38.5	36.5	40	54
Kr–N_2	185	110	92.7	86.1	96	—
Kr–O_2	178	82.9	75.0	72.2	—	—
Xe–H_2	94.5	72.3	64.8	56.0	58	90
Xe–N_2	296	193	157	131	140	—
Xe–O_2	285	143	126	111	—	—

[a] Values obtained from Eqs. (24a) and (24b) following procedures discussed in the text.

[b] Semiempirical estimates recommended by A. Dalgarno, *Advan. Chem. Phys*, **12,** 143 (1967).

[c] Values obtained from molecular beam scattering data reviewed by R. B. Bernstein and J. T. Muckerman, *Advan. Chem. Phys.* **12,** 389 (1967). Experimental uncertainties estimated to be approximately 15–20% of the values quoted.

Since the dispersion force coefficients determined by the Padé procedure depend only on the Cauchy moments, μ_k, it is of interest to examine the sensitivity of the C_{ab} results on variation of the moment values. In Table XV, we compare the dispersion force coefficients for the inert gases resulting from three different sets of moments. The first set (1) is composed of those given in Table III, constructed from dispersion data and the Stieltjes constraints; the second set (2) was constructed from dispersion data alone (Dalgarno and Kingston, 1960), and the third set (3) was constructed from absorption data alone (Barker and Leonard, 1964). We see that the lowest-order upper and lower bounds, obtained from the $[1,0]_f$ and $[1,0]_\alpha$ approximants, respectively, show little variation except in the case of Xe. The tighter bounds obtained from the $[2,1]_f$ and $[2,1]_\alpha$ approximants, however, exhibit a considerably greater sensitivity to the moment values. Semi-empirical estimates, which are also listed in Table XV, do not correspond to any of the three sets of results and are, as previously noted, closest to the

TABLE XIV

Dipole Dispersion Force Coefficients C_{ab} (*a.u.*) for Alkali Atom-Diatomic Molecule Interactions

Interacting pair	Upper bounds[a] $[1, 0]_f$	Upper bounds[a] $[2, 1]_f$	Lower bounds[a] $[2, 1]_\alpha$	Lower bounds[a] $[1, 0]_\alpha$	Semi-empirical values[b]
H-H_2	9.74	8.86	8.75	8.44	8.7
H-N_2	26.1	21.5	20.7	19.7	21
H-O_2	24.4	17.3	17.0	16.6	—
Li-H_2	148	83.1	81.4	80.5	83
Li-N_2	347	185	180	177	180
Li-O_2	317	162	159	157	—
Na-H_2	246	103	97.8	92.9	91
Na-N_2	612	236	217	205	200
Na-O_2	565	201	191	182	—
K-H_2	424	151	139	127	130
K-N_2	1,055	347	309	279	280
K-O_2	973	295	271	248	—
Rb-H_2	564	168	157	135	140
Rb-N_2	1,450	393	350	297	310
Rb-O_2	1,344	329	305	264	—
Cs-H_2	693	148	140	139	170
Cs-N_2	1,814	334	307	305	370
Cs-O_2	1,687	289	273	272	—

[a] Values obtained from Eqs. (24a) and (24b) following procedures discussed in the text.

[b] Semiempirical estimates recommended by A. Dalgarno, *Advan. Chem. Phys.* **12,** 143 (1967).

lower Padé bounds. Although it is difficult to be certain which of the three sets of coefficients used in Table XV is the most accurate, we prefer set (1) because it has been extracted from dispersion data subject to the constraint that the coefficients correspond to a convergent series of Stieltjes. Moment coefficients, such as those of set (2) for Kr and Xe, which do not satisfy the Stieltjes constraints of Eqs. (90) are obviously unsatisfactory. It is a consequence of this failure to satisfy Eq. (90f) that the set (2) Kr and Xe dispersion force coefficients of Table XV disagree with the more accurate set (1) results. Except for these and the set (3) Xe results, we see that all the other values in the table are mutually compatible. Furthermore, neglecting the Xe and set (2) Kr moment coefficients, the three sets of moments employed (Langhoff and Karplus, 1969; Dalgarno and Kingston, 1960; Barker and Leonard, 1964) exhibit a spread in values similar to the limits of uncertainty we have placed on the inert gas moments of Table III (Sect. III D). Consequenctly, Table XV indicates the spread in C-coefficient values

TABLE XV

Sensitivity of Padé Approximant Dipole Dispersion Force Coefficients C_{ab} (*a.u.*) to the Choice of Cauchy Moments[a]

Interacting pair	Upper bounds[b] $[1, 0]_f$	Upper bounds[b] $[2, 1]_f$	Lower bounds[b] $[2, 1]_\alpha$	Lower bounds[b] $[1, 0]_\alpha$	Semi-empirical values[c]
He–He (1)	1.73	1.48	1.45	1.36	—
(2)	1.73	1.50	1.46	1.38	1.47
(3)	1.74	1.49	1.43	1.37	—
Ne–Ne (1)	10.3	7.59	6.48	5.15	—
(2)	10.3	7.21	5.97	5.14	6.3
(3)	10.4	7.28	5.93	5.19	—
Ar–Ar (1)	118	75.0	63.6	57.9	—
(2)	118	81.4	68.0	58.6	65
(3)	117	75.0	62.0	59.0	—
Kr–Kr (1)	308	157	124	115	—
(2)	308	204	144	117	130
(3)	308	161	123	116	—
Xe–Xe (1)	788	489	356	272	—
(2)	786	525	334	268	270
(3)	763	362	261	245	—

[a] Choice of Cauchy coefficients: (1) this work, see Table III; (2) A. Dalgarno and A. E. Kingston, *Proc. Roy. Soc.* (*London*) **A259,** 424 (1960); (3) J. A. Barker and P. J. Leonard, *Phys. Letters* **13,** 127 (1964).

[b] Values obtained from Eqs. (24a) and (24b) as discussed in the text.

[c] Semiempirical estimates recommended by A. Dalgarno, *Advan. Chem. Phys.* **12,** 143 (1967).

induced by the uncertainties in the moment coefficients. We see that in every case the resulting variations are less than the ± 10–15% we have suggested above as suitable error limits.

E. Bounds to Other Dispersion Force Cofficients

In Sect. II.D, we showed how to use Padé approximants for determining bounds to the multipole-multipole interaction, the three-body dipole-dipole-dipole interaction, and the leading relativistic interaction for nondegenerate systems. Since the necessary quadrupole and higher-order multipole moment coefficients are, in general, not available, we illustrate the methods by presenting results for only the three-body [Eq. (42)] and relativistic terms [Eq. (46)].

Table XVI gives bounds for the dipole three-body coefficient, C_{aaa}, which involves three identical atoms. The values in Table XVI are seen to form tight bounds and should be highly accurate. Also shown in the table are accurate calculations and semiempirical estimates (Dalgarno and Davi-

TABLE XVI

Bounds to Dipole-Dipole-Dipole Dispersion Force Coefficients, C_{aaa}, Obtained from Padé Approximants[a]

System	Upper bound[b] $[1, 0]_f$	Upper bound[b] $[2, 1]_f$	Lower bounds[b] $[2, 1]_\alpha$	Lower bounds[b] $[1, 0]_\alpha$	Semi-empirical[c]
H	24.2	21.7	21.6	21.2	21.6[d]
He	1.79	1.49	1.48	1.41	1.49[d]
Ne	20.7	13.7	12.3	10.3	11.8
Ar	978	564	517	482	521
Kr	3,870	1,710	1,530	1,450	1,560
Xe	16,200	8,410	6,870	5,580	5,430
H_2	54.9	48.8	48.4	47.2	49.0
N_2	995	638	601	568	620
O_2	823	363	360	359	—
Li	3.34×10^5	1.69×10^5	1.69×10^5	1.68×10^5	—
Na	6.76×10^5	2.09×10^5	2.09×10^5	2.07×10^5	—
K	3.45×10^6	8.25×10^5	8.23×10^5	8.10×10^5	—
Rb	6.07×10^6	1.06×10^6	1.06×10^6	1.04×10^6	—
Cs	1.00×10^7	1.35×10^6	1.35×10^6	1.34×10^6	—

[a] All values in atomic units.

[b] Values obtained from Eqs. (23), (29), and (42) in the text using the moments of Table III.

[c] A. Dalgarno, I. H. Morrison, and R. M. Pengelly, *Intern. J. Quant. Chem.* **1**, 161 (1967); R. J. Bell and A. E. Kingston, *Proc. Phys. Soc.* (*London*) **88**, 901 (1966); K. T. Tang, *Phys. Rev.* **177**, 108 (1969).

[d] Accurate calculation by A. Dalgarno and G. A. Victor, *Proc. Phys. Soc.* (*London*) **90**, 605 (1967).

son, 1966; Dalgarno, 1967; Dalgarno *et al.*, 1967) obtained following procedures discussed in Sect. V.A. The latter values are seen to be in good agreement with the Padé bounds, although discrepancies are present for Ne, Xe, and N_2. Unfortunately, no experimental values are available for comparison. The C_{abc} values for all the species in the table have also been computed, and the results are given elsewhere (Langhoff and Karplus, 1970b).

In Table XVII, we show the Padé approximant values for the D_{ab} coefficients of hydrogen and the inert gases (Chang and Karplus, 1970). In most cases, reasonably tight bounds are obtained, although values for the heavier inert gases show considerable spread (Ar, Kr, Xe). Also included in the table are accurate calculations (Johnson *et al.*, 1967) for H–H, H–He, and He–He and results obtained from the semiempirical polarizability formula given in Eq. (118) in Sect. V.A. We see that the Padé values bound the accurate calculations but that the semiempirical estimates fall outside the bounds in many cases. This indicates that the semiempirical method, which yields reliable values of dipole-dipole coefficients (Dalgarno

TABLE XVII

Bounds to the Leading Relativistic Interaction Coefficient D_{ab} Obtained from Padé Approximants[a]

Interacting pair	Upper bound[b] $[1,0]_\gamma$	Upper bound[b] $[2,1]_\gamma$	Lower bound[b] $[2,1]_\beta$	Lower bound[b] $[1,0]_\beta$	Semi-empirical[c]
H–H	0.5774	0.4731	0.4575	0.4313	0.4628[d]
H–He	0.64	0.52	0.492	0.457	0.503[d]
H–Ne	2.0	1.5	1.33	1.21	1.1
H–Ar	4.8	3.0	2.57	2.40	2.0
H–Kr	7.0	4.3	3.51	3.30	2.6
H–Xe	9.8	5.7	4.49	4.25	3.4
He–He	0.93	0.69	0.640	0.581	0.6643[d]
He–Ne	3.8	2.5	1.82	1.59	1.5
He–Ar	9.1	4.5	3.07	2.84	2.3
He–Kr	14	6.6	4.07	3.81	2.9
He–Xe	20	8.9	5.02	4.73	3.7
Ne–Ne	24	13	5.28	4.39	3.7
Ne–Ar	61	23	8.54	7.67	5.6
Ne–Kr	110	38	11.3	10.2	6.9
Ne–Xe	160	54	13.7	12.6	8.9
Ar–Ar	160	43	15.3	14.3	9.5
Ar–Kr	280	74	20.6	19.4	12
Ar–Xe	410	110	25.8	24.4	16
Kr–Kr	540	140	27.9	26.3	15
Kr–Xe	810	200	35.1	33.3	20
Xe–Xe	1,200	300	44.5	42.5	26

[a] All values in atomic units.

[b] Chang and Karplus (1970). Values obtained from Eqs. (45), (46), (50), and (54) in the text using the moment coefficients of Table III.

[c] Chang and Karplus (1970). Values obtained using the semiempirical procedure of A. Dalgarno, I. H. Morrison, and R. M. Pengelly, *Intern. J. Quant. Chem.* **1**, 161 (1967).

[d] R. E. Johnson, S. T. Epstein, and W. J. Meath, *J. Chem. Phys.* **47**, 1271 (1967).

et al., 1967), can be in serious error when applied to other interactions; the approximate oscillator-strength distribution designed to provide an accurate polarizability in the normal dispersion region, does not necessarily give accurate values for other dipole properties. This emphasizes the importance and desirability of employing bounding methods to prescribe error limits for estimated values from approximate oscillator-strength distributions.

V. COMPARISONS WITH RELATED METHODS

In this section, we make some comparisons with alternative procedures for determining dispersion force coefficients. All of the methods discussed are related to the Padé procedure and make use of the polarizability as a series of Stieltjes.

A. Borel Continuation and Semiempirical Procedures

Insight into the problem of constructing dispersion force coefficients from refractivity data can be obtained by considering an alternative method for the continuation of the dynamic dipole polarizability along the imaginary frequency axis. From the power-series expansion of the polarizability given in Eq. (10),

$$\alpha(iy) = \sum_{k=0}^{\infty} \mu_k(-y^2)^k, \tag{10}$$

we construct an associated Borel series (Whittaker and Watson, 1940) in the form

$$\beta(y) = \sum_{k=0}^{\infty} (\mu_k/k!)(-y^2)^k. \tag{110}$$

Although Eq. (10) is convergent only in the interval $0 \leq |y| < \omega_{10}$, the Borel series [Eq. (110)] is convergent throughout the entire finite complex plane (Whittaker and Watson, 1940). A continuation of the polarizability expansion [Eq. (10)] within the Borel polygon (Whittaker and Watson, 1940) can be obtained from the Borel series in terms of the integral

$$\alpha(iy) = \int_0^{\infty} e^{-u}\left[\sum_{k=0}^{\infty} (\mu_k/k!)(-y^2u)^k\right] du. \tag{111}$$

However to make use of Eq. (111) by interchanging the order of summation and integration, it is necessary that the Borel series converge uniformly as a function of u. This is the case only in the interval $0 \leq |y^2u| < \omega_{10}^2$. Thus, the Borel series in Eq. (111) must be summed to, or approximated by, a specific analytical form valid over the entire range of integration. In Fig. 9, we show the Borel function, Eq. (110), calculated for atomic hydrogen (Langhoff and Karplus, 1970b) from the moments in Table I and an approximation to the Borel function in the form of a single exponential

$$\beta(y) = \mu_0 \exp(-y^2/\omega_0^2). \tag{112}$$

The parameter μ_0 is set equal to the static polarizability ($\mu_0 = 4.5$ a.u.), and ω_0 is chosen to be 0.424 (a.u.) to fit the Borel function near the origin. A good fit is obtained for small y, whereas for larger y the exponential dips below the corret result.

Substitution of Eq. (112) into Eq. (111) gives

$$\alpha(iy) = \int_0^{\infty} \mu_0 \exp\left[-\left(\frac{1+y^2}{\omega_0^2}\right)u\right] du = \frac{\mu_0\omega_0^2}{\omega_0^2 + y^2}. \tag{113}$$

Thus, $\alpha(iy)$ obtained in this way has the same analytic form as the $[1, 0]_\alpha$

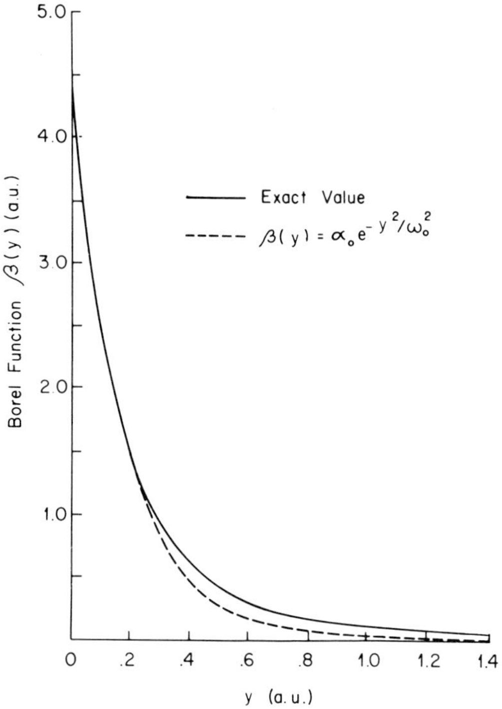

Fig. 9. Borel function for atomic hydrogen. Exact value taken from P. W. Langhoff and M. Karplus, *J. Chem. Phys.* **53**, 233 (1970). Exponential approximation obtained as discussed in the text.

Padé approximant. However, it is important to note that the Borel derivation of Eq. (113) as an approximate analytic continuation of $\alpha(iy)$ does not require any specific value for the parameters μ_0 and ω_0; their choice is dictated by the region for which we wish to have the best fit to the Borel function. By contrast, the derivation of the Padé approximant was based on the use of the Cauchy moments to evaluate the parameters. This makes it possible to obtain bounds, which are not provided by the somewhat more general, but closely related, Borel formulation.

A simple function of the form given in Eq. (113) has been used to approximate the polarizability at both real and imaginary frequencies (London, 1930; Margenau, 1939; Salem, 1960; Mavroyannis and Stephen, 1962; Tang, 1968, 1969a). However, this use has only had an *ad hoc* basis. We have here shown how such a form results for the polarizability at imaginary frequencies from a proper analytic continuation with an exponential approximation to the Borel series. Furthermore, the accuracy of the resulting dispersion force coefficient is evidently dependent upon how well a single exponential can fit the correct Borel function in the important region. For atomic hydrogen with the parameters given in the foregoing, we obtain $C_{HH} = 6.44$ (a.u.), a slight underestimate compared with the correct value of 6.499 (a.u.) (Dalgarno and Davison, 1966; Dalgarno, 1967).

Although for atomic hydrogen there are available a large number of moment coefficients, with which the Borel function may be calculated by substituting into Eq. (110), for more complex systems only the first few coefficients have been extracted from dispersion data. In these cases, an alternative approximation to the Borel function can be obtained by requiring that the power series expansion of the exponential [Eq. (112)] equal the Borel series of Eq. (110) to first order. This results in the expression

$$\omega_0^2 = \mu_0/\mu_1 , \tag{114}$$

which makes Eq. (113) identical with the $[1, 0]_\alpha$ Padé approximant. The procedure can evidently be extended to higher order by approximating the Borel function by a sum of exponentials

$$\beta(y) = \sum_{i=1}^{n} a_i \exp(-y^2/\omega_i^2) , \tag{115}$$

where the parameters a_i and ω_i are determined from the equations

$$\sum_{i=1}^{n} a_i/\omega_i^{2j} = \mu_j , \qquad j = 0, 1, \ldots, 2n-1 , \tag{116}$$

obtained by demanding that the power series expansion of Eq. (115) equal the Borel series, Eq. (110), term by term to order $2n - 1$. The resulting polarizability is

$$\alpha(iy) = \sum_{i=1}^{n} \frac{a_i\omega_i^2}{\omega_i^2 + y^2} , \tag{117}$$

which is identical with the $[n, n-1]_\alpha$ Padé approximant. Thus, the exponential approximation to the Borel function, when evaluated by Eq. (116), is an alternative way of phrasing the Padé procedure.

Prior to the introduction of rigorous bounding methods for dispersion force coefficients (Langhoff and Karplus, 1967, 1970b), semiempirical procedures which made use of approximate oscillator-strength distributions were used in such calculations (Dalgarno and Davison, 1966; Dalgarno, 1967). The distributions, although not necessarily accurate representations of the correct spectra, were constrained to satisfy known sum rules. The constraining effect of the sum rules appears to be of importance in determining the values obtained, although no quantitative test of the accuracy of the distribution was made. More recently, the dispersion data have been fitted by a sum of two Maxwell–Sellmeier type of terms (Dalgarno *et al.*, 1967)

$$\alpha(\omega) = \frac{f_1}{\omega_1^2 - \omega^2} + \frac{f_2}{\omega_2^2 - \omega^2}, \tag{118}$$

which can provide an accurate approximation to the polarizability for imaginary frequencies, although the oscillator-strength distribution is not adequately represented by $(f_1, \omega_1, f_2, \omega_2)$. The results of introducing this expression in the Casimir–Polder integral for C_{ab} have been given in Tables IX–XV for the inert gases and diatomic molecules hydrogen, nitrogen, and oxygen. A justification for using Eq. (118) and extending it to describe the analytic continuation of the polarizability along the imaginary axis is provided by the exponential approximation to the Borel function. That the Table IX semiempirical results are close to the $[2, 1]_\alpha$ Padé lower bounds is a consequence of the fact that the fitting procedure satisfies the demands of Eqs. (116). However, the inappropriateness of an approximation to the oscillator-strength distribution as simple as that appearing in Eq. (118) has been noted in attempts to use it for the relativistic corrections (Sect. IV. E).

B. Combination Rules

Dispersion force coefficients between different species can be estimated from the self-interaction coefficients. Such so-called combination rules have been used rather successfully in some recent studies (Salem, 1960; Mavroyannis and Stephen, 1960; Tang, 1968, 1969a; Hirschfelder *et al.*, 1954; Pitzer, 1959; Wilson, 1965; Mason and Monchick, 1967; Crowell, 1968). To examine the origin and accuracy of the combination rules, we make use of the Padé approximant.

The two-body combination rule can be thought of as based on the approximate polarizability expression (Salem, 1960; Mavroyannis and Stephen, 1960; Tang, 1968, 1969a)

$$\alpha_a(iy) = \frac{\mu_0^{(a)} \bar{\omega}_a^2}{\bar{\omega}_a^2 + y^2}, \tag{119}$$

where $\mu_0^{(a)}$ is the static polarizability and $\bar{\omega}_a$ is an effective excitation energy for species a. We have shown previously that the alternative choices $\bar{\omega}_a^2 = \mu_0^{(a)}/\mu_1^{(a)}$ and $\bar{\omega}_a^2 = N^{(a)}/\mu_0^{(a)}$ lead to the London (1930) and Slater–Kirwood (1931) approximations, respectively, for the dispersion force coefficient

$$C_{aa} = \tfrac{3}{4}(\mu_0^{(a)})^2 \bar{\omega}_a. \tag{120}$$

The value of C_{aa} in Eq. (120) will be greater (Slater–Kirkwood) or less

(London) than the correct dispersion force coefficient, depending upon which of the two expressions for $\bar{\omega}_a$ are used (see Sect. IV. D). The mixed dispersion coefficient is given by

$$C_{ab} = \tfrac{3}{2}\,\mu_0^{(a)}\,\mu_0^{(b)}\,\frac{\bar{\omega}_a\bar{\omega}_b}{(\bar{\omega}_a + \bar{\omega}_b)}\,, \tag{121}$$

and will again form an upper or lower bound to the correct result, depending upon the choice for $\bar{\omega}_a$ and $\bar{\omega}_b$. Alternatively, in order to establish a semiempirical combination rule, we can use Eq. (120) to define an effective excitation energy as

$$\bar{\omega}_a = \tfrac{4}{3}\,C_{aa}/(\mu_0^{(a)})^2\,, \tag{122}$$

where C_{aa} and $\mu_0^{(a)}$ are the measured or accurately calculated dispersion force coefficient and static polarizability, respectively. Making use of Eq. (122), we substitute into Eq. (121) to obtain

$$\bar{C}_{ab} = \frac{2C_{aa}C_{bb}}{(C_{aa}\mu_0^{(b)}/\mu_0^{(a)} + C_{bb}\mu_0^{(a)}/\mu_0^{(b)})}\,, \tag{123}$$

or, rewriting (Hirschfelder *et al.*, 1954; Pitzer, 1959; Wilson, 1965; Mason and Monchick, 1967; Crowell, 1968),

$$\bar{C}_{ab} = (C_{aa}C_{bb})^{1/2}\,\frac{2(\bar{\omega}_a\bar{\omega}_b)^{1/2}}{(\bar{\omega}_a + \bar{\omega}_b)} \cong (C_{aa}C_{bb})^{1/2}\,. \tag{124}$$

The last step in Eq. (124) follows from the assumption that $\bar{\omega}_a = \bar{\omega}_b$, which is approximately correct for many atomic and molecular systems.

Since the combination rule [Eq. (123)] is based on Eq. (119), the accuracy of the resulting C_{ab} coefficient depends on how well Eq. (119) can fit the polarizability at imaginary frequencies. To test this point for the inert gases, we compare the quantities $([2, 1]_\alpha + [2, 1]_f)/2\mu_0^{(a)}$ as functions of $x = y/\bar{\omega}_a$, with the function $f(x) = 1/(1 + x^2)$. The $\bar{\omega}_a$ are obtained from Eq. (122) using the average of the C_{aa} values obtained from the two [2, 1] approximants and the $\mu_0^{(a)}$ values of Table III. To the extent that $([2, 1]_\alpha + [2, 1]_f)/2$ is an accurate representation of the correct polarizability, we are thereby determining the validity of Eq. (119). The results are shown in Fig. 10, which demonstrates that rather good fits are obtained in all cases over the portions of the frequency spectrum for which the polarizabilities are large. Thus, reliable C_{ab} estimates are expected from the combination rule [Eq. (123)] for the inert gases. However, if Eq. (119) were to give a poor fit to the correct polarizabilities, even though the exact C_{aa}, C_{bb} and $\mu_0^{(a)}$, $\mu_0^{(b)}$ values were used, inaccurate C_{ab} coefficients could result. This suggests that the Padé procedure, which provides bounds on

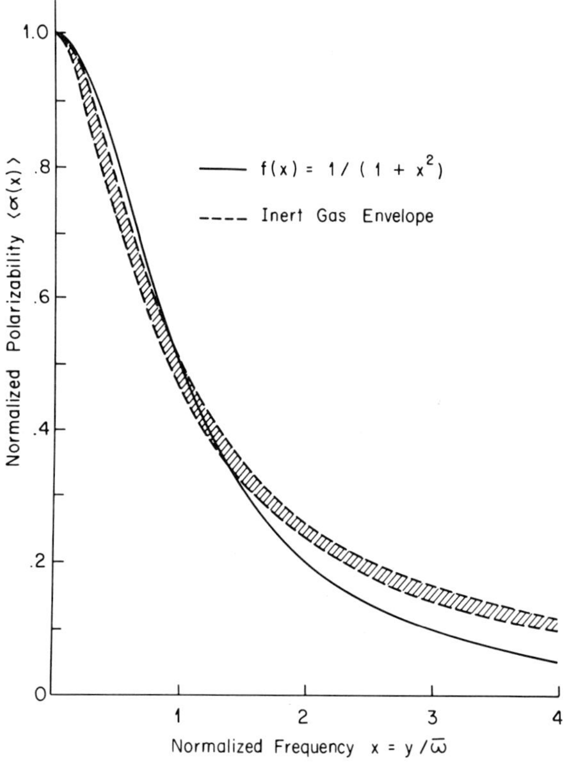

Fig. 10. Normalized inert gas polarizabilities on the imaginary axis as a function of normalized frequency determined from the Padé approximant averages.

C_{ab}, would be important to apply as well if enough information to determine the appropriate $\bar{\omega}_a$ values is available.

The combination rule given in Eq. (124) is related to a simple but important sequence of inequalities satisfied by any set of dispersion force coefficients (Weinhold, 1967, 1968, 1969). Let us consider the set of C_{ab} $(a, b = 1, N)$ as elements of a matrix. For an arbitrary real vector $|\mathbf{X}\rangle = (X_1 X_2 \cdots X_N)$, we have

$$\langle \mathbf{X} | \mathbf{C} | \mathbf{X} \rangle = \sum_{a,b=1}^{N} X_a C_{ab} X_b = \frac{3}{\pi} \int_0^\infty \left| \sum_{a=1}^{N} X_a \alpha_a(iy) \right|^2 dy\,, \tag{125}$$

where the last equality follows from the Casimir–Polder integral formula. We see that $\mathbf{C}$ must be a nonnegative matrix, since the right-hand side of Eq. (125) is nonnegative and $|\mathbf{X}\rangle$ is an arbitrary vector. Thus, the matrix

C has no negative eigenvalues, and

$$\det |\mathsf{C}| \geq 0 . \tag{126}$$

This result furnishes numerous inequalities (Weinhold, 1967, 1968, 1969). In the 2×2 case, we have

$$C_{aa}C_{bb} - C_{ab}^2 \geq 0 , \tag{127}$$

which corresponds to Eq. (124) if the equal sign in Eq. (127) is used. The higher-dimension cases furnish additional inequalities, which are somewhat unwieldy to write out but are simple to use for computations.

It has been shown (Weinhold, 1967, 1968, 1969) that a corresponding determinantal condition to Eq. (126) applies to the matrix $S_{kl} = S(k + l)$, with $S(k + l)$ the sum rule of Eq. (49). Consequently, it might appear that the Cauchy moments $[\mu_k = S(-2k - 2)]$ satisfy determinantal inequalities over and above the Stieltjes constraints of Eq. (21); i.e., the Stieltjes constraints involve sequential Cauchy moments, whereas the constraints furnished by Eq. (126) can involve nonsequential moments. Such additional constraints would be useful in a variety of applications (Weinhold, 1967, 1968, 1969; Kramer, 1969), as well as for the extraction of Cauchy moments from experimental dispersion data (Sect. III. D). However, the Stieltjes constraints, Eq. (21), are the necessary and sufficient conditions that the μ_k are moments of a nondecreasing distribution (Baker, 1965). Since the positivity of the $S(k + l)$ matrix is established directly from the nondecreasing nature of the distribution (Weinhold, 1967, 1968, 1969), the sequential constraints imply the validity of all possible nonsequential constraints. Alternately, one can demonstrate directly that the sequential Stieltjes constraints using a given number of moments imply all possible nonsequential constraints of the same order. This means that no new constraining conditions are furnished by the nonsequential constraints obtained from Eq. (126) over and above the Stieltjes constraints. Moreover, it should be noted that Eq. (21) refers to a positive definite quantity, whereas Eq. (126) requires only that the determinant be positive.

C. Alternative Bounding Methods

In addition to the Padé approximant method, a number of other procedures that result in bounds on dispersion force coefficients have been introduced in recent years. We have already mentioned the bounds obtained in the form of combination rules (Sect. V.B.) (Hirschfelder *et al.*, 1954; Tang, 1968, 1969a; Kramer, 1969; Weinhold, 1967, 1968; 1969), for two-body dispersion forces; they can be applied equally well to the three-body interaction. These techniques provide bounding relations among dispersion force coefficients, without reference to moment coefficients, and form a useful complement to the Padé procedure.

A bounding procedure intimately related to that given here has been developed by Gordon (1968). When applied to polarizabilities and dispersion forces his upper bounds are identical with those given here and his lower bounds are numerically similar but distinct, due to his explicit introduction of the resonance transition frequencies. Similar results have been obtained by Goscinski (1968), who employs the theory of operator inequalities for positive operators. Bounding procedures in the form of quantum mechanical variation techniques for the dynamic polarizability at real and imaginary frequency have been given by Epstein (1968) and Robinson (1969). Special choices for the variation functions involved result in bounds identical with those provided by the Padé method, although the variation procedures can yield other approximations, as well. In addition, Alexander (1970) has recently shown that the technique of Bell (1965), which uses moment coefficients alone in estimating dispersion force coefficients, can be modified to provide rigorous bounds. Finally, Tang (1969b) has introduced a continued factorization procedure for the polarizability which employs a number of transition frequencies to obtain more rapidly convergent bounds for both real and imaginary frequency and has applied it to the computation of dispersion force coefficients (Tang, 1970). A more detailed discussion of these methods is given separately (Langhoff, *et. al.*, 1970).

VI. CONCLUDING REMARKS

We have considered the application of the Padé approximant to the solution of a problem for which the method was designed, i.e., the moment problem for a series of Stieltjes. It has been found that, if attention is restricted to the cut complex plane $(-\omega_{10}^2 > \mathrm{Re}\, z > -\infty)$, the Padé approximant provides a useful technique for achieving rapid convergence to the correct analytic function. The function of interest in the present application is the dynamic dipole polarizability for an atomic or molecular system. The moments of the pertinent nondecreasing distribution (cumulative oscillator–strength distribution) are the Cauchy coefficients obtainable from measured dispersion and absorption data or theoretical computations. The polarizability for real frequencies is of interest because of its relation to the refractive index whereas the polarizability at imaginary frequencies can be used to determine dispersion forces. Rapid bounded convergence is obtained from low-order approximants along the imaginary frequencies axis, allowing construction of accurate bounds to dispersion force coefficients from a minimum of theoretical or experimental information in the form of Cauchy coefficients. Similar rapid bounded convergence from low-order approximants is achieved for real frequencies in the normal dispersion region. This furnishes a new method for performing bounded extrapolation of measured optical refractivity from the visible into the vacuum ultraviolet.

A variety of example involving the inert gases, alkali atoms, and simple diatomic molecules are used to illustrate the Padé approximant technique.

REFERENCES

Alexander, M. (1970a). *J. Chem. Phys.* **52,** 3354.
Alexander, M. (1970b). *Phys. Rev.* **A1,** 1397.
Axilrod, B. M., and Teller, E. (1943). *J. Chem. Phys.* **11,** 299.
Baker, G. A., Jr. (1965). "Advances in Theoretical Physics" (K. A. Brueckner, ed.), Vol. 1, pp. 1–58. Academic Press, New York.
Baker, G. A. Jr. (1969). *J. Math. Phys.* **9,** 32.
Barker, J. A., and Leonard, P. J. (1964). *Phys. Letters* **13,** 127.
Bell, R. J. (1965). *Proc. Phys. Soc.* **86,** 17.
Bernstein, R. B., and Muckerman, J. T. (1967). *Advan. Chem. Phys.* **12,** 389.
Bethe, H. A., and Salpeter, E. E. (1957). "The Quantum Mechanics of One- and Two-Electron Atoms", Academic Press, New York.
Casimir, H. B. G., and Polder, D. (1948). *Phys. Rev.* **73,** 360.
Chamberlain, G. E., and Zorn, J. C. (1963). *Phys. Rev.* **129,** 677.
Chang, T. Y., and Karplus, M. (1970). *J. Chem. Phys.* **52,** 4698.
Clementi, E. (1963). *J. Chem. Phys.* **38,** 996.
Common, A. K. (1968). *J. Math. Phys.* **9,** 32.
Crowell, A. D. (1968). *J. Chem. Phys.* **49,** 3324.
Dalgarno, A. (1962). *Advan. Phys.* **11,** 281.
Dalgarno, A. (1967). *Advan. Chem. Phys.* **12,** 143.
Dalgarno, A. (1966). Applications of time-dependent perturbation theory, *in* "Perturbation Theory and Its Applications in Quantum Mechanics" (C. H. Wilcox, ed.). Wiley, New York.
Dalgarno, A., and Kingston, A. E. (1959). *Proc. Phys. Soc.* (*London*) **73,** 455.
Dalgarno, A., and Kingston, A. E. (1960). *Proc. Roy. Soc.* (*London*) **A259,** 424.
Dalgarno, A., and Stewart, A. L. (1960). *Proc. Phys. Soc.* (*London*) **76,** 49.
Dalgarno, A., and Davison, W. D. (1966). *Advan. At. Mol. Phys.* **2,** 1.
Dalgarno, A., and Davison, W. D. (1967). *Mol. Phys.* **13,** 479.
Dalgarno, A., and Victor, G. A. (1966). *Proc. Roy. Soc.* (*London*) **A291,** 291.
Dalgarno, A., and Victor, G. A. (1967). *Proc. Phys. Soc.* (*London*) **90,** 605.
Dalgarno, A., Morrison, I. H., and Pengelly, R. M. (1967). *Intern. J. Quant. Chem.* **1,** 161.
Eisenschitz, R., and London, F. (1930). *Z. Physik* **60,** 491.
Epstein, S. T. (1968). *J. Chem. Phys.* **48,** 4718.
Fano, U., and Cooper, J. W. (1968). *Rev. Mod. Phys.* **40,** 441.
Futrelle, R. P., and McQuarrie, D. A. (1968). *Chem. Phys. Letters* **2,** 223.
Gill, P., and Heddle, D. W. O. (1963). *J. Opt. Soc. Am.* **53,** 847.

Gordon, R. G. (1968a). *J. Chem. Phys.* **48,** 3929.
Gordon, R. G. (1968b). *J. Math. Phys.* **9,** 655.
Goscinski, O. (1968). *Intern. J. Quant. Chem.* **2,** 761.
Graben, H. W., Present, R. D. (1962). *Phys. Rev. Letters* **9,** 247.
Harriman, J. M. (1956). *Phys. Rev.* **101,** 594.
Heddle, D. W. O., Jennings, R. E., and Parsons, A. S. L. (1963). *J. Opt. Soc. Am.* **53,** 847.
Hirschfelder, J. O., and Meath, W. J. (1967). *Advan. Chem. Phys.* **12,** 3.
Hirschfelder, J. O., Curtiss, C. F., and Bird, R. B. (1954). "Molecular Theory of Gases and Liquids", p. 963. Wiley, New York.
Jansen, L. (1964). *Phys. Rev.* **135,** A1292.
Johnson, R. E., Epstein, S. T., and Meath, W. J. (1967). *J. Chem. Phys.* **47,** 1271.
Karplus, M., and Kolker, H. J. (1963). *J. Chem. Phys.* **39,** 1493.
Kihara, T. (1958). *Advan. Chem. Phys.* **1,** 267.
Kirn, M. (1921). *Ann. Physik* **64,** 566.
Koch, J. (1912). *Arkiv. Math. Astron. Fysik* **8,** 20.
Korff, S. A., and Breit, G. (1932). *Rev. Mod. Phys.* **4,** 471.
Kramer, H. L. (1969). Disertation, Harvard Univ. (unpublished).
Langhoff, P. W., and Karplus, M. (1967). *Phys. Rev. Letters* **19,** 1461.
Langhoff, P. W., and Karplus, M. (1970a). *J. Chem. Phys.* **52,** 1435.
Langhoff, P. W., and Karplus, M. (1970b). *J. Chem. Phys.* **53,** 233.
Langhoff, P. W., and Karplus, M. (1969). *J. Opt. Soc. Am.* **59,** 863.
Langhoff, P. W., Karplus, M., and Hurst, R. P. (1966). *J. Chem. Phys.* **44,** 505.
Langhoff, P. W., Gordon, R. G., and Karplus, M. (1970). *J. Chem. Phys.* (to be published).
London, F. (1930). *Z. Physik Chem.* **B11,** 222.
Margenau, H. (1939). *Rev. Mod. Phys.* **11,** 1.
Mason, E. A., and Monchick, L. (1967). *Advan. Chem. Phys.* **12,** 329.
Mavroyannis, C., and Stephen, M. J. (1962). *Mol. Phys.* **5,** 629.
McQuarrie, D. A., Terebey, J. N., and Shire, S. J. (1969). *J. Chem. Phys.* **51,** 4683.
Meath, W. J., and Hirschfelder, J. O. (1966). *J. Chem. Phys.* **44,** 3197, 3210.
Meyer, L. (1969). *Advan. Chem. Phys.* **16,** 343.
Moore, C. E. (1949). Atomic Energy Levels, Vol. I. Nat. Bur. of Stds. Cir. 467.
Musulin, B., and Epstein, S. T. (1964). *Phys. Rev.* **A136,** 966.
Muto, Y. (1943). *Proc. Phys. Math. Soc. Japan* **7,** 629.
Pauly, H., and Toennies, J. P. (1965). *Advan. At. Mol. Phys.* **1,** 195.
Peck, E. R., and Khanna, B. N. (1966). *J. Opt. Soc. Am.* **56,** 1059.
Pitzer, K. S. (1959). *Advan. Chem. Phys.* **2,** 59.
Podolsky, B. (1928). *Proc. Natl. Acad. Sci. US* **14,** 253.
Robinson, P. D. (1969). *J. Phys. A.* (Proc. Phys. Soc.) **2,** 193, 295.
Salem, L. (1960). *Mol. Phys.* **3,** 441.
Salop, A., Pollack, E., and Bederson, B. (1961). *Phys. Rev.* **124,** 1431.
Samson, J. A. R. (1966). *Advan. At. Mol. Phys.* **2,** 178.
Schiff, B., and Pekeris, C. L. (1964). *Phys. Rev.* **134,** A640.
Schwartz, C. (1966). *J. Comp. Phys.* **1,** 21.
Sengupta, S., and Mukherji, A. (1967). *J. Chem. Phys.* **47,** 260.
Sherwood, A. E., and Pransnitz, J. M. (1964). *J. Chem. Phys.* **41,** 413, 429.
Slater, J. C. (1960). "Quantum Theory of Atomic Structure", Vol. I, p. 154. McGraw-Hill, New York.
Slater, J. C., and Kirkwood, J. G. (1931). *Phys. Rev.* **37,** 682.
Smith, F. J. (1965). *Mol. Phys.* **10,** 283.

Tang, K. T. (1968). *J. Chem. Phys.* **49,** 4727.
Tang, K. T. (1969a). *Phys. Rev.* **177,** 108.
Tang, K. T. (1969b). *Phys. Rev. Letters* **23,** 1271.
Tang, K. T. (1970). *Phys. Rev.* **A1,** 1033.
Weinhold, F. (1967). *J. Chem. Phys.* **46,** 2448.
Weinhold, F. (1968). *J. Phys. A.* (*Proc. Phys. Soc.*) **1,** 305, 655.
Weinhold, F. (1969a). *J. Phys. B.* (*Proc. Phys. Soc.*) **2,** 517.
Weinhold, F. (1969b). *J. Chem. Phys.* **50,** 4136.
Whittaker, E. T., and Watson, G. N. (1940). "A Course of Modern Analysis", 4th ed., p. 140. Cambridge Univ. Press, London and New York.
Wiese, W. L., Smith, M. W., and Glennon, B. M. (1968). Atomic Transition Probabilities, NBS Natl. Std. Ref. Data System 4, Vol. 1, p. 11. U.S. Government Printing Office, Washington, D. C.
Wilkinson, P. G. (1960). *J. Opt. Soc. Am.* **50,** 1002.
Wilson, J. N. (1965). *J. Chem. Phys.* **43,** 2564.

CHAPTER 3 BOUNDS FOR AVERAGES USING MOMENT CONSTRAINTS*

J. C. Wheeler[†] *and R. G. Gordon*[‡]

Department of Chemistry, Harvard University

I. INTRODUCTION

In many kinds of physical problems, one wishes to translate some information known about a system into predictions about other properties of the system. Often, of course, the known information is not sufficiently complete to determine uniquely the unknown properties. However, even in such indeterminate cases, one may still hope to determine the *allowed ranges* of predicted properties which are consistent with the known properties of the system.

In this review, we discuss some of the mathematical techniques for predicting such allowed ranges of predicted values. The properties we consider are of the form of weighted averages over some unknown, but well-defined, nonnegative distribution function that characterizes the system of interest. Certain averages over this distribution are assumed kn wn, and from these known averages we wish to predict allowed ranges for other averages over the same distribution. For example, one may know the values of an integral

$$\mu(T_i) \equiv \int_0^\infty e^{-E/kT_i} G(E)\, dE \tag{1}$$

for certain values of the parameter T_i, with some nonnegative $G(E) \geq 0$. Physically, these properties might be the partition function of some system

* Supported in part by the National Science Foundation, Grant GP 6712.

† National Science Foundation Postdoctoral Fellow. Present address: Department of Chemistry, University of California, San Diego; La Jolla, California.

‡ Alfred P. Sloan Foundation Fellow.

at temperatues T_i, and $G(E)$ is the density of states at energy E. Then we may wish to predict possible ranges for the average $\mu(T)$, at other temperatures not equal to the known ones T_i. Some other examples of physical problems of this type are summarized in Table I.

Historically, this type of problem was first posed by Tchebychev (1874) who considered the averages

$$\mu_i = \int E^i G(E)\, dE\,, \tag{2}$$

known as power *moments*. This terminology corresponds to the physical analogy of a mass distribution $G(E)$, of which μ_0 is the total mass, μ_1 is the static moment, μ_2 the *moment* of inertia, etc. Tchebychev derived his famous inequalities on the mass distribution using these moment constraints. This work was extended by Stieltjes (1884), who also used the moment constraints (2). Markov (1896) generalized the constraints to include more general conditions such as Eq. (1). Still further generalizations, such as including bounds on the distribution function itself as further constraints, have been treated by Krein (1951) and Karlin and Studden (1966).

In this review, we shall concentrate on bounds for averages, based upon the power moments (2). This case, in addition to being the most common one arising in practice, has the important advantage that the bounds can be constructed explicitly in a direct manner. For the more general constraints, such as (1), there is no general constructive method available for explicitly forming the bounds. However, existence theorems are available in these cases (Markov, 1896; Krein, 1951; Karlin and Studden, 1966), and sufficiently accurate evaluations of these bounds can be accomplished by numerical methods, such as least squares solutions (Melton and Gordon, 1969) or linear programming (Melton and Gordon, 1969; Futrelle and McQuarrie, 1968).

Section II and the Appendices simply and directly demonstrate the construction of bounds for averages, using the power moment constraints, and also certain constraints on the distribution itself. In proving results in this section, we shall consistently take the point of view that those properties of the construction which can be verified explicitly in any numerical computation may be stated with only references to proofs in the literature, whereas the proof that the numerical results obtained, on the basis of these verifiable properties, must indeed provide bounds will be given in some detail. This strategy allows us to avoid lengthy detours into the areas of continued fractions and orthogonal polynomials which are typical of most treatments of this subject, and which we feel have been, at least in part, responsible for the long delay in application of these ideas to physical and chemical problems.

Some examples and applications are given in Sect. III.

TABLE I

Examples of Bounding Linear Functionals by Knowledge of Other Linear Functionals of the Same Distribution Function

Example	Distribution function	Known averages	Averages to be bounded
Vibration of atoms in solids[a]	Density of vibration frequencies	Even powers of vibration frequency, calculated from harmonic forces between atoms	Heat capacity; energy; form factor for x-ray scattering; cumulative distribution
Long-range forces between atoms[b]	Optical absorption spectrum of separate atoms	Refractive index or polarizability of separate atoms measured at various frequencies	Potential energy of interaction between two atoms at long distances apart
Thermally averaged chemical rate constants[c]	Collision cross section for chemical reaction, as a function of energy of reactants	Thermally averaged rate constant (Laplace transforms of distribution function)	Cumulative cross section, i. e., integrated up to a given energy
Second-order perturbation sums in quantum mechanics[d]	Squares of perturbation matrix elements at various excited-state energies	Averages of powers of the excitation energy	Second-order perturbation energies
Ferromagnetic Ising models[e]	Density of zeroes of partition function in complex activity plane	Averages of powers of the activity, from virial coefficients	Thermodynamic properties of the models
Accuracy of approximate wave functions[f]	Squares of coefficients of excited state wave functions, contained in approximate ground-state function	Matrix elements of powers of the Hamiltonian	Overlap of approximate wave function with true (but unknown) wave function
Quantum mechanical scattering amplitudes[g,h]	Residues of scattering amplitude in complex energy plane	Power moments in energy or scattering amplitudes at negative energies	Scattering amplitudes at positive energies
Dielectric properties[i]	Distribution of Debye relaxation times	Real and imaginary parts of dielectric constants	Cumulative distribution of relaxation times

[a] J. C. Wheeler and R. G. Gordon, *J. Chem. Phys.* **51,** 5566 (1969).

[b] P. W. Langhoff and M. Karplus, *Phys. Rev. Letters* **19,** 26 (1967); R. G. Gordon, *J. Chem. Phys.* **48,** 3929 (1968).

[c] L. A. Melton and R. G. Gordon, *J. Chem. Phys.* **51,** 5449 (1969).

[d] R. G. Gordon, *Intern. J. Quant. Chem.* **25,** 151 (1968).

[e] G. A. Baker, *Phys. Rev.* **161,** 434 (1967).

[f] R. G. Gordon, *J. Chem. Phys.* **48,** 4948 (1968).

[g] D. Masson (this volume); R. W. Haymaker, and L. Schlessinger (this volume).

[h] R. W. Haymaker and L. Schlessinger (this volume).

[i] R. G. Gordon, *J. Chem. Phys.* (to be published).

II. THE MOMENTS PROBLEM

A. Half-Open Infinite Interval $[a, \infty)$

Let $G(x)$ be a probability density function on $[a, \infty)$ with moments μ_n defined by

$$\mu_n = \int_a^\infty x^n G(x)\, dx\,, \qquad n = 0, 1, 2, \ldots . \tag{3}$$

We wish to evalute a range of possible values for the integral

$$\mathscr{I}(F) = \int_a^\infty F(x) G(x)\, dx\,, \tag{4}$$

where $F(x)$ is a known function of x, presumed to have any required number of derivatives on the interior of $[a, \infty)$, and where $G(x)$ is unknown, but $M+1$ moments μ_n $(n = 0, 1, \ldots, M)$ are given. [$F(x)$ may contain a parametric dependence on any number of other variables z_1, z_2 ... which are held fixed throughout our arguments.] To find the desired bounds on the integral $\mathscr{I}(F)$, we construct a delta function approximation to $G(x)$ according to a prescription that depends on whether $M + 1$ is even or odd.

1. If $M + 1$ is *even* $(M + 1 = 2n)$, then construct an approximate density function

$$G^e(x) = \sum_{i=1}^{n} w_i\, \delta(x - x_i)\,, \tag{5}$$

where $\delta(x)$ is the Dirac delta function and where the n points x_i and n weights w_i are determined by the requirement that they give the first $2n$ moments correctly, that is, that

$$\sum_{i=1}^{n} w_i x_i^k = \mu_k \qquad (k = 0, 1, \ldots, M)\,. \tag{6}$$

2. If $M + 1$ is *odd* $(M + 1 = 2n + 1)$, then construct an approximate density function

$$G^o(x) = \sum_{i=0}^{n} w_i\, \delta(x - x_i)\,,$$

where $x_0 = a$, and the weights and remaining points are determined by the requirement that $G^o(x)$ give the first $2n + 1$ moments correctly.

In case 1, we are confronted $2n$ equations in $2n$ unknowns; in case 2, with $2n + 1$ equations in $2n + 1$ unknowns. Despite the formidably nonlinear appearance of these equations, they can be recast (Gordon, 1968) into the problem of diagonalizing an $n \times n$ symmetric, real, tridiagonal matrix, which is easily solved by standard techniques. The positions x_i are the eigenvalues of this matrix, and the weights w_i are given in terms of the eigenvectors. The explicit procedure for obtaining the x_i and w_i for both the even and odd cases is outlined in Appendix A.

The positions and weights of $G^e(x)$ and $G^o(x)$ are closely related, respectively, to orthogonal and quasi-orthogonal polynomials on $[0, \infty)$ defined by the exact density function, $G(x)$, through its moments. In fact, the positions x_i are just the roots of these polynomials, and the weights, w_i are functionals, defined by the moments, of related polynomials (Shohat and Tamarkin, 1950a). This relationship allows one to prove a number of theorems about $G^e(x)$ and $G^o(x)$. In particular, the theorems assure us in both cases 1 and 2 that, if $G(x)$ is indeed nonnegative, then all of the x_i are real and lie on the interval $[a, \infty)$ (Shohat and Tamarkin, 1950b), and that all of the w_i are real and positive (Shohat and Tamarkin, 1950c). Thus, $G^e(x)$ and $G^o(x)$ are reasonable approximations to $G(x)$ in that they are nonnegative and place no weight outside of $[a, \infty)$. These properties may be verified explicitly for the x_i and w_i obtained from the algorithm given in Appendix A, and so we shall not prove them here. We simply remark that a failure of these conditions indicates either that the moments one is using do not correspond to a nonnegative $G(x)$ defined on $[a, \infty)$, or that greater precision must be used in carrying through the algorithm.

A central result that we shall prove is that

$$\mathscr{I}(F) = \int_a^\infty F(x)G(x)\,dx = \begin{Bmatrix} \displaystyle\sum_{i=1}^{n} w_i^e F(x_i^e) + (K^e)^2 \frac{F^{(2n)}(\xi^e)}{(2n)!} \\ \displaystyle\sum_{i=0}^{n} w_i^o F(x_i^o) + (K^o)^2 \frac{F^{(2+1)}(\xi^o)}{(2n+1)!} \end{Bmatrix}, \quad (8)$$

where superscripts e and o refer to even and odd approximations. respectively, where ξ is some point in the interval $[a, \infty)$, and where the positive constants $(K^e)^2$ and $(K^o)^2$ are given by

$$\begin{aligned} (K^e)^2 &= \int_a^\infty \prod_{i=1}^{n} (x - x_i^e)^2 G(x)\,dx\,, \\ (K^o)^2 &= \int_a^\infty (x-a) \prod_{i=1}^{n} (x - x_i^o)^2 G(x)\,dx\,. \end{aligned} \quad (9)$$

The first term on the right-hand side of (8) is the value obtained if $G^e(x)$ is substituted for $G(x)$ in the integral, and will be referred to as $\mathscr{I}^e(F)$ or

$\mathscr{I}^o(F)$, respectively. The remaining term is the error that contains $(K)^2$, which typically becomes very small as n becomes large.

Several remarks should be made about the results implied by (8) and (9). Firstly, if the function F has the property that its derivatives, $F^{(m)}$, are of constant sign on $[a, \infty)$, then the even and odd approximations provide bounds to the integral. In particular, if successive derivatives of F are of opposite sign, then the even and odd approximations form a sequence of upper and lower bounds, for the correction terms are of opposite sign. Secondly, if these are bounds, they are the best possible bounds obtainable if only the moments are known, because we have constructed a $G^e(x)$ and $G^o(x)$ that satisfy the desired moment conditions exactly, and that *attain* these bounds. It follows immediately from this second remark that the bounds must improve as more moments are added or, at worst, possibly remain unchanged. Note also that the derivatives of $F(x)$ need not alternate in any regular order in order for both upper and lower bounds to be obtained. All that is necessary is that some derivatives be positive on $[a, \infty)$ and some negative. Finally, even if $F^{(n)}(x)$ is not of constant sign on $[a, \infty)$, if it is *bounded* there, Eqs. (8) and (9) may provide a useful bound on the error involved in replacing $\mathscr{I}(F)$ by $\mathscr{I}^e(F)$ or $\mathscr{I}^o(F)$.

Equations (8) and (9) are Gaussian integration formulas, where points x_i and weights w_i have been chosen appropriately for the weighting function $G(x)$. The derivation of (8) and (9) rests on the existence of polynomials with prescribed value and slope at certain points. For the odd case $(M + 1 = 2n + 1)$, the proof requires the existence of a polynomial $Y(x)$ of degree not exceeding $2n$ which agrees in value with $F(x)$ at each x_i $(i = 0, 1, \ldots, n)$ and whose first derivative agrees with $F'(x)$ at each x_i, except x_0; that is,

$$\begin{aligned} Y(x_i) &= F(x_i), \qquad i = 0, 1, \ldots, n, \\ Y'(x_i) &= F'(x_i), \qquad i = 1, \ldots, n. \end{aligned} \tag{10}$$

For the even case $(M + 1 = 2n)$, a polynomial of degree not exceeding $(2n - 1)$ is required which statisfies all the foregoing conditions on the x_i $(i = 1, \ldots, n)$, but which is unconstrained at x_0. Such polynomials always exist, and it is in fact easy to give an explicit construction of $Y(x)$ in a form that explicitly displays these properties and that is well suited for numerical calculations. For completeness, we exhibit a construction of these polynomials, as well as those required later for the finite interval, in Appendix B. Figure (1) shows schematically polynomials for the even and odd cases.

Given the existence of such polynomials, the proof of (8) and (9) is straightforward. It is clear, for the odd case, for example, that both $[F(x)$

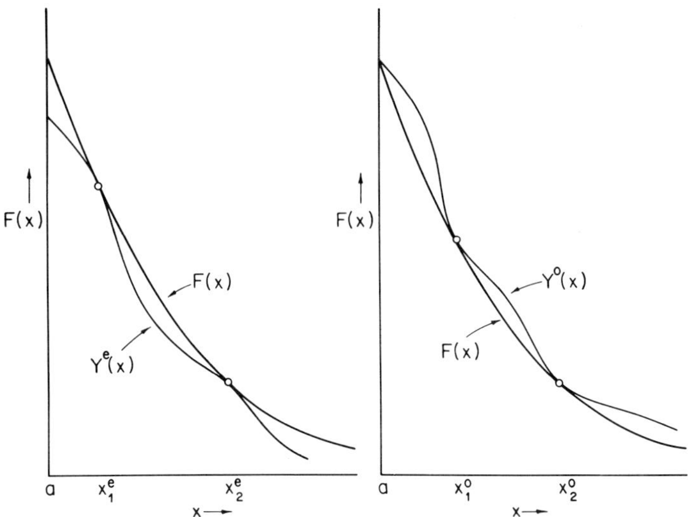

Fig. 1. Approximating polynomials $Y(x)$ described in Sect. II. A (Schematic). Figure 1a illustrates behavior of a polynomial of degree 4; 1b, that of a polynomial of degree 5, for a function $F(x)$ with the property that $F^{(4)} > 0$, and $F^{(5)}(x) < 0$. We show the behavior only schematically because the actual polynomials of this degree cannot be distinguished from the function $F(x)$ on the scale used.

$- Y(x)]$ and $(x - x_0) \Pi_{i=1}^{n} (x - x_i)^2$ vanish at least $(2n + 1)$ times on $[a, \infty)$. [Each x_i $(i = 1, \ldots, n)$ is a *double* root.] Therefore, we can choose a constant C, so that the function

$$H(x) = F(x) - Y(x) - C(x - x_0) \prod_{i=1}^{n} (x - x_i)^2 \tag{11}$$

vanishes at least $2n + 2$ times on $[a, \infty)$ with the additional root $\bar{x}$, chosen in advance. It follows that the $(2n + 1)$th derivative of H vanishes at least once, say at $\xi(\bar{x})$, on the smallest interval containing $x_0, x_1, \ldots, x_n$ and $\bar{x}$ (call it L). Since $Y(x)$ is of degree 2n or less, we have $H^{(2n+1)}[\xi(\bar{x})] = 0 = F^{(2n+1)}[\xi(\bar{x})] - (2n + 1)!\,C$, from which immediately follows the well-known result of Gaussian integration theory,

$$F(\bar{x}) - Y(\bar{x}) = \frac{F^{(2n+1)}[\xi(\bar{x})]}{(2n + 1)!} (x - x_0) \prod_{i=1}^{n} (x - x_i)^2 . \tag{12}$$

Integrating both sides of (12) with respect to the true weight function $G(x)$, we need only remark that, since $Y(x)$ is of degree $2n$ or less, its integral is expressible in terms of the first $2n$ moments, and so

$$\int_a^\infty Y(x)G(x)\,dx = \int_a^\infty Y(x)G^o(x)\,dx . \tag{13}$$

But since $Y(x) = F(x)$ at each point where $G^o(x)$ places weight,

$$\int_a^\infty Y(x) G^o(x)\, dx = \int_a^\infty F(x) G^o(x)\, dx = \sum_{i=0}^{n} w_i F(x_i) \,. \tag{14}$$

If $(x - x_0)$ is positive on $[a, \infty)$, then the second part of Eqs. (8) and (9) follows immediately by the integral mean value theorem.

The even case proceeds by the same arguments through the formula

$$F(\bar{x}) - Y(\bar{x}) = \frac{F^{(2n)}[\xi(\bar{x})]}{(2n)\,!} \prod_{i=1}^{n} (x - x_i)^2 \tag{15}$$

to the first half of Eqs. (8) and (9).

Similar arguments, again involving construction of polynomials of the type considered in Appendix B, allow one to show that, at each of their points of increase, the cumulative distributions

$$\int_{-\infty}^{x} G^e(x')\, dx' \,, \qquad \int_{-\infty}^{x} G^o(x')\, dx'$$

bound the true cumulative distributionn (Shotat and Tamarkin, 1950 d).

If the function $F(x)$ is chosen to be $(1 + xz)^{-1}$, where z is a real parameter, then the even and odd approximations $\mathscr{I}^e(F)$ and $\mathscr{I}^o(F)$ are, respectively, the $[n, n - 1]$ and $[n, n]$ Padé approximants to the function

$$g(z) = \int_a^\infty \frac{G(x)\, dx}{1 + xz} = \mu_0 + \mu_1 z + \cdots \mu_M z^M + \cdots \,. \tag{16}$$

For more general functions F, the bounds $\mathscr{I}^e(F)$ and $\mathscr{I}^o(F)$ are closely related to a generalization of the Padé approximant give by Baker (1967, 1970). For functions $F(x)$ with the property that $(-1)^n F^{(n)}(x) \geq 0$ on (a, ∞), our $\mathscr{I}^e(F)$ and $\mathscr{I}^o(F)$ are equivarent to Baker's $B_{n,-1}(z)$ and $B_{n,0}(z)$, respectively, where $F(x)$ has some parametric dependence on z. Each of these approaches has its advantages and its natural extensions to more general cases.

In the case where no lower endpoint a for $G(x)$ is known, or in which a is known to be $-\infty$, we note that the even approximation $\mathscr{I}^e(F)$ still provides bounds for $F(x)$ if $F^{(n)}(x)$ is of constant sign on $(-\infty, \infty)$. On the other hand, to obtain both upper and lower bounds, the function must have the property that $F^{(2n)}(x)$ alternates in sign with increasing n in some fashion, so that error terms of each sign can be obtained.

B. Closed Finite Interval $[a, b]$

If, in addition to the moments $(\mu_0, \mu_1, \ldots, \mu_M)$ of $G(x)$, and the knowledge of its left-hand endpoint, a, we are given the information that

$G(x) \equiv 0$ for $x > b$, we may make use of this information to improve upon the bounds to

$$\mathscr{I}(F) = \int_a^\infty G(x) F(x)\, dx = \int_a^b G(x) F(x)\, dx \tag{17}$$

obtained previously.

The bounds obtained for the infinite interval from $G^c(x)$ and $G^o(x)$, in the foregoing are still valid, and the connection with orthogonal polynomials now insures that G^e and G^o place no weight outside the interval $[a, b]$. In addition to $G^e(x)$ and $G^o(x)$, we may construct new approximate distribution functions $G_r^{\,e}(x)$ and $G_r^{\,o}(x)$ as follows:

1. If $M + 1$ is even $(M + 1 = 2n + 2)$, then construct an approximate density function

$$G_r^{\,e}(x) = \sum_{i=0}^{n+1} w_i \delta(x - x_i)\,, \tag{18}$$

where $x_0 \equiv a$, $x_{n+1} \equiv b$, and the w_i's and remaining x_i's are chosen so as to give the first $2n + 2$ moments correctly, that is,

$$\sum_{i=0}^{n+1} w_i x_i^m = \mu_m \qquad (m = 0, 1, \ldots, M)\,.$$

2. If $M + 1$ is odd $(M + 1 = 2n + 1)$, then construct an approximate density function

$$G_r^{\,o}(x) = \sum_{i=1}^{n+1} w_i \delta(x - x_i)\,, \tag{20}$$

where $x_{n+1} \equiv b$ and the weights and remaining points are determined by the moments, as usual.

The defining equations for $G_r^{\,e}(x)$ and $G_r^{\,o}(x)$ can be recast into the same form as those for G^e and G^o by the following simple transformation. The defining equation for $G_r^{\,e(o)}$ can be written as

$$\sum_{i=0(1)}^{n+1} w_i x_i^{\,m} = \mu_m \qquad [m = 0(1), \ldots, M]\,. \tag{21}$$

Multiply the mth equation by b and subtract the $(m + 1)$th, obtaining the new set of equations

$$\sum_{i=0(1)}^{n} [w_i (b - x_i)] x_i^m = (b\mu_m - \mu_{m+1}) \qquad [m = 0(1), \ldots, M - 1]\,,$$

where we have used the condition that $(x_{n+1} \equiv b)$. Now define the new positions, weights, and moments:

$$\begin{aligned} \mu_m^* &= b\mu_m - \mu_{m+1} && [m = 0(1), \ldots, M-1], \\ x_i^* &= x_i && [i = 0(1), \ldots, n), \\ w_i^* &= w_i(b - x_i) && [i = 0(1), \ldots, n]. \end{aligned} \tag{22}$$

The μ_m^* are moments of the density function

$$G^*(x) = G(x)(b - x), \tag{23}$$

which is positive on $[a, b]$, and so all of the x_i^* lie in $[a, b]$ and all of the w_i^* are positive. It follows immediately that all of the x_i lie in $[a, b]$ and that the w_i are positive $[i = 0(1), \ldots, n]$. It is not so obvious that the last weight,

$$w_{n+1} = \left(\mu_0 - \sum_{i=0(1)}^{n} \frac{w_i^*}{b - x_i}\right), \tag{24}$$

must also be positive. In fact, it can be shown from the general theory of orthogonal polynomials that, for nonnegative $G(x)$, w_{n+1} will indeed be nonnegative. It is, of course, verified explicitly that $w_{n+1} \geq 0$ in any particular case where the x_i and w_i are obtained by a numerical solution for the x_i^* and w_i^* by the method in Appendix A. We note in passing that this procedure could have been used to determine $G^o(x)$ once an algorithm for $G^e(x)$ was known. It is, in fact, more economical to use the the algorithm given in Appendix A for both G^e and G^o and then for both G_r^e and G_r^o.

The analogs of Eqs. (8) and (9) are, for $G_r^e(x)$ and $G_r^o(x)$,

$$\mathscr{I}(F) = \int_a^b F(x)G(x)\,dx = \left\{ \begin{array}{l} \mathscr{I}_r^e(F) - (K_r^e)^2 \dfrac{F^{(2n+2)}(\xi_r^e)}{(2n+2)!} \\ \mathscr{I}_r^o(F) - (K_r^o)^2 \dfrac{F^{(2n+1)}(\xi_r^o)}{(2n+1)!} \end{array} \right\}, \tag{25}$$

with

$$\left\{ \begin{array}{l} \mathscr{I}_r^e(F) = \displaystyle\sum_{i=0}^{n+1} w_i^e F(x_i^e) \\ \mathscr{I}_r^o(F) = \displaystyle\sum_{i=1}^{n+1} w_i^o F(x_i^o) \end{array} \right\}, \tag{26}$$

where superscripts e and o refer to even and odd, respectively, and it is understood that the x_i and w_i are those for $G_r^e(x)$ and $G_r^o(x)$; where ξ is some point in the interval $[a, b]$; and where the positive constants $(K_r^e)^2$

and $(K_r^o)^2$ are given by

$$(K_r^e)^2 = \int_a^b G(x)\,(x-a)\,(b-x) \prod_{i=1}^{n} (x-x_i)^2\, dx\,,$$
$$(K_r^o)^2 = \int_a^b G(x)\,(b-x) \prod_{i=1}^{n} (x-x_i)^2\, dx\,. \tag{27}$$

If the derivatives of F are of constant sign on $[a, b]$, then, for each new moment we are given, we obtain both an upper *and* a lower bound to the integral $\mathscr{I}(F)$, either from G^e and G_r^e if an even number of moments is given, or from G^o and G_r^o if an odd number of moments is given. Thus, the requirement that successive derivatives alternate in sign, which was necessary in order to obtain both upper and lower bounds in the half-open interval, is no longer necessary when an upper endpoint is given.

Again, even if $F^{(n)}(x)$ is not of constant sign on $[a, b]$, if it is bounded by a known constant, (8), (9), (25), (26), and (27) may provide useful bounds to the error incurred by approximating $G(x)$ by G^e, G^o, G_r^e, or G_r^o in the integral $\mathscr{I}(F)$.

The proof of Eqs. (25) and (27) proceeds exactly as that of Eqs. (8) and (9), and requires the existence of a polynomial $Y(x)$ of degree not exceeding $2n+1$ which agrees with $F(x)$ in value at a and b and agrees with $F(x)$ in both value and slope at x_i $(i = 1, \ldots, n)$ (even case), as well as one of degree $2n$ which agrees with F in value at b and in value and slope at x_i $(x_i = 1, \ldots, n)$ (odd case). The construction of these polynomials is included in Appendix B. Again, by similar arguments, it can be shown that

$$\int_{-\infty}^{x} G_r^e(x')\, dx'\,, \qquad \int_{-\infty}^{x} G_r^o(x')\, dx'$$

provide bounds to

$$\int_{-\infty}^{x} G(x')\, dx'$$

at each of their points of increase.

C. Other Kinds of Additional Information

1. $G(x) \equiv 0$ *in a Subinterval of* $[a, b]$

If, in addition to the moments $\mu_0, \ldots, \mu_M$ and the endpoints a and b (or just a in the $[a, \infty)$ case), one is given the information that $G(x)$ places no weight in the interval (c, d), with $a < c < d < b$, one can improve the bounds to $\mathscr{I}(F)$ by reducing the number of points x_i in the interior of the interval by one and using this information to place fixed points at c and d. The resulting equations can be recast into those for G^e and G^o by transformation analogous to (22) and (24) and the bounding theorems go through as before with the corresponding correction factors involving products of the form

$$(x - c)(x - d) \prod_i (x - x_i)^2 ,$$

which is nonnegative except in the interval (c, d), where $G(x)$ places no weight. The correction term to $\mathscr{I}(F)$ is thus of a definite sign if the appropriate derivative of $F(x)$ is of constant sign on $[a, c]$ and on $[d, b]$. Of course, if one is able to calculate moments for $G(x)$ on the intervals $[a, c]$ and $[d, b]$ *separately*, even further improvement be obtained by constructing bounds for the intervals separately.

2. *Lower Bounds to $G(x)$ Known*

Suppose there is a known function $L(x)$ that satisfies the inequality

$$0 \leq L(x) \leq G(x) \qquad \text{on} \qquad [c, d] \subseteq [a, b] \tag{28}$$

and for which the moments

$$\mu_m^L = \int_c^d L(x) x^m \, dx , \qquad m = 0, 1, \ldots , \tag{29}$$

and the integral

$$\mathscr{I}_L(F) = \int_c^d L(x) F(x) \, dx \tag{30}$$

can be calculated. [The function $L(x)$ may arise, for example, because a series expansion can be found for $G(x)$ about some point in $[c, d]$ and the first few terms are known to provide a lower bound, or it may simply be a constant that it is known $G(x)$ exceeds.] Subtracting the moments $\mu_m{}^L$ from the μ_m for $G(x)$, we obtain the moments $\mu_m = (\mu_m - \mu_m{}^L)$ for the (positive) density function $\hat{G}(x)$ defined by

$$\hat{G}(x) = \left. \begin{cases} G(x) - L(x) , & x \in [c, d] \\ G(x) , & x \in [a, b] - [c, d] \end{cases} \right\} . \tag{31}$$

If we now apply the bounding procedure described previously to $\hat{G}(x)$ and calculate the contribution $\mathscr{I}_L(F)$ separately, we shall obtain improved bounds to $\mathscr{I}(F)$. Indeed, it is easily seen that these are the best bounds possible, given only the moments $(\mu_0, \mu_1, \ldots, \mu_m)$, the left and right endpoints a and b, and the lower bound $L(x)$ to $G(x)$, for the density function consisting of $L(x)$ plus the delta-function approximation $\hat{G}^e(x)$, $\hat{G}^o(x)$,

$\hat{G}_r^e(x)$, or $\hat{G}_r^o(x)$ is everywhere as great as the lower bound to $G(x)$, and it satisfies the required moment conditions. The same approach can be applied, of course, when the right-hand endpoint is not known or is infinite. In that case, only $\hat{G}^e(x)$ and $\hat{G}^o(x)$ can be constructed, but the bounds obtained are again optimal for the given information.

There is a very general and elegant theory (Krein, 1951; Karlin and Studden, 1966) concerning the optimal bounds to $\mathscr{I}(F)$ which can be obtained when either upper or lower bounds, or both, are known for $G(x)$. Unfortunately, the construction of the optimal bounds is, in general, very difficult when an upper bound to $G(x)$ is known. In the case when only a lower bound is known, the construction given here provides the optimal bounds, and is equivalent to the prescription given by the more general theory.

3. *An Accurate Expression for $G(x)$ Is Known in a Subinterval of $[a, b]$*

It frequently occurs in applications that the function $F(x)$ diverges or that it has singular derivatives at one end of the interval $[a, b]$. If $F(x)$ diverges at $x = a$, then the bounds involving $G^o(x)$ and $G_r^e(x)$ which place a weight at $x_0 = a$ are themselves infinite. If $F(x)$ is finite but has singular derivatives, the bounds will be finite, but their convergence may be rather poor.

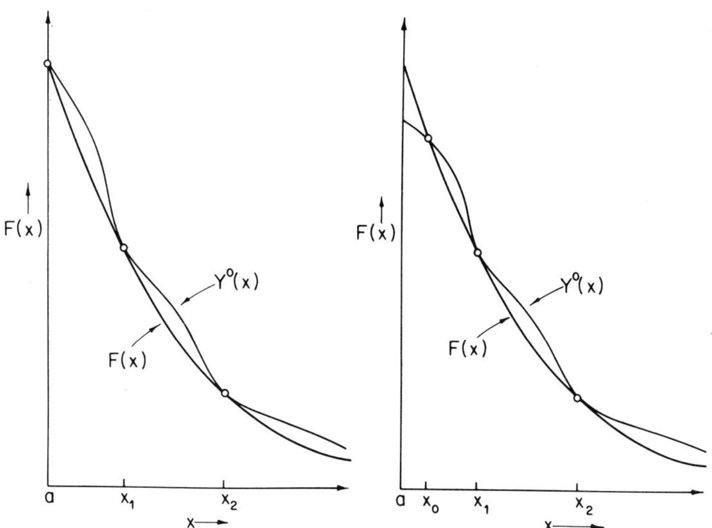

Fig. 2. Schematic illustration of the effect on the polynomial $Y(x)$ described in Sect. II. C, of shifting the point x_0 from (Fig. 1a) to $c > a$ (Fig. 1b) for a function $F(x)$ with the property $F^{(5)}(x) < 0$.

Under such circumstances, it would be useful to have a bounding procedure in which the fixed point x_0 was chosen to be some number $c > a$. Figure 2 shows the effect on the polynomial $Y(x)$, introduced in Eqs. (10), of moving x_0 from a to $c > a$. According to Eq. (12), if $F^{(2n+1)}(x)$ is of constant sign on $[a, b]$, then the difference $[F(x) - Y(x)]$ is of one sign on $[a, c)$ and of the other sign on $(c, b]$. In addition, it is usually found that the magnitude of $F(x) - Y(x)$ is larger in $[a, c]$ than its typical values in $[c, b]$ and that $|F(a) - Y(a)|$ increases rapidly as c is increased. [Of course, if $F(a)$ is infinite, then $|F(x) - Y(x)|$ will become infinite as $x \to a$.]

Suppose that an approximate density function $G_A(x)$ is known to agree with $G(x)$ to within some specified fractional error on $[a, c]$, where $a < c < b$, and assume that the moments of $G_A(x)$ on $[a, c]$ and the integral

$$\mathscr{I}_A(F) = \int_a^c F(x)G_A(x)\,dx \tag{32}$$

can be calculated explicitly, either in closed form or numerically. Consider the function

$$\hat{\hat{G}}(x) = \begin{cases} G(x) - G_A(x), & a \leq x \leq c \\ G(x), & c < x \leq b \end{cases}. \tag{33}$$

If $\hat{\hat{G}}(x)$ were *identically* zero in $[a, c]$, we could simply apply the bounding methods described previously with $x_0 = c$ to calculate bounds to the contribution of $\hat{\hat{G}}(x)$ to $\mathscr{I}(F)$. Of course $\hat{\hat{G}}(x)$ is not identically zero in $[a, c]$ and may even be negative there. As a consequence, there is no *guarantee* that the delta function densities $\hat{\hat{G}}^e(x)$, $\hat{\hat{G}}^o(x)$, $\hat{\hat{G}}_r^{\,e}(x)$, $\hat{\hat{G}}_r^{\,o}(x)$ have the properties that all x_i lie in $[c, b]$ (or even in $[a, b]$!) and all w_i are positive. Nevertheless, in practice it is found that, if $G_A(x)$ represents $G(x)$ sufficiently accurately, then the x_i's do lie in $[c, d]$ and the w_i's are positive. If $G_A(x)$ arises from a power series for $G(x)$ about $x = a$, for example, then one will generally be able to choose c small enough so that the x_i's and w_i's behave properly.

In place of Eqs. (8) and (25), we now write the equivalent but more useful relation

$$\mathscr{I}(F) = \int_a^b \hat{\hat{G}}(x)F(x)\,dx = \begin{cases} \mathscr{I}^o(F) + E^o \\ \mathscr{I}_r^e(F) + E_r^e \end{cases}, \tag{34}$$

where

$$\begin{aligned} E^o &= \int_a^b \hat{\hat{G}}(x)[F(x) - Y^o(x)]\,dx\,, \\ E_r^e &= \int_a^b \hat{\hat{G}}(x)[F(x) - Y_r^e(x)]\,dx\,, \end{aligned} \tag{35}$$

and where $\mathscr{I}^o(F)$ and $\mathscr{I}_r^e(F)$ are the values obtained from the delta function approximations $\hat{\hat{G}}^o(x)$ and $\hat{\hat{G}}_r^e(x)$ to $\hat{G}(x)$, and $Y^o(x)$ and $Y_r^e(x)$ are the corresponding polynomials that agree in value with $F(x)$ at $x = c$ and in value and slope at the x_i corresponding to $\hat{\hat{G}}_o(x)$ and $\hat{\hat{G}}_r^e(x)$. Since the differences $[F(x) - Y^o(x)]$ and $[F(x) - Y_r^e(x)]$ can be calculated explicitly at any point in $[a, b]$ (Appendix B), it may be possible to verify that the contribution to E^o and E_r^e from $[a, c]$ is necessarily dominated by that from $(c, b]$, so that $\mathscr{I}^o(F)$ and $\mathscr{I}_r^e(F)$ are indeed bounds.

Two additional requirements must be met in order to accomplish this. First, one must know something of the general shape of $G(x)$ in $[c, d]$ in order to assert that $\int_c^d [F(x) - Y(x)]G(x)\,dx$ is of some minimum size. [For example, if $G(x)$ were, in fact, a point distribution that placed weight only where $Y(x) = F(x)$, then the integral would be zero.] Secondly, either the singularity in $F(x)$ at a must be integrable, or else the difference $G_A(x) - G(x)$ must tend to zero as $x \to a$ sufficiently rapidly that

$$\int_a^{a+\varepsilon} F(x)\hat{G}(x)\,dx$$

exists. Despite these limitations, this technique has been found useful, and is also illustrated in Sect. III.

III. APPLICATIONS

A. Inequalities in Closed Form

When only the first few moments are known, the bounds can be calculated explicitly and expressed in terms of the moments. Let

$$\langle F \rangle = \frac{\int_a^b F(x)G(x)\,dx}{\int_a^b G(x)\,dx} = \left(\frac{\mathscr{I}(F)}{\mu_0}\right), \tag{36}$$

so that

$$\left.\begin{cases} \langle x \rangle = \mu_1/\mu_0, \\ \langle (\Delta x)^2 \rangle = \langle (x - \langle x \rangle)^2 \rangle = (\mu_2 - \mu_1^2)/\mu_0, \\ \langle (\Delta x)^3 \rangle = \langle (x - \langle x \rangle)^3 \rangle = (\mu_3 - 3\mu_2\mu_1 + 2\mu_1^3)/\mu_0, \\ \text{etc.} \end{cases}\right\} \tag{37}$$

Then the first few moment inequalities become

$F'(x) \geq 0$:

$$F(a) \leq \langle F(x) \rangle \leq F(b); \tag{38}$$

$F''(x) \geq 0$:

$$F(\langle x\rangle) \leq \langle F(x)\rangle \leq \left(\frac{b-\langle x\rangle}{b-a}\right)F(a) + \left(\frac{\langle x\rangle - a}{b-a}\right)F(b)\,; \tag{39}$$

$F^{(3)}(x) \geq 0$:

$$\frac{\langle(\Delta x)^2\rangle}{\langle(x-a)^2\rangle}F(a) + \left(\frac{\langle x-a\rangle^2}{\langle(x-a)^2\rangle}\right)F\left(\frac{\langle x(x-a)\rangle}{\langle x-a\rangle}\right)$$

$$\leq \langle F(x)\rangle$$

$$\leq \left(\frac{\langle b-x\rangle^2}{\langle(b-x)^2\rangle}\right)F\left(\frac{\langle x(b-x)\rangle}{\langle b-x\rangle}\right) + \frac{\langle(\Delta x)^2\rangle}{\langle(b-x)^2\rangle}F(b)\,; \tag{40}$$

$F^{(4)}(x) \geq 0$:

$$W_-F(x_-) + W_+F(x_+) \leq \langle F(x)\rangle$$

$$\leq W_aF(a) + W_bF(b) + W_1F(x_1)\,, \tag{41}$$

where

$$\left.\begin{aligned}
x_\pm &= \langle x\rangle + \frac{\langle(\Delta x)^3\rangle}{2\langle(\Delta x)^2\rangle} \pm \left\{\langle(\Delta x)^2\rangle + \left(\frac{\langle(\Delta x)^3\rangle}{2\langle(\Delta x)^2\rangle}\right)^2\right\}^{1/2}, \\
W_+ &= \left(\frac{\langle x\rangle - x_-}{x_+ - x_-}\right) \qquad W_- = \left(\frac{x_+ - \langle x\rangle}{x_+ - x_-}\right), \\
x_1 &= \frac{\langle x(x-a)(b-x)\rangle}{\langle(x-a)(b-x)\rangle}, \\
w_1 &= \frac{\langle(x-a)(b-x)\rangle^3}{\langle(x-a)^2(b-x)\rangle\langle(b-x)^2(x-a)\rangle}, \\
w_a &= \frac{\langle(b-x)^3\rangle\langle(b-x)\rangle - \langle(b-x)^2\rangle^2}{(b-a)\langle(b-x)(x-a)^2\rangle}, \\
w_b &= \frac{\langle(x-a)^3\rangle\langle x-a\rangle - \langle(x-a)^2\rangle^2}{(b-a)\langle(b-x)^2(x-a)\rangle}.
\end{aligned}\right\} \tag{42}$$

The first pair of inequalities is trivial. The first of the second pair is well known as Jensen's inequality (Hardy *et al.*, 1964) for convex functions. The second of this pair is an obvious consequence of the convexity of F, and holds even with the $\langle\ \rangle$ signs removed. The higher-order inequalities rapidly become more complex, and less obvious.

In each case, if the sign of $F^{(n)}$ is reversed, the order of the inequality is simply reversed. The first few inequalities can be used to provide bounds on higher moments of $G(x)$. For example, the positivity of w_a and w_b for the case $F^{(4)} > 0$ is easily verified by applying the inequality for $F^{(3)} > 0$ to $\langle(b-x)^3\rangle$ and $\langle(x-a)^3\rangle$.

B. Numerical Example: Simple Harmonic Solid

We now illustrate the procedures of Sect. II by applying them to a specific problem: that of obtaining the equilibrium thermal properties of a simple harmonic solid. We shall consider a face-centered cubic crystal with nearest-neighbor central force constants. The thermal properties can be expressed as averages of functions of frequency and temperature over the distribution of normal mode frequencies for the crystal. The convenient variables in which to express these thermal properties are the dimensionless quatities

$$x = (\omega/\omega_{\max})^2, \qquad \tau = (kT/\hbar\omega_{\max}),$$

where ω is the frequency of a normal mode, $\omega_{\max}$ is the largest normal mode frequency that is determined by the force constant and the mass of the particles, T is the absoulte temperature, k is Boltzman's constant, and $\hbar$ Planck's constant divided by 2π. If we let $G(x)\,dx$ be the *fraction* of normal modes in the interval $[x, x + dx]$, then the vibrational free energy F, internal energy U, and heat capacity C_v are given by (Wheeler and Gordon, 1969)

$$\left.\begin{aligned} f(\tau) &= \left(\frac{F}{3N\hbar\omega_{\max}}\right) = \tau \int_0^1 G(x) \ln\left[2 \sinh \frac{x^{1/2}}{2\tau}\right] dx\,, \\ u(\tau) &= \left(\frac{U}{3N\hbar\omega_{\max}}\right) = \tau \int_0^1 G(x)\left[\frac{x^{1/2}}{2\tau} \cosh\left(\frac{x^{1/2}}{2\tau}\right)\right] dx\,, \\ c(\tau) &= \left(\frac{C_v}{3Nk}\right) = \tau \int_0^1 G(x)\left[\frac{x^{1/2}/2\tau}{\sinh (x^{1/2}/2\tau)}\right] dx\,. \end{aligned}\right\} \tag{44}$$

Each of the functions of $(x^{1/2}/2\tau)$ appearing in the integrals in Eq. (44) has the property that, for all $\tau > 0$, as a function of x, its derivatives are constant in sign on $x \in [0, 1]$, and that successive derivatives alternate in sign (Wheeler and Gordon, 1969) so that we can use this example to illustrate the techniques on the infinite half-interval $[0, \infty)$ as well as on the closed interval $[0, 1]$. In particular, the zero-point vibrational energy, obtained by taking the limit $\tau \to 0$ in the equation for U, is just

$$u_0 \equiv \frac{U_0}{3N\hbar\omega_M} \int_0^1 G(x)\left(\frac{x^{1/2}}{2}\right) dx\,,$$

where the function $x^{1/2}$ obviously has the required properties.

An exact expression for $G(x)$ has not been obtained, even for the relatively simple case of nearest-neighbor interactions through central forces, but the *moments* of $G(x)$ can be calculated exactly, and a large number are available (Isenberg, 1963). The first few moments, together with the re-

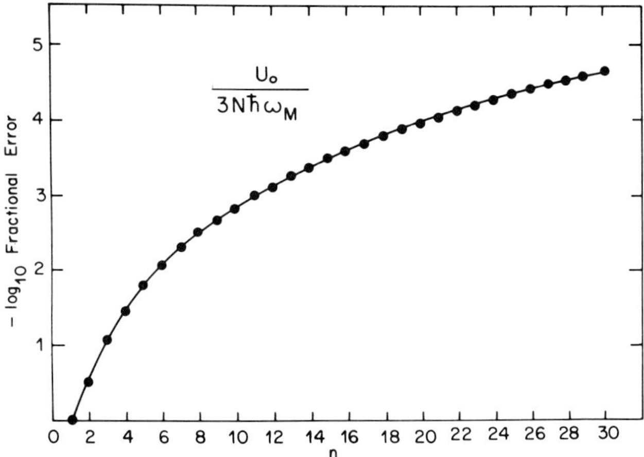

Fig. 3. Accuracy of bounds to the vibrational zero-point energy as a function of the number of moments used. The ordinate is the negative logarithm of the fractional difference of successive upper and lower bounds; the abscissa is the number of moments used. With 20 moments, we obtain the zero-point energy to about a part in 10^4; with 30 moments, to about two parts in 10^5.

sulting bounds on u_0, are

n	μ_n	Lower bound	Upper bound
0	1	*0*	$< u_0 < 0.5$
1	$\frac{1}{2}$	0.25	$< u_0 <$ *0.354*
2	$\frac{5}{16}$	*0.316*	$< u_0 < 0.345$
3	$\frac{57}{256}$	0.3300	$< u_0 <$ *0.3423*

Thus the first three moments already restrict u_0 to a range of values less than 4% of its smallest possible value. The italicized numbers are those that would be avilable if the upper endpoint were not known. Using only these values, three moments only restrict u_0 to a range of about 8% of its smallest possible value. Figure 3 gives the dependence of the fractional error on the number of moments, n. The ordinate is minus the logarithm (base 10) of the fractional difference between the bounds. With 10 moments, we obtain a spread of about 1×10^{-3}; with 20 moments of about 1×10^{-4}; and with 30 moments of about 2×10^{-5}. The bounds using 30 moments are

$$u_0 = 0.34088^{83}_{07}\,. \tag{47}$$

1. *Lower Bound to G(x) Known*

Additional information about $G(x)$ is known. Its general shape (Fig. 4) is known from root-sampling techniques (Isenberg, 1963; Leighton,

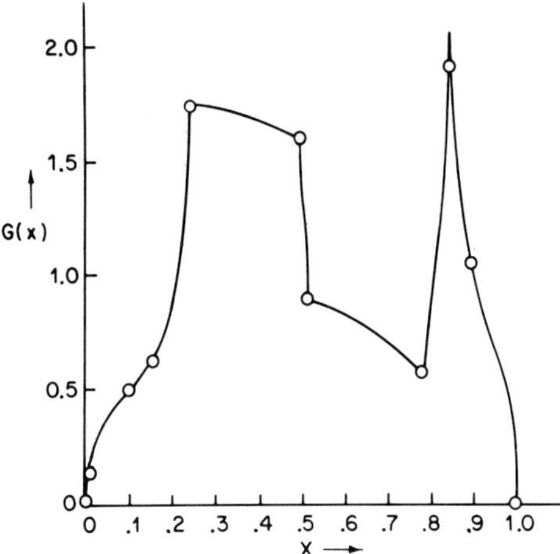

Fig. 4. Qualitative behavior of $G(x)$ for fcc lattice with nearest-neighbor interactions after Leighton (1948). $G(x)$ is known to have infinite derivative at $x = \frac{1}{4}, \frac{1}{2}$, and $[(2 + \sqrt{2}/4)] = 0.853 \cdots$.

1948), and it is known from very general topological theorems (Van Hove, 1953) that $G(x)$ has singularities with divergent slope at certain points on the interval $[0, 1]$, the nearest to the origin being at $x = \frac{1}{4}$ for the lattice under consideration. In addition, $G(x)$ may be expanded in a power series about $x = 0$ of the form

$$2x^{1/2}G(x) = A_1x + A_2x^2 + A_3x^3 + \cdots, \tag{48}$$

where the coefficients A_j can be calculated with great precision, and are found to be positive (Isenberg, 1966) We have argued elsewhere (Wheeler and Gordon, 1969) that, on the basis of the known properties of $G(x)$, the function

$$L(x) = 2.60345x + 4.6349x^2 + 9.722x^3 \tag{49}$$

is surely a lower bound to $2x^{1/2}G(x)$, on the interval $0 \leq x \leq 0.1$. In fact, this bound could be improved noticeably by (1) extending its range to $x = \frac{1}{4}$, (2) using, a larger number of coefficients, A_j, which are known, and (3) using more significant figures in the coefficients, but the bound (49) will serve to illustrate the technique described in Sect. II. If we now define $\hat{G}(x)$ as in Eq. (31) and proceed to apply our moment procedure to it, we obtain improved bounds to u_0. The improvement when only a few moments are known is only slight, but, when more moments are known, the effect on the bounds is quite striking. Figure 5 shows the behavior of the fractional deviation of the bounds as a function of n. With 10 moments, we now obtain a fractional spread of about 3×10^{-4}; with 20

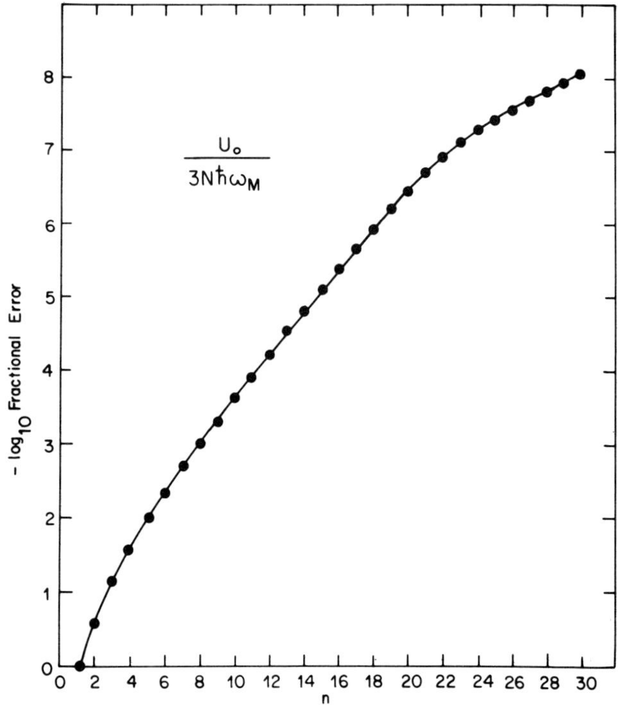

Fig. 5. Accuracy of bounds to vibrational zero-point energy using moments and three-term low-freqnency series [Eq. (49)] as a function of the number of moments used. Ordinate and abscissa are as in Fig. 1. We now obtain an accuracy of better than one in 10^6 with 20 moments, and of about a part in 10^8 with 30 moments.

moments of about 3×10^{-7}; and with 30 moments of about 1×10^{-8}. The bounds using 30 moments are

$$u_0 = 0.3408872^{204}_{177}. \tag{50}$$

Note that we are obtaining bounds to u_0 using Eqs. (8), (9), (25), and (27), even though the function $F(x) = x^{1/2}/2$ is singular at the origin with derivatives that diverge there. We need not worry about whether $F^{(M)}(\xi)$ is large: since the sign of the correction is known, and because we have both upper and lower bounds, we can simply take their agreement as conclusive evidence that ξ does not, in fact, lie too close to the origin.

The singularity in $x^{1/2}$ at $x = 0$ does undoubtedly slow down the convergence, however; and we might expect that the convergence would be better at nonzero temperatures where $u(\tau)$ has finite derivatives at the origin. This is indeed the case: the bounds improve monotonically with increasing temperature. At high temperatures, our bounds may be compared with those from the high-temperature expansion obtained by ex-

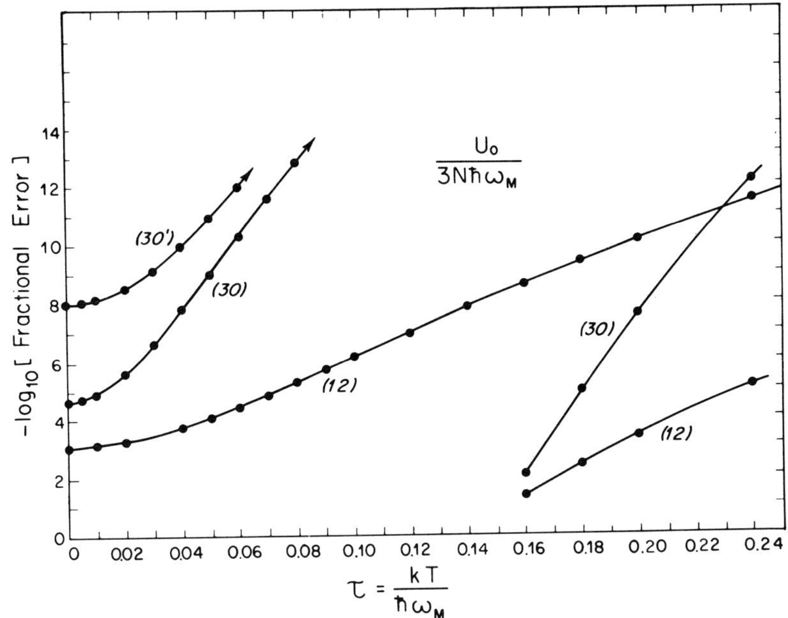

Fig. 6. Accuracy of bounds to the internal energy as a function of the normalized temperature, $\tau = kT/\hbar\omega_M$. The ordinate is the negative logarithm of the fractional difference of successive bounds. The curves at the left are obtained using our method, and those on the right by the high-temperature series. The curves are labeled with numbers indicating the number of moments used. The curve labeled (30′) was obtained using Eq. (49). The high-temperature series fail below $\tau \cong 0.16$.

panding the integrand for $u(\tau)$ in Eq. (44) in powers of inverse temperature, the coefficients being moments over $G(x)$. One obtains an alternating series that converges for $\tau > 1/2\pi \cong 0.159$. Figure 6 compares the accuracy of the moments method and the high τ series as a function of τ for 12 and 30 moments. Note that both techniques improve with increasing temperature. Note also that, at higher temperatures, where the integrand for $u(\tau)$ is less singular, using the lower bound (49) to $G(x)$ causes relatively little improvement.

2. *Accurate Approximation to G(x) Known*

The integrand for the free energy $f(\tau)$ in Eq. (44) diverges to $-\infty$ as $x \to 0$. The free energy can be convenietly rewritten in the form

$$f(\tau) = -\tau \ln \tau + \left(\frac{\tau}{2}\right)\int_0^1 \ln(x)G(x)\,dx + \tau\int_0^1 G(x)\ln\left[\frac{\sinh(x^{1/2}/2\tau)}{x^{1/2}/2\tau}\right]dx\,, \tag{51}$$

which isolates the singular part of the integrand. The third term on the right-hand side of (51) is well behaved for all x, and has derivatives of constant sign on [0, 1], and its successive derivatives alternate in sign (Wheeler and Gordon, 1969). The accuracy of the bounds obtained for it behaves similarly, as a function of n and τ, to that for the bounds for $u(\tau)$. The second term in Eq. (51), however, involves the function $F(x) = \frac{1}{2}\ln(x)$, which is singular at $x = 0$. In order to obtain lower bounds to this term, we must move x_0 away from the origin.

The three-term series for $L(x)$ in Eq. (49) is not only a lower bound to $2x^{1/2}G(x)$: sufficiently near the origin, it is a very good *approximation* as well. For $x \leq 0.02$, it should give $2x^{1/2}G(x)$ to better than one part in 10^4. Accordingly, we have chosen

$$\hat{G}(x) = \begin{cases} G(x) - \dfrac{L(x)}{2x^{1/2}} & (0 \leq x \leq 0.02)\,, \\ G(x) & (0.02 < x \leq 1.0)\,, \end{cases} \tag{52}$$

and applied the technique described in Sect. II when an approximation to $G(x)$ is known. In order to estimate the error term in Eq. (35), we must now calculate $F(x) - Y(x)$ explicitly, where $F(x) = \frac{1}{2}\ln x$. Using the points x_i from the 30 moment approximation $G_r^e(x)$, we have found that $F(x) - Y(x)$ is greater than 10^{-7} over the middle half of each interval $[x_i, x_{i+1}]$, and ranges from -7.8×10^{-4} at $x = 0.18$ to -2.5×10^{-3} at 0.16; -5.9×10^{-3} at 0.14; -1.2×10^{-2} at 0.12; and -2.4×10^{-2} at 0.1. The fraction of the total weight which $G(x)$ places in [0, 0.02] is estimated from $L(x)$ to be less than (3×10^{-3}). On the basis of the general shape of $G(x)$ given in Fig. 4, we therefore conclude that the error term E_r^e in Eqs. (34) and (35) surely satisfies the inequality

$$\begin{aligned} E > {} & \tfrac{1}{2}(0.9)(10^{-7}) \\ & - (10^{-4})(3 \times 10^{-3})\left[\left(\frac{1}{10}\right)(8 \times 10^{-4} + 9 \times 10^{-3} + 4 \times 10^{-2}) \right. \\ & \left. - 2\left(\frac{A_1 \ln(0.01)}{6}(0.01)^{3/2}\right)\right] > 0\,, \end{aligned} \tag{53}$$

so that the $\mathcal{I}_r^e(F)$ does indeed provide a lower bound. A similar procedure confirms the bounds using 20 moments. We thus have rigorous lower as well as upper bounds on the integral

$$\mathcal{L} = \int_0^1 \ln x^{1/2} G(x)\, dx\,. \tag{54}$$

The results are

$$\begin{aligned} \mathcal{L} &= -0.428^{901}_{847} && (20 \text{ moments})\,, \\ \mathcal{L} &= -0.4288^{706}_{687} && (30 \text{ moments})\,, \end{aligned} \tag{55}$$

with an accuracy of about a part in 10^4 and a few parts in 10^6, respectively.

Salsburg and Huckaby (1968) have estimated this integral by another method, involving calculation of $\mathscr{L}$ for finite crystals and extrapolating to infinite size, and have estimated

$$\mathscr{L} = -0.42887\,, \tag{56}$$

in excellent agreement with our results.

If, instead of using the three-term series (49), we use the full 30-term series for $G(x)$ calculated by Isenberg (1966), we can approximate $G(x)$ to better than a part in 10^{10} on $(0 \leq x \leq 0.1)$. Carrying out the same kind of analysis as in Eq. (53), we find improved bounds for $\mathscr{L}$:

$$\begin{aligned} \mathscr{L} &= -0.428869^{66}_{52} \qquad \text{(20 moments)}\,, \\ \mathscr{L} &= -0.4288695 66^{29}_{15} \qquad \text{(30 moments)}\,, \end{aligned} \tag{57}$$

with a precision of a few parts in 10^7 and 10^{10}, respectively. Using the same procedure, we also obtain improved bounds on u_0:

$$\begin{aligned} u_0 &= 0.3408872^{22}_{10} \qquad \text{(20 moments)}\,, \\ u_0 &= 0.3408872202^{93}_{89} \qquad \text{(30 moments)}\,, \end{aligned} \tag{58}$$

giving an accurracy of a few parts in 10^8 and about one in 10^{11}.

3. *Bounds as Approximations*

The heat capacity $c(\tau)$ defined by Eq. (44) provides an illuminating example of the techniques we have been using. The function

$$F(x) = \left[\frac{x^{1/2}/2\tau}{\sinh\,(x^{1/2}/2\tau)}\right]$$

appearing in the integrand varies drastically in shape as the parameter τ is varied. At high temperatures ($\tau \gtrsim 1$), $F(x)$ varies smoothly over the region $0 < x < 1$, and the bounds obtained from the moments alone are extremely precise. For $\tau \ll 1$, however, $F(x)$ decays very rapidly from its value, 1, at $x = 0$, toward zero with increasing x. This rapid change causes the error terms in (8) and (25) to be large, and the bounds fail to converge well. This failure can be understood from another point of view. Each of the bounds to $c(\tau)$ consists of a sum of weighted Einstein heat capacity functions with weights and characteristic frequencies chosen to satisfy the moments. Thus, at very low temperatures, the behavior of $c(\tau)$ is dominated by the Einstein function with the lowest characteristic frequency: if this frequency is zero, $c(\tau)$ tends to the corresponding weight; if it is nonzero, $c(\tau)$ vanishes as a typical Einstein heat capacity function as $\tau \to 0$. We remark, in passing, that a consequence of this interpretation is that the usual Einstein heat capacity law with a single characteristic frequency,

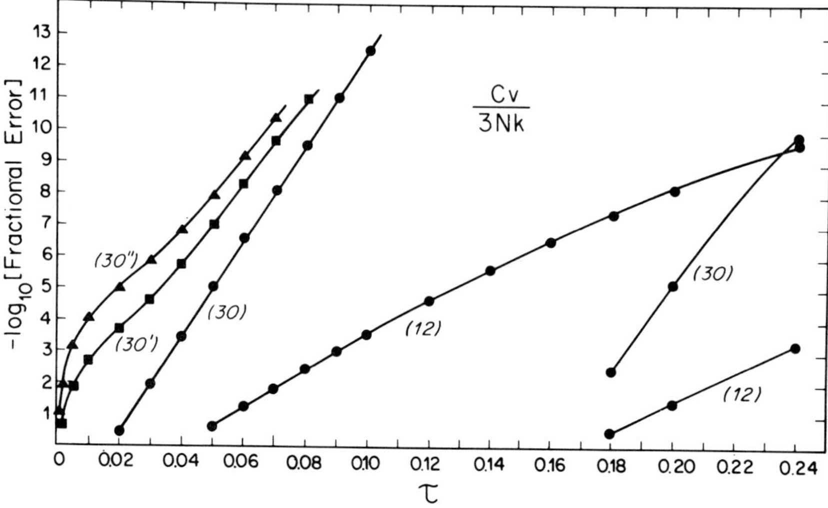

Fig. 7. Accuracy of the bounds to C_v as a function of reduced temperature τ. The ordinate is minus the logarithm of the fractional difference between upper and lower bounds. Curves on the left were obtained using our bounds, and those on the right using the high-temperature series. The curves are labeled with numbers indicating the number of moments used. The curve marked (30′) was obtained using a three-term low-frequency series, Eq. (49), as a lower bound for $2x^{1/2}G(x)$; the curve marked (30″) was obtained using a 15-term low-frequency series to bound $G(x)$ from below. The high-temperature series fail below $\tau \cong 0.16$.

chosen to give μ_1 correctly, provides a lower bound to the correct heat capacity at all temperatures.

If we now make use of Eq. (49) to provide a lower bound to $\hat{G}(x)$ and obtain bounds to $c(\tau)$ using $G(x)$ in Eq. (31), the bounds converge down to lower temperatures than before, but still fail at sufficiently low τ. Figure 7 shows the dependence of the accuracy of our bounds on temperature. The curves at the right give the accuracy obtained using the high-temperature expansion with 12 and 30 moments; those at the left give the accuracy of the bounds obtained here. With the moments in the foregoing, $c(\tau)$ is determined to an accuracy of 10^{-4} or better down to $\tau \cong 0.04$, using 30 moments. With the three-term series (49) and 30 moments [curve (30′)] this accuracy is maintained down to about $\tau = 0.02$. Using a 15-term series lower bound for $G(x)$ and 30 moments [curve (30″)], an accuracy of 1×10^{-4} is maintained down to $\tau = 0.01$. Similarly, if 1% accuracy is required, these three sets of bounds can be used down to $\tau = 0.03$, $\tau = 0.005$, and $\tau = 0.002$, respectively.

Although making use of a lower bound to $G(x)$ improves the bounds to $c(\tau)$ and extends the range of τ in which they converge rapidly, the bounds still fail to converge at $\tau = 0$. This is because $G^o(x)$ and $G_r^e(x)$ still place a weight at zero, so that the corresponding bounds for $c(\tau)$ still tend to a constant (smaller now) as $\tau \to 0$. The *lower* bounds to $c(\tau)$, however, are now dominated by the contribution from $G_L(x)$ at low τ, and must approach the correct τ^3 behavior for $c(\tau)$ to within the accuracy to which A_1 in Eq. (48) is known. This suggests that the lower bounds to $c(\tau)$ may also be quite accurate approximations to $c(\tau)$ over the *whole* temperature range.

That this suggestion is born out in fact can be shown by using the full 30-term series for $G(x)$ as an accurate approximation to it. Using the technique described earlier, which led to the bounds (57) and (58), one obtains $c(\tau)$ to one part in 10^7 over the entire temperature range. Using this information, we have verified that the lower bound to $c(\tau)$, obtained using $L(x)$ in Eq. (49) and 30 moments, in fact approximates $c(\tau)$ to better than one part in 10^5 over the entire temperature range.

A similar, but less drastic sort of asymmetry in the accuracy of upper and lower bounds is found for the zero-point energy. The integrand for U_0, $F(x) = x^{1/2}/2$, also has the characteristic that it is (mildly) singular at $x = 0$. As a result, the upper bounds in Eqs. (46), (47), and (50) are all closer than the lower bounds to the common value of the bounds in Eq. (58). In Eq. (46), there is a spread of about 4% in the bounds using three moments, whereas the upper bound is less than 1% away from the correct value. The bounds in Eq. (47) differ by about two parts in 10^5, whereas the upper bound is correct to about three parts in 10^6. The bounds in Eq. (50) differ by one part in 10^8, whereas the upper bound is correct to better than one part in 10^9.

It may generally be expected that, when the function $F(x)$ in the integrand of (17) has a singularity, even of finite nature, at an endpoint of the interval, a or b, those bounds which place a weight at that endpoint will tend to be less accurate than those which do not. On the other hand, when the function $F(x)$ is smoothly varying on $[a, b]$, one may expect the upper and lower bounds to be of comparable accuracy.

IV. CONCLUSION

When moments, expansion coefficients, or other pieces of precise but partial information are known about a probability density function, it is possible to use these pieces of information to place bounds on the possible range of values of certain integrals over the density function. Each new piece of information allows one to decrease the range of values still further, and different kinds of information may often be incorporated into the same bound. When a substantial number of moments is known, the precision of the bounds can be remarkably high. Two major advantages of the procedures we have discussed here are, one, that an explicit, economical, and numerically stable procedure for *constructing* the bounds is available and, two, that, since the procedure involves the construction of a density function that satisfies the given constraints, one knows that the bounds are the best bounds obtainable with the given information.

APPENDIX A

We present here an algorithm for obtaining the x_i's and w_i's required for $G^e(x)$ and $G^o(x)$ in Sec. II. A. We do so for the interval $[0, \infty]$: any other choice of the left endpoint a may be transformed to this case by a simple change in variable and corresponding transformation of the μ_m. The construction of the x_i and w_i takes place in two main stages: (1) transformation of the moments $\mu_0, \mu_1, \ldots, \mu_m$ into a set of coefficients $\alpha_1, \alpha_2, \ldots, \alpha_{m+1}$ according to the product-difference algorithm, given below, and (2) use of these coefficients to construct a tridiagonal matrix (one for G^e, and another for G^o) which is then *diagonalized* to give the x_i as the eigenvalues, and the w_i as squares of elements of the eigenvector matrix. The proof that this construction *must* yield the correct x_i and w_i has been given elsewhere (Gordon, 1968) and will not be repeated here. In any numerical application, the construction may be verified explicitly by computing the moments of G^e and G^o from the x_i and w_i and comparing with $\mu_0, \mu_1, \ldots, \mu_m$.

A. Product Difference Algorithm

The coefficients α_k are found by constructing a triangular table of numbers $P_{i,j}$ according to the following prescription illustrated in Table II. The first column is initialized to zero, with the exception of $P_{1,1}$, which is taken as unity. The second column lists the moments as in Table II:

$$P_{j,2} = (-1)^{j-1}\mu_{j-1} \qquad (j = 1, \ldots, M+1). \tag{A1}$$

The table is then filled up by columns, proceeding to the right from column 3 according to a "product-difference" (PD) recursion relation

$$P_{i,j} = P_{1,j-1}P_{i+1,j-2} - P_{1,j-2}P_{i+1,j-1}. \tag{A2}$$

Within each column, one starts at the top and works downward. When a triangular portion of the table is complete, then the α_k are given by

$$\alpha_k = P_{1,k+1}/(P_{1,k} \cdot P_{1,k-1}). \tag{A3}$$

This scheme has many points in common with Rutishauser's quotient-difference (QD) algorithm (Rutishauser, 1957) which starts from ratios, rather than the moments themselves, and produces the α_n by alternating

TABLE II

Product-Difference Table for the Corresponding Continued Fraction, Including All Those Terms Which Can Be Calculated Using μ_0, μ_1, μ_2, μ_3, and μ_4

1	2	3	4	5	6
1 1	μ_0	μ_1	$(\mu_0\mu_2 - \mu_1^2)$	$\mu_0(\mu_3\mu_1 - \mu_2^2)$	$(\mu_0^2\mu_1\mu_2\mu_4 - \mu_0\mu_1^3\mu_4 - \mu_0\mu_1\mu_2^3$ $- \mu_0^2\mu_1\mu_3^2 + 2\mu_0\mu_1^2\mu_2\mu_3)$
2 0	$-\mu_1$	$-\mu_2$	$-(\mu_0\mu_3 - \mu_2\mu_1)$	$\mu_0(\mu_2\mu_3 - \mu_1\mu_4)$	
3 0	μ_2	μ_3	$(\mu_0\mu_4 - \mu_1\mu_3)$		
4 0	$-\mu_3$	$-\mu_4$			
5 0	μ_4				
6 0					

quotients and differences. Essentially the same results are obtained by the PD and QD algorithms. The main advantage of the PD method is that it saves all divisions until the end, whereas the QD algorithm may break down during iteration because of trying to divide by zero or a very small number. Both the PD and QD algorithms are rather sensitive to roundoff error, and must be carried out with double-precision arithmetic. In this respect, the PD algorithm has an additional advantage in that the entire recursion may be carried out in the field of integer arithmetic, completely avoiding the roundoff error, provided that the moments are rational numbers. On the other hand, the use of the PD algorithm does require scaling of the moments by an apporpriate factor γ according to

$$\mu_k' = \mu_k \gamma^k, \qquad (k = 0, \ldots, M) \tag{A4}$$

in order to keep the elements of the PD table of the same order of magnitude.

The α_n can also be expressed in terms of Hankel determinents, but, since the expressions for the determinants needed requires $\sim n!$ multiplications and additions, compared to the $\sim n^2$ operations required to construct the PD table, it is clear that the recursion schemes are more suitable for calculations.

B. Construction and Diagonalization of Matrix

Once the α_k are determined, we construct a real, symmetric, tridiagonal matrix $\mathbf{M}^e$ for the even case $(M + 1 = 2n)$, the elements of which are given by

$$\begin{aligned} M_{1,1}^e &= \alpha_2, \\ M_{j,j}^e &= (\alpha_{2j-1} + \alpha_{2j}), \qquad j = 2, \ldots, n, \\ M_{j,j+1}^e &= M_{j+1,j}^e = -(\alpha_{2j}\alpha_{2j+1})^{1/2}, \qquad j = 1, \ldots, n-1, \end{aligned} \tag{A5}$$

with all other elements zero; that is,

$$M \equiv \begin{pmatrix} \alpha_2 & -(\alpha_2\alpha_3)^{1/2} & 0 & 0 & 0 & & \\ -(\alpha_2\alpha_3)^{1/2} & (\alpha_3+\alpha_4) & -(\alpha_4\alpha_5)^{1/2} & 0 & 0 & \cdots & \\ 0 & -(\alpha_4\alpha_5)^{1/2} & (\alpha_5+\alpha_6) & -(\alpha_6\alpha_7)^{1/2} & 0 & & \\ & \cdots & & \ddots & & -(\alpha_{2n-2}\alpha_{2n-1})^{1/2} & (\alpha_{2n-1}+\alpha_{2n}) \end{pmatrix}. \tag{A6}$$

(The matrix is real because the α_k must all turn out to be nonnegative if the μ_k are moments of a nonnegative density function on $[0, \infty]$.) Let $\mathbf{U}^e$ be the $n \times n$ matrix of eigenvectors of $\mathbf{M}^e$, and λ_j^e its eigenvalues, so that

$$\lambda_j^e = (\mathbf{U}^{e^{-1}}\mathbf{M}^e\mathbf{U}^e)_{jj} \qquad (j = 1, \ldots, n)\,. \tag{A7}$$

Then the positions x_i^e and weights w_i^e are then given by

$$x_i^e = \lambda_i^e\,, \qquad w_i^e = \alpha_1(\mathrm{U}_{1i}^e)^2\,.$$

Since the U_{1i}^e are real (they are elements of the transformation matrix of a real, symmetric matrix), it is obvious that the w_i are positive. From the form of (A6) and the positivity of the α_k, it is easy to show that $\mathbf{M}$ is positive definite, so that all the x_i lie in $[0, \infty]$.

The positions and weights for the odd case, say $(M + 1 = 2n - 1)$, are determined by the same procedure from the matrix $\mathbf{M}^o$ obtained from M^e by changing only one element:

$$M_{nn}^o = \alpha_{2n-1}\,, \qquad M_{ij}^o = M_{ij}^e \text{ (otherwise)}\,. \tag{A9}$$

Only the first $2n - 1$ moments are required for its construction. The positions x_i^o and weights w_i^o are now determined using (A7) and (A8) with $\mathbf{M}^o$, $\mathbf{U}^o$, and λ_j^o in place of $\mathbf{M}^e$, $\mathbf{U}^e$ and λ_j^e. One of the eigenvalues λ_j^o will now be zero.

APPENDIX B

We give here an explicit construction of the polynomials required in Sec. II. We first construct a polynomial $Y(x)$ of degree exactly $2n + 1$ which agrees with $F(x)$ in value and in slope at n points $x_1, \ldots, x_n$ and agrees with $F(x)$ in value alone at *two* additional points, say a, b. We then show how the construction may be modified to produce a polynomal of degree $2n$ which satisfies only the conditions at $x_1, \ldots, x_n$ and the auxiliary condition at a, and a polynomial of degree $2n - 1$ which satisfies only the conditions at $x_1, \ldots, x_n$.

Let

$$S_k = \left(\frac{1}{x_k - a}\right) + \left(\frac{1}{x_k - b}\right) + \sum_{\substack{j=1 \\ (j \neq k)}}^{n} \left(\frac{2}{x_k - x_j}\right), \qquad (k = 1, \ldots, n),$$

$$\begin{aligned} Q_a(x) &= \left(\frac{b - x}{b - a}\right) \prod_{j=1}^{n} \left(\frac{x - x_j}{x_j - a}\right)^2, \\ Q_b(x) &= \left(\frac{x - a}{b - a}\right) \prod_{j=1}^{n} \left(\frac{x - x_j}{b - x_j}\right)^2, \end{aligned} \tag{B1}$$

$$Q_k(x) = \left(\frac{x - a}{x_k - a}\right)\left(\frac{x - b}{x_k - b}\right) \prod_{\substack{j=1 \\ (j \neq k)}}^{n} \left(\frac{x - x_j}{x_k - x_j}\right)^2, \qquad (k = 1, \ldots, n).$$

Then the polynomial $Y(x)$ that satisfies the equations

$$\left.\begin{aligned} Y(x_i) &= F(x_i) \\ Y'(x_i) &= F'(x_i) \end{aligned}\right\} \qquad (i = 1, \ldots, n), \tag{B2a}$$

$$Y(a) = F(a), \tag{B2b}$$

$$Y(b) = F(b) \tag{B2c}$$

is given by

$$\begin{aligned} Y(a) = {} & F(a)Q_a(x) + F(b)Q_b(x) \\ & + \sum_{k=1}^{n} F'(x_k)(x - x_k)Q_k(x) \\ & + \sum_{k=1}^{n} F(x_k)Q_k(x)[1 + (x_k - x)S_k] \end{aligned} \tag{B3}$$

This polynomial is explicitly of degree $2n + 1$ and and is easily seen to satisfy Eqs. (B2a)–(B2c), since $S_k = Q_k'(x_k)$ and $Q_k'(x_j) = 0$ if $j \neq k$.

The polynomial of degree $2n$ which satisfies (B2a) and (B2b) but not (B2c) may be obtained from (B1) and (B3) by formally taking the limit $b \to +\infty$ in each equality in Eq. (B1) and dropping the term involving $F(b)$ from Eq. (B3). Similarly, the polynomial of degree $2n - 1$ which satisfies just Eq. (B2a) is obtained by taking the limits $b \to \infty$ $a \to -\infty$ in Eq. (B1) and dropping the terms involving $F(a)$ and $F(\mathrm{b})$ in Eq. (B3).

These polynomials can also be determined by a set of linear equations for the unknown coefficients of the powers of x, but the equations are rather ill conditioned and do not determine the value of $Y(x)$ as precisely as (B3). We have verified that, with 28-figure arithmetic, the two procedures yield essentially identical results for the difference $F(x) - Y(x)$, for polynomials $Y(x)$ of degree up to 30, whenever this difference is greater than about one part in 10^{16} of either $F(x)$ or $Y(x)$.

REFERENCES

Baker, G. A., Jr. (1967). *Phys.* Rev. **161,** 434.

Baker, G. A., Jr. (1970). The Padé approximant method and some related generalizations (*in* this volume).

Futrelle, R. P., and McQuarrie, D. A. (1968). *Chem. Phys. Letters* **2,** 223.

Gordon, R. G. (1968). *J. Math. Phys.* **9,** 655.

Hardy, G. H., Littlewood, J. E., and Polya, G. (1964). "Inequalities", Sec. 3.5. (Cambridge Univ. Press, London and New York.

Isenberg, C. (1963). *Phys. Rev.* **132,** 2427.

Isenberg, C. (1966). *Phys. Rev.* **150,** 712.

Krein, M. G. (1951). *Usp. Mat. Nauk.* **6,** 3. [English transl. in Am. Math. Soc. Trasl. Ser. 2, Vol. 12 (1959)].

Karlin, S., and Studden, W. J. (1966). "Tchebycheff Systems: With Applications in Analysis and Statistics". Wiley (Interscience), New York.

Leighton, R. B. (1948). *Rev. Mod. Phys.* **20,** 165.

Markov, A. (1896). *Mem. Acad. Sci. St.-Petersburg Class Phys. Meth.* **3.**

Melton, L. A., and Gordon, R. G. (1969). *J. Chem. Pheys.* **51,** 5449.

Rutishauser, H. (1957). "Der Quotienten-Differenzen-Algorithmus". Birkhäuser, Basel/ Stuttgart.

Salsburg, Z., and Huckaby, D. (1968). Dept. of Chem. Rice Univ., Houston, Texas (private communication).

Shohat, J. A., and Tamarkin, J. D. (1950a). The Problem of Moments, 2nd ed., Vol. 1. Am. Math. Surveys, Am. Math. Soc., Providence Rhode Island, See especially, p. 34–39.

Shohat, J. A., and Tamarkin, J. D. (1950b). The Problem of Moments, 2nd ed., Vol. 1, p. 36, Lemma 2.8. Am. Math. Surveys, Am. Math. Soc. Providence, Rhode Island.

Shohat, J. A., and Tamarkin, J. D. (1950c). The Problem of Moments, 2nd ed., Vol. 1, p. 38. Am. Math. Surveys, Am. Math. Soc., Providence, Rhode Island.

Shohat, J. A., and Tamarkin, J. D. (1950d). The Problem of Moments, 2nd ed., Vol. 1, p. 43. Am. Math. *Soc.* Surveys, Am. Math. Soc., Providence, Rhode Island.

Stieltjes, T. J. (1884). *Compt. Rend.* **97,** 740, 798 (1884); **99,** 508; (1894). *Ann. Fac. Sci. Toulouse* **8,** 1; (1895). **9,** 5.

Tchebychev, P. (1874). *J. Math. Phys. Appl.* **19,** 157. Markov, A. (1884). *Math. Ann.* **24,** 172.

Van Hove, L. (1953). *Phys. Rev.* **89,** 1189.

Wheeler, J. C., and Gordon, R. G. (1969). *J. Chem. Phys.* **51,** 5566.

CHAPTER 4 TURBULENT DIFFUSION: EVALUATION OF PRIMITIVE AND RENORMALIZED PERTURBATION SERIES BY PADÉ APPROXIMANTS AND BY EXPANSION OF STIELTJES TRANSFORMS INTO CONTRIBUTIONS FROM CONTINUOUS ORTHOGONAL FUNCTIONS

Robert H. Kraichnan

Dublin, New Hampshire

I. INTRODUCTION

This work started with an exploration of Padé approximants as tools for summing some divergent series that arise in the theory of diffusion of fluid particles by a random, or turbulent, velocity field. The turbulent diffusion can be described by a perturbation expansion with a diagram structure very similar to that of some quantum field theories. The effective expansion, or coupling, parameter is the ratio of turbulent to molecular diffusivity. An essential feature is that the limit of infinite coupling strength (pure turbulent diffusion) is physically well posed and interesting. This suggests working with the irreducible (renormalized) diagram expansions, whose terms remain finite in the limit, as well as the primitive expansion in powers of coupling constant.

A type of series which arises can be expressed as a generalized Stieltjes integral,

$$F(z) = \int_A^B \frac{\rho(a)\,da}{1 - az} = \sum_{n=0}^{\infty} c_n z^n . \tag{1}$$

Here $\rho(a)$ and a are real, z is complex, and the limits A and B ($B > A$) may be finite or infinite. In our applications $A = -\infty$, $B = \infty$. In the case of the primitive perturbation expansion, $\rho(a)$ is a smooth, continuous function, positive for all a. If the coupling constant is finite, we find that z is complex, or pure imaginary, so that Padé approximants to (1) converge. However, the limit of pure turbulent diffusion corresponds to z approaching the real axis, and the Padé approximants fail to converge in this, the case of principal interest. In this limit, $\rho(a)$ must be uniformly approximated as a function of a. In particular, $\rho(0)$ determines the effective turbulent diffusivity at large times. Padé approximants replace $\rho(a)$ by a sum of Dirac functions and give either zero or infinity as the approximation at any given value of a.

In the case of the irreducible expansions, we are unable to prove that (1) is always a valid representation. If it is valid, examples show that $\rho(a)$ cannot be positive for all a. Thus, we must deal with cases that exceed the proved domain of convergence of Padé approximants.

These difficulties with the Padé approximants have suggested an alternative procedure in which $\rho(a)$ is expanded in a set of continuous orthogonal functions:

$$\rho(a) = \sum_{n=0}^{\infty} b_n P_n(a) w(a) , \tag{2}$$

where the $P_n(a)$ are a complete set of polynomials orthogonal with respect to the positive weight $w(a)$ in the interval (A, B). We shall see that each b_n can be expressed as a linear function of $c_0, c_1, \ldots, c_n$. The two most prominent contrasts with Padé approximation are, first, the representation of $\rho(a)$ by a sum of continuous functions rather than a sum of Dirac functions, and, second, the fact that, once a suitable $w(a)$ is chosen, (2) involves only linear operations on the c_n. Thus, if $F(z) = F_1(z) + F_2(z)$, each approximant to $F(z)$ is the sum of the corresponding approximants to $F_1(z)$ and $F_2(z)$. Moreover, each successive approximant is obtained by simply adding one more term to the previous approximant, in contrast to the complete change of the Dirac-function sum between successive Padé approximants, The convergence theory of the continuous orthogonal function expansion method is largely unexplored. We shall see, however, that it can be shown to converge in some cases where Padé approximants either do not converge or have unknown convergence properties.

The material that follows is arranged with the hope that readers without interest in turbulence will find, at the beginning, the features that may be of general interest. We shall start with the convergence theory of the expansion of Stieltjes transforms in continuous orthogonal functions and compare this procedure with Padé approximation by working some examples. The diffusion problem is introduced in Sect. IV and developed as a specimen of stochastic field theory, without assuming any prior knowledge of turbulence theory.

II. CONTINUOUS ORTHOGONAL EXPANSION OF STIELTJES TRANSFORMS

Suppose that the polynomials $P_n(a) = \sum_{m=0}^{n} d_{nm} a^m$ ($d_{nn} \neq 0$) satisfy

$$\int_A^B w(a) P_n(a) P_m(a)\, da = \delta_{nm} N_n \,, \tag{3}$$

where $w(a) > 0$ for $A < a < B$ and the normalization factors N_n are all finite. Expanding the integrand of (1), we have

$$c_n = \int_A^B \rho(a) a^n\, da \,, \tag{4}$$

whence the b_n in the formal expansion (2) are given by

$$N_n b_n = \int_A^B \rho(a) P_n(a)\, da = \sum_{m=0}^{n} d_{nm} c_m \,. \tag{5}$$

Thus,

$$F(z) = \sum_{n=0}^{\infty} b_n I_n(z) \,, \tag{6}$$

where

$$I_n(z) = \int_A^B P_n(a) w(a) (1 - az)^{-1}\, da \,, \tag{7}$$

We now seek conditions under which the formal expansion (6) converges to $F(z)$.

Dividing (2) by $[w(a)]^{1/2}$, squaring the equation, and using (1), we find

$$\int_A^B [\rho(a)]^2 [w(a)]^{-1}\, da = \sum_{n=0}^{\infty} N_n |b_n|^2 \,. \tag{8}$$

Thus $\sum_{n=0}^{\infty} N_n |b_n|^2$ converges if the left side of (8) exists. Next, consider the expansion

$$[w(a)]^{1/2}/(1 - az) = \sum_{n=0}^{\infty} p_n(z) P_n(a) [w(a)]^{1/2} \,, \tag{9}$$

where the left side is considered as a function of a and the $p_n(z)$ are coefficients to be determined. Squaring (9) and integrating gives

$$\int_A^B w(a) |1 - az|^{-2}\, da = \sum_{n=0}^{\infty} N_n |p_n(z)|^2 \,, \tag{10}$$

whereas multiplication of (9) by $[w(a)]^{1/2}$ and integration gives $p_n(z) = I_n(z)/N_n$. Hence, if the left side of (10) exists,

$$\sum_{n=0}^{\infty} N_n |p_n(z)|^2 = \sum_{n=0}^{\infty} |I_n(z)|^2/N_n \tag{11}$$

converges. But if $\sum_n |I_n(z)|^2/N_n$ and $\sum_n N_n |b_n|^2$ both converge, then $\sum_n I_n(z) b_n$ converges, by Schwarz' inequality. Thus, the existence of the integrals on the left sides of both (8) and (10) is a sufficient condition for the convergence of the right side of (6).

But does (6) converge to the function $F(z)$ defined by (1)? Consider

$$\begin{aligned} F_r'(z) &= \int_A^B \rho_r'(a)(1 - az)^{-1}\, da\,, \\ \rho_r'(a) &= \rho(a) - \sum_{n=0}^{r} b_n P_n(a) w(a)\,. \end{aligned} \tag{12}$$

By Schwarz' inequality,

$$|F(z)| \leq \left[\int_A^B |\rho_r'(a)|^2 [w(a)]^{-1}\, da \int_A^B w(a)\,|1 - az|^{-2}\, da\right]^{1/2}, \tag{13}$$

If the $P_n(a)[w(a)]^{1/2}$ are a complete set of orthogonal functions, the existence of the left side of (8) implies that the first integral on the right side of (13) goes to zero as $r \to \infty$, whereas the second integral is finite if $|1 - az| > 0$ everywhere in (A, B). Under these conditions,

$$F_r'(z) \to 0, \qquad r \to \infty\,. \tag{14}$$

So far, we have established that the continuous orthogonal expansion (6) converges to $F(z)$ if the denominator of (1) has no zero in the range of integration and if (2) converges in the sense of mean-square approximation of $\rho(a)[w(a)]^{-1/2}$. If (A, B) is infinite, this requires that $w(a)$ fall off faster than algebraically at large a, for the P_n(a) to be normalizable, but more slowly than $a\,|\rho(a)|^2$.

The smoothing effect of the integration in (1) extends the domain of convergence of (6) to some cases where the preceding criteria are violated. Suppose first that z has the real value x and $1 - ax$ has a zero in (A, B). In this case, we define $F(x)$ and the $I_n(x)$ by the principal-part integrals. Assume that $\rho(a)$ is smooth in the neighborhood of $a = 1/x$ and that the mean-square convergence of (2) is accompanied also by uniform convergence to $\rho(a)$ and $d\rho(a)/da$ in the neighborhood of $a = 1/x$. This is sufficient to insure convergence of the principal-part integral over the neighborhood, and we conclude that, in such cases, (6) converges to $F(x)$.

Next, we shall consider some cases where $\rho(a)$ is not square-integrable. Inserting the explicit value of $p_n(z)$ into (9), we have

$$(1 - az)^{-1} = \sum_{n=0}^{\infty} I_n(z) P_n(a)/N_n , \tag{15}$$

upon dividing by $[w(a)]^{1/2}$. We have noted that, if $|1 - az| > 0$ everywhere in (A, B), (9) converges in mean square provided only that $w(a)$ yields a complete set of normalizable orthogonal functions $P_n(a)[w(a)]^{1/2}$. Now suppose that $w(a)$ is such that (9) and (15) are uniformly convergent in the neighborhood of the point $a = u$, in (A, B). If we write $1/(1 - uz)$ as a formal integral,

$$(1 - uz)^{-1} = \int_A^B \delta(a - u)(1 - az)^{-1}\, da , \tag{16}$$

then (6) gives us back precisely (15), at $a = u$. We conclude that, with such a $w(a)$, if it exists, (6) converges to $F(z) = 1/(1 - uz)$ provided u is in (A, B) and $|1 - az| > 0$ for all a in (A, B).

Suppose next that (A, B) includes $a = 0$ and that it is possible to find a $w(a)$ such that the sth derivative of (15) with respect to a is convergent at $a = 0$. Carrying out the differentiation, we have that

$$z^s = \sum_{n=0}^{\infty} I_n(z)[d^s P_n(a)/da^s]_{a=0}/(N_n s!) = \sum_{n=s}^{\infty} I_n(z)\, d_{ns}/N_n \tag{17}$$

is convergent. But this is identical with the formal expansion (5), (6) for $F(z) = z^s$. Two additional conditions are now sufficient to insure convergence of (6) to $F(z)$ for any $F(z)$ analytic at $z = 0$ and with z inside the radius of convergence of $\sum_n c_n z^n$. They are, first, that (17) converge for all z and s; and second, that the normalized errors

$$\left| z^s - \sum_{x=s}^{r} I_n(z)\, d_{ns} \middle/ N_n \right| \Big/ |z^s|$$

have a bound independent of r and s. On the basis of some examples (see Sect. III), we conjecture that $w(a) = \exp(-a^2)$ satisfies all the conditions of this paragraph in the domain $(-\infty, \infty)$. Here the condition $|1 - az| > 0$ excludes all z on the real axis.

Now let us return to square-integrable $\rho(a)$. If (A, B) is a finite interval, any $w(a)$ that is positive throughout (A, B) yields a complete set of orthogonal functions and satisfies the sufficient conditions for convergence of (6) to $F(z)$. If (A, B) is infinite, the situation is less clear. Suppose that $\rho(a)$ goes to zero like $\exp(-|a^{1/q}|)$ as $a \to \infty$, which implies $c_n \sim$

$q\Gamma(nq+q)$ for large n. We can always find a $w(a)$ such that

$$\int_A^B |\rho(a)|^2 [w(a)]^{-1}\, da$$

exists, and, consequently, (6) converges. But whether it converges to $F(z)$ depends on the completeness properties of the functions $P_n(a)[w(a)]^{1/2}$. It is not necessary that this set be complete for all continuous functions, but only that they be complete with respect to $\rho(a)$, and this depends on the closeness of match between $\rho(a)$ and $w(a)$. It is not clear to what extent Carleman's criterion for determinancy of the moment problem (Wall, 1948) is relevant to the continuous orthogonal expansion method. If we expand $\rho(a)$ about a $w(a)$ chosen to match the c_n at large n, we are implicitly using the knowledge that $\rho(a)$ is smooth at large a, a fact that may alter the conditions for determinancy of $\rho(a)$ from its moments. A *sufficient* condition for completeness, and, hence, for convergence to the correct function when the left-hand side of (8) exists, is that $w(a)$ fall off at least exponentially fast at large a (Kraichnan, 1970a).

A number of generalizations of the continuous orthogonal expansion method are possible. Equation (1) is a particular case of the integral transform

$$F(z) = \int_A^B \rho(a) f[ag(z)]\, da = \sum_{n=0}^{\infty} c_n z^n . \tag{18}$$

If $f(z)$ and $g(z)$ have known Taylor series with finite coefficients, the b_n in (2) can again be found from the c_n. If $\rho(a)$ is square integrable, the convergence theory given previously extends easily to (18). The choice $f[ag(z)] = e^{iaz}$ provides a way of finding the Fourier transform of $F(z)$ from the coefficients of its Taylor series. It is also easy to extend the convergence theory to the case where z and a are N-dimensional vectors, f in (18) is replaced by $f\,(\sum_{r,s=1}^N C_{rs} a_r z_s)$, and da is replaced by $d^N a$. This generalization suggests itself as a means of approximating the functional power series that characterize problems with an infinite number of degrees of freedom. Several generalizations, and a detailed treatment of the application to Fourier transforms, are discussed by Kraichnan (1970a).

III. EXAMPLES COMPARING CONTINUOUS ORTHOGONAL EXPANSION AND PADÉ APPROXIMANTS

In all the examples that follow, $\rho(a)$ is an even function of a, with the domain of integration $(-\infty, \infty)$, a specialization suggested by the later applications to the turbulent diffusion problem. We may therefore rewrite (1) as

$$F(z) = 2\int_0^\infty \rho(a)/(1 - a^2z^2)\, da\,. \tag{19}$$

We shall take

$$w(a) = (2\pi)^{-1/2}q^{-1}\exp(-a^2/2q^2)\,, \tag{20}$$

with q a positive scaling parameter, so that the $P_n(a)$ are the Hermite polynomials

$$He_0(\bar{a}) = 1\,, \qquad He_2(\bar{a}) = \bar{a}^2 - 1\,, \qquad He_4(\bar{a}) = \bar{a}^4 - 6\bar{a}^2 + 3\,,$$
$$He_6(\bar{a}) = \bar{a}^6 - 15\bar{a}^4 + 45\bar{a}^2 - 15, \ldots, \bar{a} = a/q,$$

with orthogonality relation $\int_{-\infty}^{\infty} w(a)He_n(\bar{a})He_m(\bar{a})\,da = n!\,\delta_{nm}$. The $I_n(z)$ can be computed by direct quadrature or with the aid of the recursion relation

$$z^2\int_{-\infty}^{\infty} w(a)a^{2n+2}(1 - a^2z^2)^{-1}\,da$$
$$= -\int_{-\infty}^{\infty} w(a)a^{2n}\,da + \int_{-\infty}^{\infty} w(a)a^{2n}(1 - a^2z^2)^{-1}\,da\,. \tag{21}$$

For the first example, take $\rho(a) = (2\pi)^{-1/2}\exp(-a^2/2)$, so that

$$F(z) = \sum_{n=0}^{\infty} (2n!/2^n n!)z^{2n} = 1 + z^2 + 3z^4 + 15z^6 + \cdots\,. \tag{22}$$

The Padé approximant sequences $[n, n]$ and $[n + 1, n]$ for this case are the approximants of the continued fraction

$$F(z) = \frac{1|}{|1} - \frac{z^2|}{|1} - \frac{2z^2|}{|1} - \frac{3z^2|}{|1} - \frac{4z^2|}{|1} - \cdots, \tag{23}$$

a special case of Gauss's continued fraction, and are known to converge if $\mathrm{Im}(z) \neq 0$ (Wall, 1948). We label Padé approximants throughout this article according to powers of z^2 rather than z. For $z = i$, the first seven Padé approximants are

$[n, m]$	[0, 0]	[1, 0]	[1, 1]	[2, 1]	[2, 2]	[3, 2]	[3, 3]
$F_{[m,n]}$	1.	0.5	0.75	0.6	0.692707	0.631579	0.672414

The exact value is $F(i) = 0.655680$. The $[n, n]$ and $[n + 1, n]$ sequences bound the exact value from above and below, respectively.

If we take $q = 1$, $w(a)$ is identical with $\rho(a)$ and (6) is trivial. For $q \neq 1$, (6) is an infinite series. Consider first $q^2 = 2$, which makes $w(a)$

fall off more slowly than $\rho(a)$. The first seven values of b_{2n} and $b_{2n}I_{2n}(i)$ in (6), together with the errors $\delta F_n = F - \sum_{m=0}^{n} b_{2n}I_{2m}$ are

n	0	1	2	3	4	5	6
$2n!\, b_{2n}$	1	−0.5	0.75	−1.875	6.5625	−29.53125	162.421875
$b_{2n}I_{2n}$	0.545641	0.079616	0.020633	0.006383	0.002161	0.000774	0.000288
δF_n	−0.110038	−0.030423	−0.009790	−0.003407	−0.001246	−0.000471	−0.000183

For $q^2 = 4$, a stronger mismatch of $w(a)$ and $\rho(a)$, the corresponding values are

n	0	1	2	3	4	5	6
$2n!b_{2n}$	1	−0.75	1.6875	−6.328125	33.2227	−224.253	1850.09
$b_{2n}I_{2n}$	0.438182	0.116479	0.048284	0.024427	0.013369	0.007677	0.004557
δF_n	−0.217497	−0.105849	−0.057565	−0.033138	−0.019769	−0.012092	−0.004765

As $q \to \infty$, any given $b_{2n}I_{2n}(z)$ is $\propto (q\,|\,z\,|)^{-1}$, and the number of $b_{2n}I_{2n}(z)$ which must be summed to maintain a given accuracy is $O(q)$.

For $q^2 = 1/2$, a mismatch in the other direction, we find

n	0	1	2	3	4	5	6
$2n!b_{2n}$	1	1	3	15	105	945	10395
$b_{2n}I_{2n}$	0.757872	−0.136808	0.049946	−0.023164	0.012216	−0.006996	0.004247
δF_n	0.102193	−0.034616	0.015331	−0.007833	0.004383	−0.002613	0.001634

This choice violates the condition that $|\rho(a)|^2/w(a)$ be integrable, and the expansion of $\rho(a)$, in fact, does not converge. Equation (6) converges to $F(z)$ despite this, because of the effect of $1/(1 - az)$ in reducing the contributions at large a, where the high-order polynomials contribute the most. For $q^2 < 1/2$, (6) no longer converges. In the limit $q \to 0$, it can be shown that $b_{2n}I_{2n}(z) \to c_{2n}z^{2n}$, so that the continuous orthogonal expansion reproduces the original power series.

Next, consider

$$\rho(a) = -\,(2\pi)^{-1/2}He_2(a)\exp(-\,a^2/2), \tag{24}$$

which yields the power series

$$F(z) = 0 + 2z^2 + 12z^4 + 90z^6 + 840z^2 + 9450z^{10}\cdots, \tag{25}$$

For $z = i$, the exact value is $F(\mathrm{i}) = 0.319502$. The Padé approximants are

$[n, m]$	[1, 0]	[1, 1]	[2, 1]	[2, 2]	[3, 2]	[3, 3]
$F_{[n,m]}$	0	0.285714	0.723404	0.321101	0.384849	0.325898

In contrast to the first example, these approximants no longer form upper and lower bounds, and their convergence properties are not clear from the values given. The continuous orthogonal expansion with $q^2 = 2$ gives

n	0	1	2	3	4	5	6
$2n!b_{2n}$	0	−1	3	−11.25	52.5	295.313	1949.06
$b_{2n}I_{2n}$	0	0.159231	0.082532	0.038295	0.017291	0.007743	0.003460
δF_n	−0.319502	−0.160271	−0.077738	−0.039443	−0.022152	−0.014409	−0.010949

A difference in behavior between the continuous orthogonal expansion and Padé approximants is brought out sharply by considering now the linear combination

$$\rho(a) = (2\pi)^{-1/2} \exp(-a^2/2)[1 + AHe_2(a)], \tag{26}$$

which yields

$$F(z) = 1 + (1 - 2A)z^2 + 3(1 - 4\mathrm{A})z^4 + 15(1 - 6A)z^6 + \cdots. \tag{27}$$

Since the continuous orthogonal expansion involves only linear operations on the c_n, it yields approximants to (27) whose errors are simply the linear weightings of the errors from the two preceding examples. On the other hand, the error of a given Padé approximant depends nonlinearly on the power-series coefficients. The approximant [1, 1] to (27) blows up at $z = i$ for $A = 2/7$, whereas, for $0.25 < A < 0.5$, this approximant qualitatively misrepresents $\rho(a)$ by yielding a Dirac function at imaginary a. Nevertheless, it is likely that the $[n, n]$ Padé approximants do converge to $F(z)$ as $n \to \infty$. We should also note that there is a technique for splitting an indefinite $\rho(a)$ into an all-positive and an all-negative part, working only with the c_n, and then constructing Padé approximants to each part separately (Baker, 1970).

Now let us take z real, interpret (19) and the $I_n(z)$ as principal-part integrals, and return to the example (22). Here it is known that the Padé approximants do not converge to the correct answer. The value of a high-order approximant depends on the accidental circumstance of how close the nearest δ-function in the approximation to $\rho(a)$ happens to be to the pole of $(1 - az)^{-1}$. For $z = 1$, the first few Padé approximants to (22) are $[0, 0] = 1$, $[1, 0] = 0$, $[1, 1] = 0.5$, $[2, 1] = -0.25$, $[2, 2] = 0$, $[3, 2] = -0.277027$, $[3, 3] = -1$. The exact value is $F(1) = 0.724778$. The continuous orthogonal expansion is again trivially exact at lowest order for $q = 1$. For $q^2 = 2$, it yields

n	0	1	2	3	4	5	6
$2n!b_{2n}$	1	−0.5	0.75	−1.875	6.5625	−29.5313	162.422
$b_{2n}I_{2n}$	0.424436	0.25	0.067223	0.021378	0.003088	−0.000416	−0.00125
δF_n	−0.300342	−0.050342	0.016881	0.038258	0.041347	0.040931	0.039684

Notice that the error changes sign at $n = 2$ and reaches a maximum at $n = 4$, whereas the behavior thereafter suggests a very slow eventual decline to zero, possibly with further slow oscillations. According to Sect. II, convergence is assured if the expansion uniformly approximates $\rho(a)$ in the neighborhood of $a = 1/z = 1$.

In the examples so far given, the continuous orthogonal expansion method is better than Padé approximation with regard to quantitative accuracy in low orders, faithfulness of qualitative representation of $\rho(a)$, and domain of convergence. Now we shall turn to examples of singular $\rho(a)$, where Padé approximants are superior. First consider $F(z) = (1 - z^2)^{-1}$. Here all Padé approximants $[n, m]$ give $F(z^2)$ exactly, for $n \geq 1$. The continuous orthogonal expansion with $q = 1$ gives, for $z = i$,

n	0	1	2	3	4	5	6
$2n!b_{2n}$	1	0	−2	16	−132	1216	−12440
$b_{2n}I_{2n}$	0.655680	0	−0.046400	−0.040703	−0.029265	−0.019547	−0.012435
δF_n	0.155680	0.155680	0.109280	0.068577	0.039312	0.019764	0.007330

According to Sect. II, this expansion converges to $F(i) = 0.5$ provided that the expansion of $(1 + a^2)^{-1}$ in Hermite polynomials $He_{2n}(a)$ converges uniformly at $a = 1$.

As a final example, let us take $F(z) = \exp(-z^2)$ at $z = i$. The Padé approximants are $[0, 0] = 1$, $[1, 0] = \infty$, $[1, 1] = 3$, $[2, 1] = 2.66667$, $[2, 2] = 2.71429$, $[3, 2] = 2.71875$, $[3, 3] = 2.71831$, whereas the exact value is $e = 2.71828$. The continuous orthogonal expansion in Hermite functions, with $q = 1$, gives

n	0	1	2	3	4	5	6
$2n!b_{2n}$	1	−2	9.5	−67.6667	634.708	−7351.88	101124.9
$b_{2n}I_{2n}$	0.655680	0.311359	0.220398	0.172141	0.140718	0.118181	0.101082
$\sum_{m=0}^{n} b_{2n}I_{2n}$	0.655680	0.967039	1.187437	1.359578	1.500296	1.618477	1.719560

For $z^2 = -0.5$, where the exact value is $F(z) = 1.64872$, we find

n	0	1	2	3	4	5	6
$b_{2n}I_{2n}$	0.757872	0.273616	0.158164	0.104496	0.073846	0.054430	0.041318
$\sum_{m=0}^{n} b_{2n}I_{2n}$	0.757872	1.031489	1.189652	1.294149	1.367995	1.422425	1.463743

It is unclear from these tables whether there is convergence to the correct values of $F(z)$ or not. Shank's "iterated e_1" algorithm for accelerating convergence of a sequence (Shanks, 1955) applied to the $\sum_{m=0}^{n} b_{2n}I_{2n}$ for $n = 0$ through $n = 7$ yields the projections 2.66644 for $z^2 = -1$ and 1.64432 for $z^2 = -0.5$. These values are sufficiently close to correct to be consistent with convergence, but we have not verified that the Hermite functions satisfy the conditions for convergence of the continuous orthogonal expansion to analytic $F(z)$ that were stated in Sect. II. At any rate, it is clear that this expansion is not a practicable way to approximate the exponential function.

IV. DIFFUSION BY A RANDOM VELOCITY FIELD

Let $\hat{\mathbf{v}}(\mathbf{x})$ be a time-independent, incompressible $[\nabla \cdot \hat{\mathbf{v}}(\mathbf{x}) = 0]$ velocity field in three dimensions. A particle started at $\mathbf{x} = 0$, $t = 0$ which moves with the fluid has a trajectory $\mathbf{y}(t)$ determined by

$$d\mathbf{y}/dt = \hat{\mathbf{v}}(\mathbf{y}), \qquad \mathbf{y}(0) = 0 . \tag{28}$$

If the motion is averaged over a statistical ensemble of velocity fields, the probability density that the particle reaches $\mathbf{x}$ at time t is

$$G(\mathbf{x}, t) = \langle \delta^3[\mathbf{x} - \mathbf{y}(t)] \rangle , \tag{29}$$

where $\langle \; \rangle$ denotes ensemble average. We shall assume that the ensemble is isotropic, homogeneous, and multivariate normal, and that the root-mean-square value of the velocity component in any direction is v_0. Then,

$$\langle \hat{\mathbf{v}}(\mathbf{x} + \mathbf{r}) \cdot \hat{\mathbf{v}}(\mathbf{x}) \rangle = 2 \int_0^\infty E(k) \frac{\sin(kr)}{kr} \, dk , \tag{30}$$

where $r = |\mathbf{r}|$ and $E(k)$ is called the energy spectrum function of the velocity field. It is normalized by $\int_0^\infty E(k)\, dk = 3v_0^2/2$. The multivariate normal property, together with isotropy, means that the average of any odd-order product of velocity amplitudes vanishes, whereas the average of any even-order product equals the sum of the products of second-order averages (covariances) obtained by pairing the amplitudes in all possible ways. Thus,

$$\begin{aligned}\langle \hat{v}_i(\mathbf{x})\rangle &= 0\,,\\ \langle \hat{v}_i(\mathbf{x})\hat{v}_j(\mathbf{x}')\rangle &\equiv V_{ij}(\mathbf{x},\mathbf{x}')\,,\\ \langle \hat{v}_i(\mathbf{x})\hat{v}_j(\mathbf{x}')\hat{v}_k(\mathbf{x}'')\rangle &= 0\,,\\ \langle \hat{v}_i(\mathbf{x})\hat{v}_j(\mathbf{x}')\hat{v}_k(\mathbf{x}'')\hat{v}_l(\mathbf{x}''')\rangle &= V_{ij}(\mathbf{x},\mathbf{x}')V_{kl}(\mathbf{x}'',\mathbf{x}''')\\ &\quad + V_{ik}(\mathbf{x},\mathbf{x}'')V_{jl}(\mathbf{x}',\mathbf{x}''')\\ &\quad + V_{il}(\mathbf{x},\mathbf{x}''')V_{jk}(\mathbf{x}',\mathbf{x}'')\,,\quad \text{etc.}\end{aligned} \tag{31}$$

Since the sum of a set of normal variables is normal, a multivariate normal velocity field may be realized in the form

$$\hat{\mathbf{v}}(\mathbf{x}) = \sum_{n=1}^{N} [\mathbf{v}(\mathbf{k}_n)\cos(\mathbf{k}_n\cdot\mathbf{x}) + \mathbf{w}(\mathbf{k}_n)\sin(\mathbf{k}_n\cdot\mathbf{x})]\,, \tag{32}$$

where the Fourier coefficients $\mathbf{v}(\mathbf{k}_n)$ and $\mathbf{w}(\mathbf{k}_n)$ have normal distributions over the ensemble. Incompressibility requires $\mathbf{k}_n\cdot\mathbf{v}(\mathbf{k}_n) = \mathbf{k}_n\cdot\mathbf{w}(\mathbf{k}_n) = 0$, whereas isotropy and homogeneity require that the $\mathbf{k}_n$ be densely and isotropically distributed in $\mathbf{k}$ space and that the coefficients for distinct wave vectors $\mathbf{k}_n$ and $\mathbf{k}_m$ be uncorrelated. In the limit $N\to\infty$, $\hat{\mathbf{v}}(\mathrm{x})$ is multivariate normal provided only that the Fourier coefficients for distinct wave vectors are statistically independent, without restriction on their univariate distributions.

If the correlation (30) falls off sufficiently fast for r greater than some characteristic length l, we expect that the evolution of $G(\mathbf{x}, t)$ should resemble a molecular diffusion process with mean free path l and thermal velocity v_0, provided $t \gg l/v_0$. If we define an effective "turbulent diffusivity" by

$$\kappa(t) = \tfrac{1}{3}\langle \mathbf{y}(t)\cdot d\mathbf{y}(t)/dt\rangle\,, \tag{33}$$

then $\kappa(\infty)$ should have a finite limit value $\sim l v_0$.

$G(\mathbf{x}, t)$ is the Green's function for the evolution of an initial probability distribution of particles:

$$\phi(\mathbf{x}, t) = \int G(\mathbf{x}-\mathbf{x}', t)\phi(\mathbf{x}', 0)\, d^3x'\,, \tag{34}$$

where $\phi(\mathbf{x}, t)$ is the probability density at time t. The function

$$\begin{aligned} g_k(t) &= \int G(\mathbf{x}, t)\exp(-i\mathbf{k}\cdot\mathbf{x})\, d^3x\\ &= 4\pi k^{-2}\int_0^\infty G(\mathbf{x}, t)(kx)\sin(kx)\, dx \end{aligned} \tag{35}$$

satisfies $g_r(0) = 1$ and describes the decay of the Fourier components of the initial distribution. We shall also introduce the Laplace transform,

$$G_k(u) = \int_0^\infty e^{-ut} g_k(t)\, dt\,. \tag{36}$$

$G_k(0)$ is the relaxation time for an initial distribution $\cos(\mathbf{k}\cdot\mathbf{x})$.

Provided that a finite fraction of the particles are not trapped near their starting places, we anticipate that $G_k(0)$ is finite for all finite k. If so, and if the motion does act like a molecular diffusion process for large space and time intervals, we expect the behavior

$$g_k(t) \approx \exp[-k^2\kappa(\infty)t] \qquad (kl \ll 1,\ t \gg l/v_0)\,, \tag{37}$$

which implies

$$G_k(0) \approx 1/k^2\kappa(\infty) \qquad (kl \ll 1)\,. \tag{38}$$

On the other hand, for $v_0t \ll l$, $G(\mathbf{x}, t)$ should differ inappreciably from what it would be if $\hat{\mathbf{v}}(\mathbf{x})$ were independent of $\mathbf{x}$. Thus, since $\hat{\mathbf{v}}(0)$ is normal with zero mean (Gaussian),

$$G(x, t) \approx (2\pi v_0^2t^2)^{-3/2} \exp[-x^2/2v_0^2t^2] \qquad (t \ll lv_0)\,, \tag{39}$$

when

$$g_k(t) \approx \exp(-v_0^2k^2t^2/2) \qquad (t \ll l/v_0)\,. \tag{40}$$

This implies

$$G_k(0) \approx [(\pi/2)/(v_0k)]^{1/2} \qquad (kl \gg 1)\,, \tag{41}$$

since, for such k, $g_k(t)$ is already so small at $t \sim l/v_0$ that (40) holds for all t that contribute appreciably to $G_k(0)$.

In Sect. VI we shall develop approximants to $G_k(u)$ and $\kappa(\infty)$ by applying Padé approximation and continuous orthogonal expansions to series representations of these functions. First, we shall examine the results of computer simulation of the diffusion, which confirm the conjectures on qualitative behavior stated previously and provide a quantitative basis for assessing the approximants. The velocity field was represented in the form (32). For each realization in the ensemble, a set of N vector Fourier coefficient pairs $\mathbf{v}(\mathbf{k}_n)$, $\mathbf{w}(\mathbf{k}_n)$ was constructed by taking

$$\mathbf{v}(\mathbf{k}_n) = \boldsymbol{\zeta}_n \times \mathbf{k}_n\,, \qquad \mathbf{w}(\mathbf{k}) = \boldsymbol{\xi}_n \times \mathbf{k}_n\,,$$

where the vectors $\boldsymbol{\zeta}_n$ and $\boldsymbol{\xi}_n$ were picked independently from a three-dimensional Gaussian distribution. The vectors $\mathbf{k}_n$ were picked from a statistically isotropic distribution so shaped that the desired $E(k)$ would be realized in the limit $N \to \infty$. In each realization, the Fourier coefficients were stored and (28) was solved forward in time. The velocity field was synthesized at each needed point by (32). This process was repeated for R realizations, and $\kappa(t)$ and $g_k(t)$ were computed as

$$\kappa(t) = (3R)^{-1} \sum_{n=1}^{R} \mathbf{y}_r(t) \cdot d\mathbf{y}_r(t)/dt \,,$$

$$g_k(t) = R^{-1} \sum_{n=1}^{R} \frac{\sin(k\,|\,\mathbf{y}_r(t)\,|)}{k\,|\,\mathbf{y}_r(t)\,|} \,, \tag{42}$$

where $\mathbf{y}_r(t)$ is the particle trajectory in the rth realization. The results shown in Figs. 1 and 2 were obtained with $N = 100$, $R = 2000$ for the spectrum shapes

$$E_1(k) = \tfrac{3}{2}\, v_0^2\, \delta(k - k_0) \,,$$

$$E_2(k) = 16(2/\pi)^{1/2} v_0^2 k^4 k_0^{-5} \exp(-\,2k^2/k_0^2) \tag{43}$$

Each of these spectra peaks at $k = k_0$. Repetitions with $N = 50$ showed no statistically significant deviations.

The calculations were continued to $v_0 k_0 t = 15$, by which time the quantities $\int_0^t g_k(s)\, ds$ appeared to have reached asymptotic values, to within the statistical fluctuation, for all $k/k_0 \geq 0.75$. The points in Fig. 2 for $k/k_0 = 0.5$ and $k/k_0 = 0.25$ were obtained by fitting $g_k(t)$ to an exponential decay over the interval $10 \leq v_0 k_0 t \leq 15$ and then extrapolating to $t = \infty$.

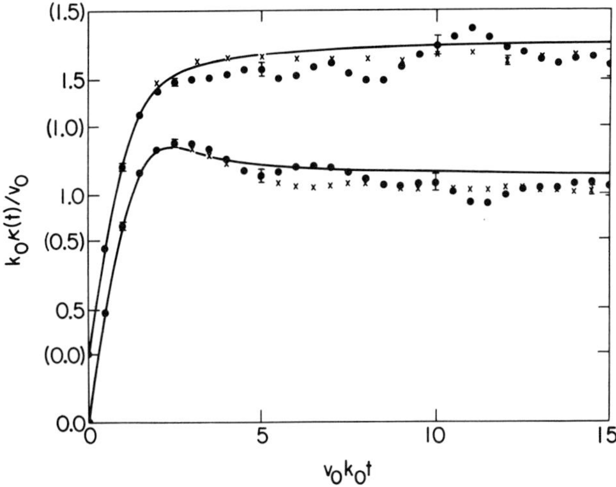

Fig. 1. Values of $\kappa(t)$ from computer experiments. Lower solid dots, spectrum E_1, $N = 100$, $R = 2000$. Lower crosses, spectrum E_1, $N = 50$, $R = 23{,}750$. Upper solid dots, spectrum E_2, $N = 100$, $R = 2000$. Upper crosses, spectrum E_2, $N = 50$, $R = 21{,}000$. Bars show probable error. Curves are the direct-interaction approximations for the two spectra, discussed in Sect. VII. [Scale ( ) for E_2.]

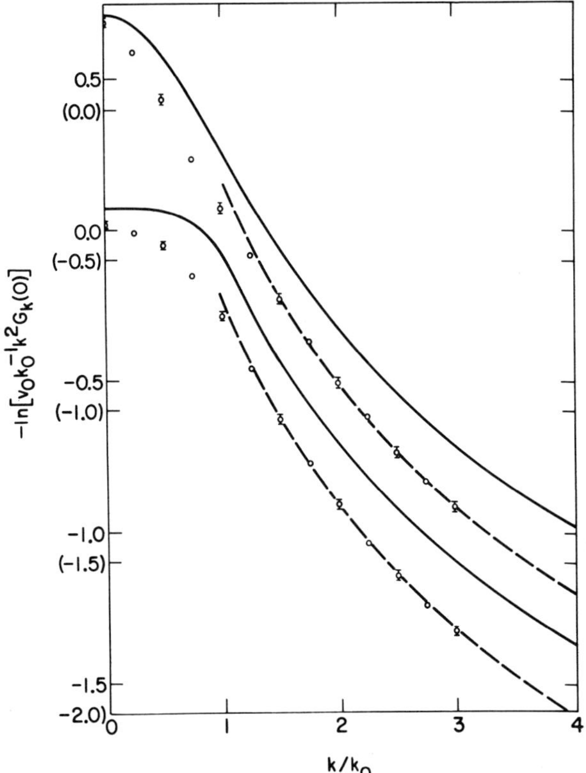

Fig. 2. $G_k(0)$ from computer experiments. Upper points and curves, spectrum E_2. Lower points and curves, spectrum E_1. Points are for $N = 100$, $R = 2000$. Dashed curves give the asymptotic relation (41). Solid curves are the direct-interaction approximation, discussed in Sect. VII. [Scale ( ) for E_2.]

The points at $k = 0$ are the values of $\kappa(\infty)$, which are estimated as the average value of $\kappa(t)$ over the interval $10 \leq v_0k_0t \leq 15$.

The leveling off of $\kappa(t)$ at large v_0k_0t and the agreement with (37)–(41) evident in Figs. 2–4 are consistent with our conjecture that the dispersion of particles by the random velocity field acts like a classical diffusion process at long times and like a spatially uniform random motion at short times. Figure 4 shows that (40) is a good approximation even when the validating inequality is substantially violated. (The results for spectrum E_2 are similar.) The computer experiment gives no evidence of trapping of particles, which would show up by leveling off of $g_k(t)$ at nonzero values. Absence of trapping also is indicated by the fact that $\langle|\mathbf{y}(t)|^4\rangle/\langle|\mathbf{y}(t)|^2\rangle^2$ was found to stay close to the value 5/3, characteristic of a Gaussian $G(\mathbf{x}, t)$, throughout the calculation.

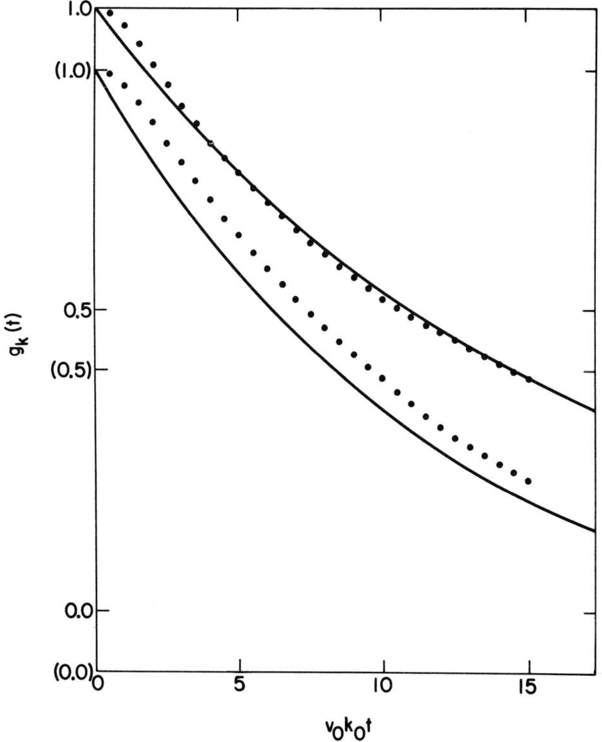

Fig. 3. Test of Eq. (37). Upper points and curve, spectrum E_1. Lower points and curve, spectrum E_2. Points give $g_k(t)$ from computer experiments ($N = 100$, $R = 2000$) with $k = 0.25k_0$. Curves give Eq. (37), with $k(\infty)$ taken from the experiments. [Scale ( ) for E_2.]

In contrast, repetition of the experiment in two dimensions gave clear evidence of trapping, a phenomenon also indicated by simple theory in two dimensions: the streamlines close on themselves in the neighborhood of each minimun of the stream function. Trapping poses severe additional difficulties in constructing approximants. It disappears if the velocity field varies randomly in time as well as in space (Kraichnan, 1970b).

V. PERTURBATION EXPANSIONS FOR THE DIFFUSION PROBLEM

The simple turbulent diffusion problem formulated in Sect. IV leads to perturbation expansions with a diagram structure analogous to that encountered in quantum electrodynamics. A feature of special interest is that the limit of infinite coupling strength is the physical case. If

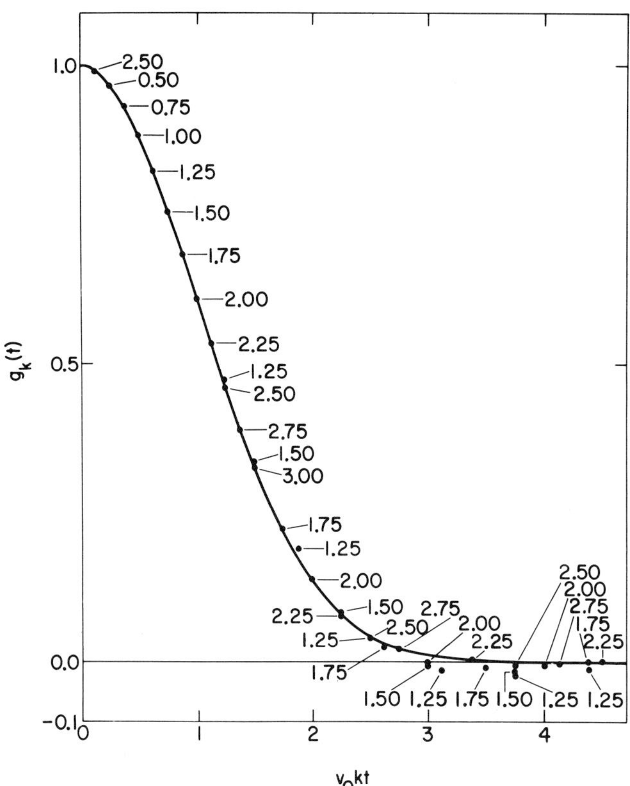

Fig. 4. Test of the asymptotic form (40) for $g_k(t)$, with spectrum E_1 ($N = 100$, $R = 2000$). The curve is Eq. (40). The points are labeled by the values of k/k_0.

$\psi(\mathbf{x}, t) = \delta^3[\mathbf{x} - \mathbf{y}(t)]$, then (28) gives

$$\partial\psi(\mathbf{x}, t)/\partial t = -\hat{\mathbf{v}}(\mathbf{x}) \cdot \nabla\psi(\mathbf{x}, t)\,, \qquad \psi(\mathbf{x}, 0) = \delta^3(\mathbf{x})\,, \tag{44}$$

which states that $\psi(\mathbf{x}, t)$ is constant along the trajectory. In order to develop the desired expansions, we generalize (44) to

$$(\partial/\partial t + \nu)\psi(\mathbf{x}, t) = -\lambda\hat{\mathbf{v}}(\mathbf{x}) \cdot \nabla\psi(\mathbf{x}, t)\,, \qquad \psi(\mathbf{x}, 0) = \delta^3(\mathbf{x})\,, \tag{45}$$

where ν is a positive damping factor and λ is an ordering parameter. The physical case is then $\nu \to 0$, $\lambda = 1$. The solutions of (45) with $\lambda = 1$ differ from those of (44) only by a factor $\exp(-\nu t)$, which may be interpreted as a decay of the total probability of finding a particle.

The Fourier transform of (45) is

$$(\partial/\partial t + \nu)\psi_k(t) = -i\lambda \sum_{\mathbf{q}} \mathbf{k} \cdot \mathbf{v}(\mathbf{q})\psi_{\mathbf{k}-\mathbf{q}}(t)\,, \qquad \psi_{\mathbf{k}}(0) = 1\,, \tag{46}$$

where

$$\hat{\mathbf{v}}(\mathbf{x}) = \sum_{\mathbf{q}} \mathbf{v}(\mathbf{q}) \exp(i\mathbf{q}\cdot\mathbf{x}) , \qquad \mathbf{v}(-\mathbf{q}) = [\mathbf{v}(\mathbf{q})]^* , \qquad \mathbf{q}\cdot\mathbf{v}(\mathbf{q}) = 0 ,$$

and the sum is over all allowed wave vectors in a cyclic box of side L $(L \to \infty)$. We have $g_{\mathbf{k}}(t) = \langle \psi_{\mathbf{k}}(t) \rangle$. The Laplace transform of (46) is

$$(\nu + u)\psi_k(u) = 1 - i\lambda \sum_{\mathbf{q}} \mathbf{k}\cdot\mathbf{v}(\mathbf{q})\psi_{\mathbf{k}-\mathbf{q}}(u) , \tag{47}$$

which may be solved by iteration in the form of an infinite perturbation series:

$$\begin{aligned}\psi_{\mathbf{k}}(u) = G_{\mathbf{k}}{}^0(u) &- i\lambda \sum_{\mathbf{q}} \mathbf{k}\cdot\mathbf{v}(\mathbf{q}) G_{\mathbf{k}}{}^0(u) G^0_{\mathbf{k}-\mathbf{q}}(u) \\ &- \lambda^2 \sum_{\mathbf{q},\mathbf{q}'} \mathbf{k}\cdot\mathbf{v}(\mathbf{q})\, (\mathbf{k}-\mathbf{q})\cdot\mathbf{v}(\mathbf{q}') G_{\mathbf{k}}{}^0(u) G^0_{\mathbf{k}-\mathbf{q}}(u) G^0_{\mathbf{k}-\mathbf{q}-\mathbf{q}'}(u) + \cdots ,\end{aligned} \tag{48}$$

where $G_{\mathbf{k}}{}^0(u) = 1/(\nu + u)$.

$G_{\mathbf{k}}(u)$ is obtained by averaging (48) and reducing the averages over products of $\mathbf{v}$ factors to products of covariances, as in (31). The isotropy incompressibility and homogeneity properties imply

$$\langle v_i(\mathbf{q}) v_j(\mathbf{q}') \rangle = \tfrac{1}{2}(2\pi/L)^3 P_{ij}(\mathbf{q})\, U(q)\, \delta_{\mathbf{q},-\mathbf{q}'} , \tag{49}$$

where $P_{ij}(\mathbf{q}) = \delta_{ij} - q_i q_j/q^2$ and $U(q) = E(q)/(2\pi q^2)$ $[3v_0{}^2 = \int U(q)\, d^3q]$. The normal-distribution pairing rule leads to a simple diagram representation of the final result. Terms with an odd power of λ vanish. If the power of λ is $2n$, draw a horizontal line with $2n$ marks on it, corresponding to the $2n$ factors $\mathbf{v}$ as they occur, from left to right, in (48). Then connect the marks into n pairs with curved lines, corresponding to a breakup of the product of $\mathbf{v}$ factors into covariances. There are $R_{2n} = 2n!/n!2^n$ ways of doing this, giving R_{2n} diagrams, and R_{2n} terms, of order $2n$. The series written explicitly through fourth order, with the wave vector sums re-

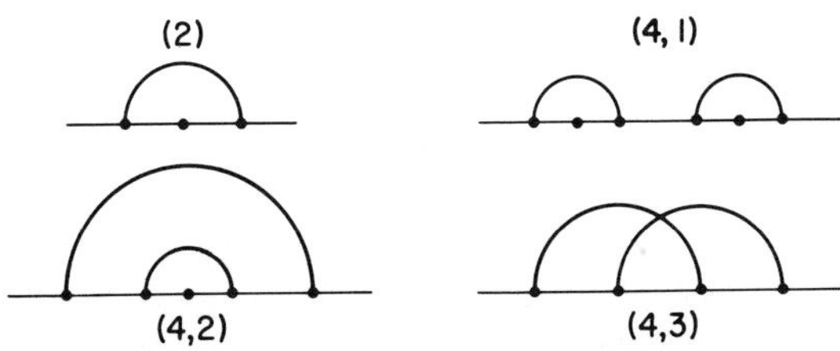

Fig. 5. All second- and fourth-order diagrams.

expressed as Fourier integrals, is

$$G_{\mathbf{k}}(u) = G_{\mathbf{k}}{}^0(u) + \lambda^2 G_{\mathbf{k}}{}^2(u) + \lambda^4[G_{\mathbf{k}}^{4,1}(u) + G_{\mathbf{k}}^{4,2}(u) + G_{\mathbf{k}}^{4,3}(u)] + \cdots ,$$

$$G_{\mathbf{k}}{}^2(u) = -\int d\mathbf{q}[\mathbf{k}\cdot\mathsf{P}(\mathbf{q})\cdot\mathbf{k}]U(q)G_{\mathbf{k}}{}^0(u)G_{\mathbf{k}-\mathbf{q}}^0(u)G_{\mathbf{k}}{}^0(u) ,$$

$$\begin{aligned} G_{\mathbf{k}}^{4,1}(u) &= \int d\mathbf{q}\, d\mathbf{q}'[\mathbf{k}\cdot\mathsf{P}(\mathbf{q})\cdot\mathbf{k}][\mathbf{k}\cdot\mathsf{P}(\mathbf{q}')\cdot\mathbf{k}]U(q)\,U(q') \\ &\quad\times G_{\mathbf{k}}{}^0(u)G_{\mathbf{k}-\mathbf{q}}^0(u)G_{\mathbf{k}}{}^0(u)G_{\mathbf{k}-\mathbf{q}'}^0(u)G_{\mathbf{k}}{}^0(u) , \end{aligned} \tag{50}$$

$$\begin{aligned} G_{\mathbf{k}}^{4,2}(u) &= \int d\mathbf{q}\, d\mathbf{q}'[\mathbf{k}\cdot\mathsf{P}(\mathbf{q})\cdot\mathbf{k}][(\mathbf{k}-\mathbf{q})\cdot\mathsf{P}(\mathbf{q}')\cdot(\mathbf{k}-\mathbf{q})]U(q)\,U(q') \\ &\quad\times G_{\mathbf{k}}{}^0(u)G_{\mathbf{k}-\mathbf{q}}^0(u)G_{\mathbf{k}-\mathbf{q}-\mathbf{q}'}^0(u)G_{\mathbf{k}-\mathbf{q}}^0(u)G_{\mathbf{k}}{}^0(u) , \end{aligned}$$

$$\begin{aligned} G_{\mathbf{k}}^{4,3}(u) &= \int d\mathbf{q}\, d\mathbf{q}'[\mathbf{k}\cdot\mathsf{P}(\mathbf{q})\cdot(\mathbf{k}-\mathbf{q}')][(\mathbf{k}-\mathbf{q})\cdot\mathsf{P}(\mathbf{q}')\cdot\mathbf{k}]U(q)\,U(q') \\ &\quad\times G_{\mathbf{k}}{}^0(u)G_{\mathbf{k}-\mathbf{q}}^0(u)G_{\mathbf{k}-\mathbf{q}-\mathbf{q}'}^0(u)G_{\mathbf{k}-\mathbf{q}'}^0(u)G_{\mathbf{k}}{}^0(u) , \end{aligned}$$

$$\cdot \quad \cdot \quad \cdot$$

The second- and fourth-order diagrams are shown in Fig. 5. In writing (50), the identity $\mathbf{q}\cdot\mathsf{P}(\mathbf{q}) = 0$ has been used.

The function $\Gamma_{\mathbf{k}}(u)$, defined by

$$G_{\mathbf{k}}(u) = [\nu + u + \lambda^2\Gamma_{\mathbf{k}}(u))]^{-1} , \tag{51}$$

is an effective dynamical damping factor, analogous to the mass operator of quantum field theory. The expansion of $\Gamma_k(u)$ is easily found from that of $G_k(u)$. $\Gamma_k^{m,n}(u)$ vanishes if the corresponding diagram for $G_k^{m,n}(u)$ can be broken into two parts (each containing at least two vertices) by a single cut of the horizontal line. We shall call such diagrams disconnected. For all other diagrams, $\Gamma_k^{n,m}(u)$ is obtained from $G_k^{n,m}(u)$ by removing the two factors $G_k{}^0(u)$. Thus,

$$\Gamma_{\mathbf{k}}(u) = \Gamma_{\mathbf{k}}{}^2(u) + \lambda^2[\Gamma_{\mathbf{k}}^{4,2}(u) + \Gamma_{\mathbf{k}}^{4,3}(u)] + \cdots ,$$

$$\Gamma_{\mathbf{k}}{}^2(u) = \int dq[\mathbf{k}\cdot\mathsf{P}(\mathbf{q})\cdot\mathbf{k}]U(q)G_{\mathbf{k}-\mathbf{q}}^0(u) ,$$

$$\begin{aligned} \Gamma_{\mathbf{k}}^{4,2}(u) &= -\int dq\, dq'[\mathbf{k}\cdot\mathsf{P}(\mathbf{q})\cdot\mathbf{k}][(\mathbf{k}-\mathbf{q})\cdot\mathsf{P}(\mathbf{q}')\cdot(\mathbf{k}-\mathbf{q})]U(q)\,U(q') \\ &\quad\times G_{\mathbf{k}-\mathbf{q}}^0 G_{\mathbf{k}-\mathbf{q}-\mathbf{q}'}^0 G_{\mathbf{k}-\mathbf{q}} , \end{aligned} \tag{52}$$

$$\begin{aligned} \Gamma_{\mathbf{k}}^{4,3}(u) &= -\int dq\, dq'[\mathbf{k}\cdot\mathsf{P}(\mathbf{q})\cdot(\mathbf{k}-\mathbf{q}')][(\mathbf{k}-\mathbf{q})\cdot\mathsf{P}(\mathbf{q}')\cdot\mathbf{k}]U(q)\,U(q') \\ &\quad\times G_{\mathbf{k}-\mathbf{q}}^0 G_{\mathbf{k}-\mathbf{q}-\mathbf{q}'}^0 G_{\mathbf{k}-\mathbf{q}'}^0 , \end{aligned}$$

$$\cdot \quad \cdot \quad \cdot$$

where the argument u is implied. This series for $\Gamma_{\mathbf{k}}(u)$ can also be obtained directly, by inversion of (51) and expansion of $1/G_{\mathbf{k}}(u)$, a procedure that remains valid when the $\mathbf{v}$ distribution is not normal and there is no diagram representation.

A further reworking gives $\Gamma_{\mathbf{k}}(u)$ as a series in G functions rather than G^0 functions. Here, only those terms in (52) are retained whose diagrams cannot be broken into two parts (each containing at least two vertices) by two cuts of the horizontal line. In the terms retained, each factor $G_{\mathbf{p}}{}^0(u)$ is replaced by $G_{\mathbf{p}}(u)$. This corresponds in quantum field theory to renormalization by elimination of all self-energy parts. The resulting series is

$$\begin{aligned}
\Gamma_k(u) &= \bar{\Gamma}_k{}^2(u) + \lambda^2\bar{\Gamma}_k{}^4(u) + \lambda^4\bar{\Gamma}_k{}^6(u) + \cdots, \\
\bar{\Gamma}_k{}^2(u) &= \int d\mathbf{q}[\mathbf{k}\cdot\mathsf{P}(\mathbf{q})\cdot\mathbf{k}]U(q)G_{\mathbf{k}-\mathbf{q}}(u)\,, \\
\bar{\Gamma}_k{}^4(u) &= \bar{\Gamma}_k^{4,3}(u)\,, \\
\bar{\Gamma}_k{}^6(u) &= \bar{\Gamma}_k^{6,1}(u) + \bar{\Gamma}_k^{6,2}(u) + \bar{\Gamma}_k^{6,3}(u) + \bar{\Gamma}_k^{6,4}(u), \ldots .
\end{aligned} \tag{53}$$

The diagrams for the four sixth-order terms are shown in Fig. 6. Equation (52) can be recovered from (53) by expanding each G factor according to (50). We shall call (53) the irreducible series for $\Gamma_k(u)$. It can be obtained without diagrammatics, and generalized to arbitrary statistical distribution of $\mathbf{v}$, by reverting (50) to give $G_k{}^0(u)$ as a series in $G_p(u)$ functions and substituting into (52).

If $k \gg k_0$ holds with sufficient strength that $k \gg q$ for all q where $U(q)$ is appreciable, then $G_{\mathbf{k}}(u) \approx G_{\mathbf{k}-\mathbf{q}}(u) \approx G_{\mathbf{k}-\mathbf{q}-\mathbf{q}'}(u)$, etc., and the integrations over wave vectors in (50), (52), and (53) are easily performed. The results are

$$G_k(u) = (\nu + u)^{-1}\left\{1 + \sum_{n=1}^{\infty}(-1)^n R_{2n}\lambda^{2n}[v_0 k/(\nu + u)]^{2n}\right\} \qquad (k \gg k_0)\,, \tag{54}$$

$$\Gamma_k(u) = (v_0 k)^2(\nu + u)^{-1}\sum_{n=1}^{\infty}(-1)^{n-1}R'_{2n}\lambda^{2n-2}[v_0 k/(\nu + u)]^{2n-2} \qquad (k \gg k_0)\,, \tag{55}$$

$$\Gamma_k(u) = (v_0 k)^2 G_k(u)\sum_{n=1}^{\infty}(-1)^{n-1}S_{2n}\lambda^{2n-2}[v_0 k G_k(u)]^{2n-2} \qquad (k \gg k_0)\,, \tag{56}$$

where R'_{2n} is the number of connected diagrams of order $2n$ and S_{2n} is the number of irreducible diagrams of order $2n$. In this limit, each surviving diagram of a given order makes the same contribution.

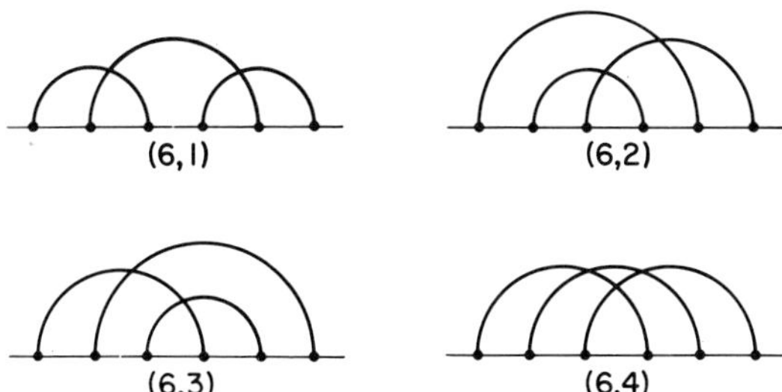

Fig. 6. All irreducible sixth-order diagrams.

Equation (57) is the Laplace transform of the expansion, in powers of λ, of $g_k(t) = \exp(-\nu t - v_0^2 k^2 \lambda^2 t^2/2)$, so that there is agreement with (40). Equations (54)–(56) are identical with the corresponding series for the function $\int_0^\infty e^{-ut}\langle q(t)\rangle\, dt$, where $q(t)$ is the amplitude of a random-frequency oscillator,

$$(d/dt + \nu)q(t) = -ia\lambda q(t)\,, \qquad q(0) = 1\,, \tag{57}$$

and the probability distribution of a is

$$\rho(a) = (v_0 k)^{-1}(2\pi)^{-1/2}\exp(-a^2/2v_0^2k^2)\,. \tag{58}$$

The R_{2n} satisfy $R_{2n+2} = (2n+1)R_{2n}$, so that the first few values, starting with $n = 1$, are 1, 3, 15, 105, 945, 10395, 135135. The first few values of the R'_{2n}, starting with $n = 1$, are 1, 2, 10, 74, 706, 8162, 110410, 1708994. We shall see in Sect. VI that $R'_{2n}/R_{2n} \to 1$ as $n \to \infty$. The S_{2n} satisfy $S_{2n+2} = n\sum_{m=1}^{n} S_{2m}S_{2n+2-2m}$ (Orszag, 1969), and the first few values starting with $n = 1$ are 1, 1, 4, 27, 248, 2830, 38232, 593859, 10401712. It can be shown from the diagram topology (Orszag, 1969) that $S_{2n}/R_{2n} \to 1/e$ as $n \to \infty$. Thus all three series (54)–(56) are strongly divergent, and the elimination of disconnected or reducible diagrams in (55) or (56) does not affect the rate of growth of the coefficients at very large n.

Reversion of (56) gives

$$G_k(u) = (v_0k)^{-2}\Gamma_k(u)\left(1 + \sum_{n=1}^{\infty}(-1)^{n-1}V_{2n+2}\lambda^{2n}[\Gamma_k(u)/v_0k]^{2n}\right), \tag{59}$$

where it can be shown that V_{2n} is the number of diagrams of order $2n$ without vertex parts: that is, diagrams that cannot be cut into two parts, each containing at least two vertices, by cutting the horizontal line twice and a curved line once. The first three V_{2n}, starting with $n = 2$, are 1, 1, 7. It can be shown (Orszag, 1969) that $V_{2n}/R_{2n} \to 1/e^2$ as $n \to \infty$, so that, again, the elimination of diagrams does not affect the asymptotic rate of growth of the coefficients.

VI. APPROXIMANTS TO THE PRIMITIVE PERTURBATION EXPANSIONS

If $\nu = 0$, (46) conserves $\sum_{\mathbf{k}} |\psi_{\mathbf{k}}(t)|^2$. Therefore, the frequencies of the normal modes are all real, in this case, and clearly proportional to λ. For $\nu > 0$, each normal-mode amplitude is multiplied by $\exp(-\nu t)$. It follows that $G_k(u)$ satisfies the dispersion relation

$$G_k(u) = \int_{-\infty}^{\infty} \rho_k(a)\,(\nu + u + ia\lambda)^{-1}\,da\,, \tag{60}$$

where

$$\rho_k(a) = \pi^{-1}\,\mathrm{Re}[G_k(-ia)]_{\nu=0,\,\lambda=1} \geq 0\,. \tag{61}$$

Moreover, $\rho_k(a)$ is an even function of a, by symmetry, so that (60) can be written as

$$\begin{aligned} G_k(u) &= (\nu + u)^{-1} \int_{-\infty}^{\infty} \rho_k(a)[1 + ia\lambda/(\nu + u)]^{-1}\,da \\ &= 2(\nu + u)^{-1} \int_0^{\infty} \rho_k(a)[1 + a^2\lambda^2/(\nu + u)^2]^{-1}\,da\,. \end{aligned} \tag{62}$$

Comparing (62) and (50), we find $\int_{-\infty}^{\infty} \rho_k(a)\,da = 1$. It follows from (62), (51), and a theorem on the reciprocal of a Stieltjes integral (Masson, 1970) that

$$\begin{aligned} \Gamma_k(u) &= (\nu + u)^{-1} \int_{-\infty}^{\infty} \rho_k'(a)[1 + ia\lambda/(\nu + u)]^{-1}\,da \\ &= 2(\nu + u)^{-1} \int_0^{\infty} \rho_k'(a)[1 + a^2\lambda^2/(\nu + u)^2]^{-1}\,da\,, \end{aligned} \tag{63}$$

where $\rho_k'(a)$ is real, nonnegative, even in a, and independent of ν and λ.

Explicit expressions for $\rho_k(a)$ and $\rho_k'(a)$ in terms of each other follow immediately from (62), (63), and (51). Putting $\lambda = 1$, $u = -ia$, and taking the real part of (51) in the limit $\rightarrow 0$, we find

$$\rho_k(a) = \rho_k'(a)\left\{\left[a + \int_{-\infty}^{\infty} \rho_k'(a')\,(a - a')^{-1}\,da'\right]^2 + [\pi\rho_k'(a)]^2\right\}^{-1}. \tag{64}$$

Rewriting (51) as $\Gamma_k(u) = [G_k(u)]^{-1} - (\nu + u)$ and repeating the procedure, we find

$$\rho_k'(a) = \rho_k(a)\left\{\left[\int_{-\infty}^{\infty} \rho_k(a')\,(a - a')^{-1}\,da'\right]^2 + [\pi\rho_k(a)]^2\right\}^{-1}. \tag{65}$$

The coefficients in (50) are all finite if $U(q)$ is integrable over $\mathbf{q}$ space, which shows that $\rho_k(a)$ falls off at large a faster than any power. Therefore, since $\int_{-\infty}^{\infty} \rho_k(a)\,da = 1$, the principal-part integral in (65) approaches

a^{-1} at large a, and we have

$$a^2\rho_k(a)/\rho_k'(a) \to 1 , \qquad a \to \infty , \tag{66}$$

which, in turn, implies $(R'_{2n}/R_{2n})_{n\to\infty} \to 1$.

The qualitative properties conjectured in Sect. IV and confirmed by the computer experiment imply a substantial difference in character between $\rho_k(\mathrm{a})$ and $\rho_k'(a)$ at low k. If (37) is correct,

$$\rho_k(a) \approx \pi^{-1}\{a^2 + [k^2\kappa(\infty)]^2\}^{-1} \qquad (kl \ll 1, |a| \ll v_0/l) . \tag{67}$$

This functional form implies that the moment integrals $\int_0^\infty \rho_k(a)a^{2n}\,da$ converge increasingly poorly as $kl \to 0$. On the other hand, $k^{-2}\Gamma_k(u)$ should approach a limit as $kl \to 0$ if the motion acts like a classical diffusion process. In particular, $k^{-2}\Gamma_k(0) \to \kappa(\infty)$ as $k \to 0$. We shall therefore confine our efforts to approximating $\Gamma_k(u)$. The high k limit series (54) and (55) are qualitatively similar in that R_{2n} and R'_{2n} never differ by a factor of more than 1.5, and we do not find there the difference in behavior between $\rho_k(a)$ and $\rho_k'(a)$ that is expected at low k.

The Padé approximants to the positive-density Stieltjes integral (63) converge for all parameter values except $\lambda/(\nu + u)$ pure imaginary. This is an embarrassment, since it excludes the physical case $\nu = 0$, $\lambda = 1$, $u = -ia$. Let us take $\lambda = 1$ and consider the approximants to $\Gamma_k(0)$, the quantity of greatest interest, as a function of ν. Equation (63) shows that $\nu\Gamma_k(0)$ is a power series in ν^{-2}. Since $\rho_k'(a)$ is nonnegative, the $[n, n]$ Padé approximants all upper-bound $\nu\Gamma_k(0)$, whereas the $[n + 1, n]$ approximants all lower-bound that quantity. [We label the approximants to $\nu\Gamma_k(0)$ by the powers of ν^{-2} in denominator and numerator.] In the limit $\nu \to 0$, each $[n, n]$ approximant yields $\Gamma_k(0) \to \infty$, whereas each $[n + 1, n]$ approximant gives $\Gamma_k(0) \to 0$. This is not a shortcoming of the Padé method. We can easily see that no better bounds are possible if our *only* information is a finite number of moments of $\rho_k'(a)$, or a finite number of moments of the velocity field. Thus, suppose we alter $\rho_k'(a)$ in the neighborhood of $a = 0$ to any value desired. If the alteration is made in a sufficiently small interval, the change in any finite-order moment of $\rho_k'(a)$ can be made as small as desired. But, in the limit $\nu \to 0$, $\Gamma_k(0) = \pi\rho'(0)$. In order to obtain finite bounds on $\Gamma_k(0)$, some information or assumption about smoothness must be used, in addition to the values of the low-order moments. The alterations in $\rho_k'(0)$ just appealed to can be realized physically as follows. First, let $\hat{\mathbf{v}}(\mathbf{x})$ be zero everywhere in a finite fraction of the realizations. Then $G_k(0)$ is infinite, and, by (51), $\Gamma_k(0) = 0$ if $\nu = 0$. But if the fraction of such realizations is small enough, we can make any finite-order moment of the velocity field, and, hence, any moment of $\rho_k'(a)$, as close to the multivariate-normal value as desired. On the other hand, let $\hat{\mathbf{v}}(\mathbf{x})$ be constant in space and nonzero, in a finite fraction of the realizations, at $t = 0$. Switch $\hat{\mathbf{v}}(\mathbf{x})$ off in these realizations at a later time t such that $\psi_k(t)$ is negative for a given k. Thereafter these realizations will make a growing, negative contribution to $\int_0^t g_k(s)\,ds$, which, finally, can be limited by changing $\hat{\mathbf{v}}(\mathbf{x})$ to a typical, random field at a third time. In this way, the positive contributions to $G_k(0)$ from the typical realizations can be balanced by negative contributions from the anomalous realizations so as to make $G_k(0)$ as small as desired and, hence, $\Gamma_k(0)$ as large as desired, again with an arbitrarily small change in any finite-order moment.

It is possible to obtain crude upper and lower bounds on $\Gamma_k(0)$ for $\nu = 0$ from the Padé approximants for $\nu > 0$, provided that two physically plausible conjectures are valid. First, we assume that the extra damping ν should decrease the relaxation time of each k mode; that is, $[\nu + \Gamma_k(0)]_{\nu>0} > [\Gamma_k(0)]_{\nu=0}$. On the other hand, $\nu > 0$ should decrease the turbulent diffusivity because it damps fluctuations in $\psi(\mathbf{x})$ and, in fact, damps the total probability of finding a particle (cf. Saffman, 1960). Therefore, we expect $[\Gamma_k(0)]_{\nu>0} < [\Gamma_k(0)]_{\nu=0}$. Under these assumptions, every $[n, n]$ approximant, in our previous notation, should yield a $\nu + \Gamma_k(0)$ value that upper-bounds $[\Gamma_k(0)]_{\nu=0}$, whereas each $[n + 1, n]$ approximant should yield a $\Gamma_k(0)$ that lower-bounds $[\Gamma_k(0)]_{\nu=0}$. We obtain a least upper bound by minimizing as a function of ν and a greatest lower bound by maximizing as a function of ν.

Evaluation of (52) at $\lambda = 1$ for the spectra (43) involves only elementary integrals. The result, through sixth-order diagrams, is

$$\begin{aligned}\Gamma_k(u) = (v_0 k)^2(\nu + u)^{-1}\{1 &+ [v_0 k_0/(\nu + u)]^2(A + 2k^2/k_0^2) \\ &+ [v_0 k_0/(\nu + u)]^4(B + Ck^2/k_0^2 + 10k^4/k_0^4) + \cdots\}\,, \end{aligned} \qquad (68)$$

where

$$\begin{aligned} &A = 1.25\,, \quad B = 7.34375\,, \quad C = 23.75 \text{ (spectrum } E_2)\,, \\ &A = 1\,, \quad B = 3.5\,, \quad C = 19 \text{ (spectrum } E_1)\,. \end{aligned} \qquad (69)$$

The $[0, 0]$ and $[1, 0]$ approximants are

$$\begin{aligned} &[\Gamma_k(0)]_{0,0} = v_0^2 k^2/\nu\,, \\ &[\Gamma_k(0)]_{1,0} = v_0^2 k^2 \nu^{-1}/[1 + (2k^2/k_0^2 + A)v_0^2 k_0^2 \nu^{-2}]\,. \end{aligned} \qquad (70)$$

They yield, respectively, a least upper bound for $\nu + \Gamma_k(0)$ at $\nu = v_0 k$ and a greatest lower bound for $\Gamma_k(0)$ at $\nu^2 = v_0^2(2k^2 + Ak_0^2)$. Under the assumptions stated, this implies

$$\Gamma_k(0) < 2v_0 k\,, \qquad \Gamma_k(0) > \tfrac{1}{2}\, v_0 k^2/(Ak_0^2 + 2k^2)^{1/2} \qquad (\nu = 0)\,. \qquad (71)$$

In the limit $k \gg k_0$, (71) gives the bounds 2 and 0.35355 for $\Gamma_k(0)/(v_0 k)$, whereas from (41) the exact value is $(2/\pi)^{1/2} = 0.79788$. If

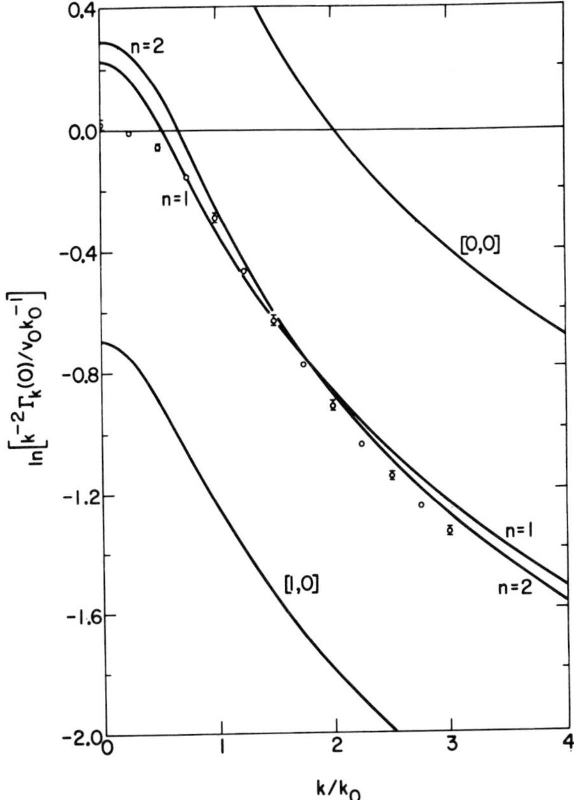

Fig. 7. Conjectured Padé bounds [0, 0] and [1, 0] for $\Gamma_k(0)$, together with approximants $n = 1$ and $n = 2$ from the continuous orthogonal expansion for (68). Spectrum E_1. Points are computer experiment ($N = 100$, $R = 2000$).

the procedure is extended to the [1, 1] and [2, 1] approximants found from (55), we obtain the improved bounds 1.604 and 0.4446. The conjectures can be verified from the exact solution in this high k limit, but the indicated rate of convergence is so poor that there is no incentive to compute the higher approximants for the whole k spectrum.

The conjectured bounds (71) are compared with the computer experiment in Figs. 7 and 8. It is easy to see from the form of (52) that all the least upper bounds from higher approximants yield infinite $k^{-2}\Gamma_k(0)$ at $k = 0$, whereas all the greatest lower bounds are finite, as with (71).

Next, we shall turn to constructing approximants to (52) by the continuous orthogonal expansion method. Here an anticipation that $\rho_k'(a)$ is a smooth function is implied. With the normal velocity distribution, there seems no way in which a particular $a \neq 0$ can be singled out, so that the

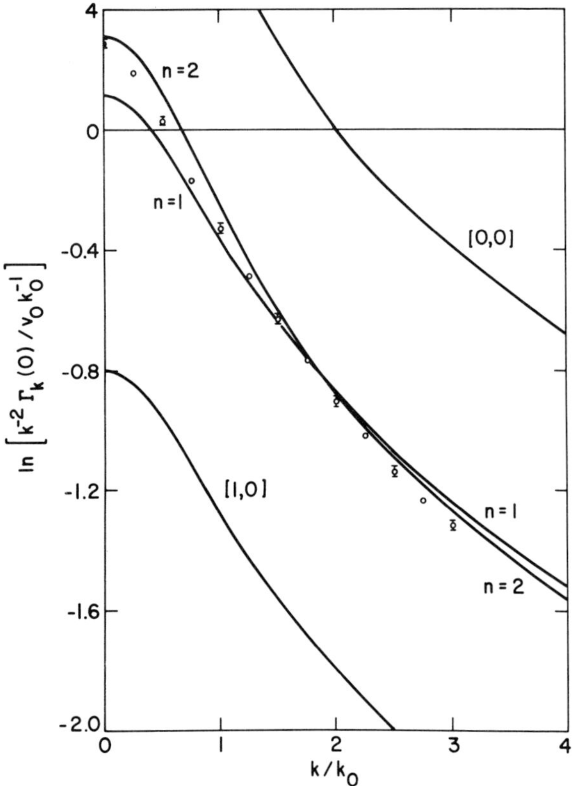

Fig. 8. Figure 7 repeated for spectrum E_2.

only possibility of singular behavior is at $a = 0$. Trapping would imply that $\rho_k(0)$ is infinite, and, therefore, $\rho_k'(0) = 0$. We expect, therefore, that in any case $\rho_k'(a)$ is square integrable, so that, if $w(a)$ falls off slowly enough at large a, we shall successfully approximate ρ_k' in mean square.

Let us start with the $k \gg k_0$ limit, where it is easy to carry the approximations to relatively high orders. As a trial choice, subject to correction if necessary to assure convergence, we take Gaussian $w(a)$, choosing q in (20) so as to make $b_2 = 0$ in (2). Multiplying (63) and (55) by $\nu + u$, taking $z = \lambda/(\nu + u)$, and comparing with (1), we have $c_{2n} = (v_0k)^2 R'_{2n+2}(v_0k)^{2n}$ $(n = 0, 1, 2 \ldots)$. Carrying out the expansion as in Sects. II and III, we have $q^2 = 2v_0^2k^2$ and find, for the first few b_{2n}, using the values of R'_{2n} from Sect. V,

n	0	1	2	3	4	5	6
$\dfrac{2n!\,b_{2n}}{(v_0k)^2}$	1	0	−0.5	1.75	−4.875	1.9375	174.156

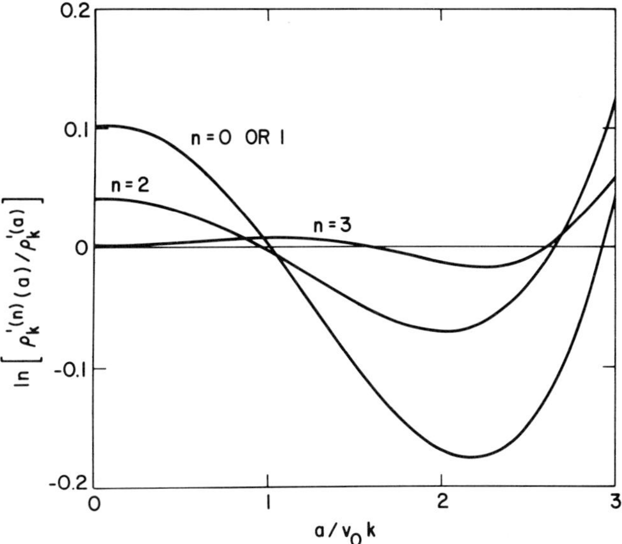

Fig. 9. Logarithmic error in continuous orthogonal expansion approximants to $\rho_k'(a)$ in the high k limit.

The approximants $\rho_k'^{(n)}(a) = \sum_{m=0}^{n} b_{2m}(a) P_n(a)$ at $a = 0$ are

n	0	1	2	3	4	5	6
$\dfrac{\rho_k'^{(n)}(0)}{v_0 k}$	0.282095	0.282095	0.264464	0.254179	0.250598	0.250456	0.251522

whereas the exact value is $\rho_k'(0)/v_0 k = (2/\pi)^{1/2}/\pi = 0.253975$. The approximants to $\Gamma_k(a)$ at $\nu = 0$, $\lambda = 1$ are given by $\Gamma_k(a) = \pi\rho_k'(a)$.

These results show rapid convergence toward the exact value through $n = 3$, then a much slower improvement, accompanied by slow oscillations of the sign of the error, as n increases further. This behavior is insensitive to changes of width of $w(a)$ over a moderate range. The choice $q = v_0 k$ gives 0.398942, 0.199471, 0.249339, 0.257650, 0.256611, 0.254845, 0.253919, for the approximants $n = 0, \ldots, 6$. It follows from (66), the exact function (58), and Sect. II that there will be convergence in mean square for any $q > v_0 k\sqrt{2}$. Figure 9 shows the logarithmic error in $\rho_k'^{(n)}(a)$ as a function of a for the original choice $q^2 = 2v_0^2k^2$. The exact values were computed from (58) and (65). The fractional errors are surprisingly small out to $a = 3v_0 k$, where the absolute value of $\rho_k'(a)$ is already small. The accuracy of any given approximant eventually deteriorates at large enough a, where the information needed to determine $\rho_k'(a)$ comes from high orders of perturbation theory.

The extension of the approximants $n = 1$ and $n = 2$ to the entire k range is compared with values of $\Gamma_k(0)$ from the computer experiments in Figs. 7 and 8. These curves were obtained using a Gaussian $w(a)$, with width chosen at each k to fit (68) with $b_2 = 0$. In contrast to the high k limit, we cannot assert that this $w(a)$ falls off slowly enough at large a to assure convergence. In order to extend the approximants reliably to high orders, we must either estimate the asymptotic rate of growth of the coefficients in (68) or rely on some automatic scheme for correcting $w(a)$, if necessary. A crude scheme of this kind is to take $w(a)$ of the form $\exp(-|a|^p/q)$ and adjust p and q as n rises so as to maximize the projection of $\rho_k'^{(n)}(a)[w(a)]^{-1/2}$ on $P_0(a)[w(a)]^{1/2}$, relative to the projection on higher functions. Probably a better scheme for the present problem, where Gaussian $w(a)$ is specially indicated bacause of the normal velocity distribution, is to take

$$w(a) = \sum_{i=1}^{S} C_i \exp(-a^2 q_i) , \tag{72}$$

where, say, $S > (n + 1)/2$, and, again, the parameters are chosen to maximize the projection on the lowest orthogonal function. The convergence theory of these schemes remains to be investigated.

The possibility that $w(a)$ may have to be corrected at high orders is of rather academic interest because of the labor of computing higher coefficients in (68)

We note in Figs. 7 and 8 that the approximations to the computer-experiment values are fairly uniform over k in the case of spectrum E_2 but deteriorate for $k < k_0$ in the case of the more sharply peaked spectrum E_1. In either case, they are a very great improvement over the conjectured Padé bounds. The relatively poor low-k approximation with spectrum E_1 probably is connected with the fact that $\kappa(t)$ overshoots its asymptotic value before leveling off (Fig. 1). That phenomenon is a consequence of regions of negative correlation exhibited by the velocity covariance (30). It can be shown that the quantity $d\kappa(t)/dt$ is equal to the correlation of the particle's current and initial velocities (Lagrangian velocity correlation) and that its frequency spectrum is just $[k^{-2}\rho_k'(a)]_{k\to 0}$. Thus, the negative region in $d\kappa(t)/dt$ indicates a peculiarity in the shape of $\rho_k'(a)$ at low k, and it is not surprising that two Hermite functions are an inadequate approximation. Probably a better choice for $w(a)$ at low k would be the direct-interaction approximation for $\rho_k'(a)$, discussed in Sect. VII.

VII. APPROXIMANTS TO THE IRREDUCIBLE EXPANSION

Since the physical case is $\nu \to 0$, it seems natural to seek approximants to the irreducible expansion (53), whose terms are finite in the limit, rather than work with the primitive perturbation expansions. We shall take $\nu \to 0$ throughout this section. The simplest way to treat (53) is as a power series in λ^2, ignoring, for the purpose of classifying terms, the implicit dependence on λ^2 through the G factors. Alternatively, and probably better, we may take $\lambda = 1$ at the start and consider (53) to be a functional power series in $G_k(u)$. A sufficient condition for Padé approximants or a continuous orthogonal expansion to converge is the existence of well-behaved integral representations analogous to (63).

A special role is played by the lowest-order truncation of (53), which is also the [0, 0] Padé approximant. Inserting this truncation in (51), with $\lambda = 1$, we have

$$G_k(u) = \left\{u + \int d\mathbf{q}[\mathbf{k}\cdot\mathsf{P}(\mathbf{q})\cdot\mathbf{k}]U(q)G_{\mathbf{k}-\mathbf{q}}(u)\right\}^{-1} \tag{73}$$

The equivalent time-domain equation is

$$dg_k(t)/dt = -\int_0^t \int d\mathbf{q}[\mathbf{k}\cdot\mathsf{P}(\mathbf{q})\cdot\mathbf{k}]U(q)g_{\mathbf{k}-\mathbf{q}}(t-s)g_k(s)\,ds\,,$$
$$g_k(0) = 1\,. \tag{74}$$

These equations can be shown to represent exactly the statistical dynamics of a model dynamical system that has an infinite number of degrees of freedom and also conserves $\sum_\mathbf{k} |\psi_\mathbf{k}(t)|^2$ (Kraichnan, 1961). As a result, it can be asserted that the solutions have important consistency properties. In particular, $\rho_k(a) = \pi^{-1}\,\mathrm{Re}[G_k(-ia)]$ as computed from (73) is nonnegative for all a. Equations (73) and (74) have been called the direct-interaction approximation.

Figures 1 and 2 show that the direct-interaction approximation gives fairly good values for $\kappa(t)$ over the entire time range and for $G_k(0)$ over the entire k range, with both spectra E_1 and E_2. It also yields fairly faithful curves for $g_k(t)$ over the whole k range (Kraichnan, 1970b).

The success of the direct-interaction approximation raises the hope that higher Padé approximants to (53) will provide a usefully convergent sequence. Let us start again with the high k limit. Comparison of (55) and (56) suggests that we seek an integral representation of the form

$$\Gamma_k(u) = G_k(u)\int_{-\infty}^{\infty} \frac{\rho_k''(a)\,da}{1 + ia\lambda G_k(u)} = 2G_k(u)\int_0^{\infty} \frac{\rho_k''(a)\,da}{1 + a^2\lambda^2[G_k(u)]^2}\,, \tag{75}$$

wherein $1/G_k(u)$ plays the role that $\nu + u$ does in (63). Here λ and $G_k(u)$ occur only as $\lambda G_k(u)$ in the integrand so that the distinction between series in λ and functional power series is empty in the high k limit.

The first few Padé approximants to (56), labeled by powers of λ^2, are

$\Gamma_k(u) = v_0 k G_k(u)[n, m]$, where

$$
\begin{aligned}
&[0, 0] = 1\,, \qquad [1, 0] = 1/(1 + x^2)\,, \qquad [1, 1] = (1 + 3x^2)/(1 + 4x^2)\,, \\
&[2, 1] = (3 + 20x^2)/(3 + 23x^2 + 11x^4)\,, \\
&[2, 2] = (11 + 129x^2 + 167x^4)/(11 + 140x^2 + 263x^4)\,, \\
&[3, 2] = (167 + 2916x^2 + 8919x^4)/(167 + 3083x^2 + 11334x^4 + 3551x^6)\,, \\
&[3, 3] = \frac{3511 + 86065x^2 + 477870x^4 + 375899x^6}{3511 + 89576x^2 + 553402x^4 + 665784x^6}\,, \ldots,
\end{aligned}
\tag{76}
$$

where $x = v_0 k \lambda G_k(u)$, and the numerical values of the S_{2n} have been used. If we take $\lambda = 1$ and insert the exact value $G_k(0) = (\pi/2)^{1/2}/v_0 k$ into (76), we obtain for $G_k(0)\Gamma_k(0)$ the successive approximants 1.57080, 0.61102, 1.23203, 0.81576, 1.11706, 0.90216, 1.06560. To the orders calculated, the $[n, n]$ approximants form a decreasing sequence bounding the exact value $G_k(0)\Gamma_k(0) = 1$ from above, whereas the $[n + 1, n]$ approximants are an increasing sequence bounding this value from below. This is consistent with a representation (75) with nonnegative $\rho_k''(a)$; moreover, the poles of (76) are all at imaginary x, and have residues of correct sign to give a positive δ-function contribution to $\rho_k''(a)$.

In order to obtain approximants to $G_k(u)$ itself, let us substitute the successive approximants (76) into (51) and solve for $G_k(u)$ at $\lambda = 1$, $\nu = 0$, taking always the branch that yields $G_k(0) \to \nu^{-1}$ as $\lambda \to 0$. This gives for $v_0 k G_k(0)$ the sequence of approximants 1, ∞, 1.12417, 1.53729, 1.17963, 1.36237, 1.20876. Again the $[n, n]$ and $[n + 1, n]$ sequences appear to bound the exact value $\Gamma_k(0) = 1/G_k(0) = v_0 k/1.25331$ from above and below, respectively.

We shall call the approximations just formed *internal* Padé approximants, since they involve approximation of the functional form of $\Gamma_k(u)$ within the integral equation (51). Their behavior for $u = -ia$, $a \neq 0$, is complicated. Through the orders studied, they yield sets of branch points along the real a axis, so that $\rho_k(a)$ consists of a finite number of finite-width passbands separated by stop bands. In all the approximants $\rho_k(a)$ is finite everywhere within the passbands, except for [1, 0], where $\rho_k(a)$ has a singularity of type $|a|^{-1/3}$. We conjecture that as $n \to \infty$ the primary passbands of the approximants $[n, n]$ and $[n + 1, n]$ spread to include any given a and that $\rho_k(a)$ within that band converges to the exact $\rho_k(a)$. Figure 10 shows $\rho_k(a)$ as given by [0, 0] and [3, 3], together with the exact $\rho_k(a)$. These solutions were continued through the branch points by requiring $\rho_k(a) \geq 0$, and they were checked by verifying $\int_{-\infty}^{\infty} \rho_k(a)\, da = 1$. For later comparison, Fig. 10a shows approximants constructed by continuous orthogonal expansion.

The internal Padé approximations can be extended to the full k range

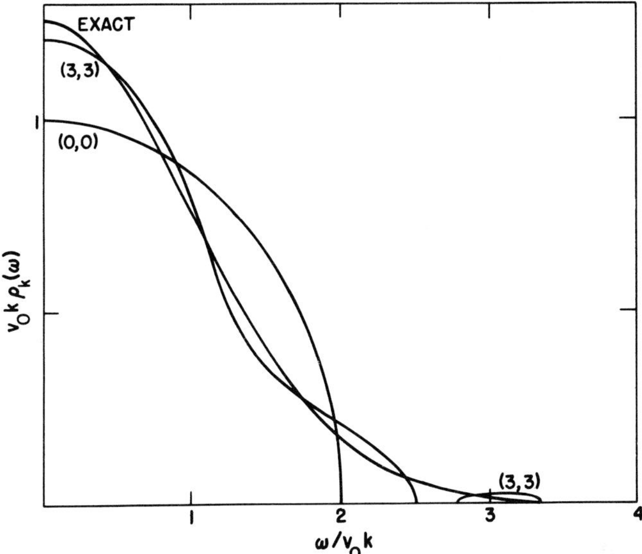

Fig. 10. Approximations to high-k $\rho_k(a)$ from internal Padé approximants.

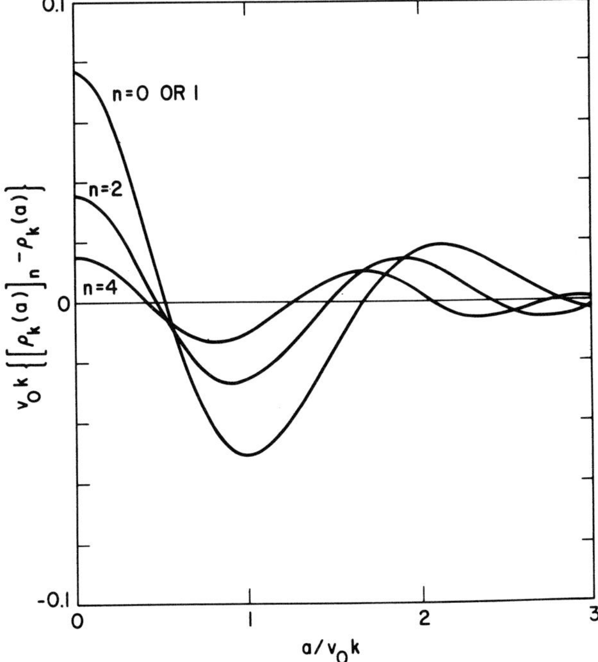

Fig. 10a. Approximants to high-k $\rho_k(a)$ from using (78) in (51).

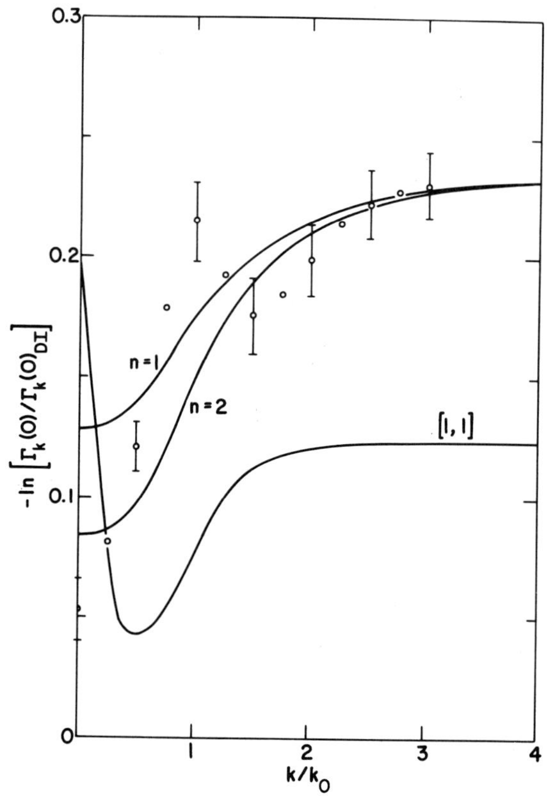

Fig. 11. Logarithmic deviation of $\Gamma_k(0)$ from direct-interaction values, for spectrum E_1. Points are computer experiment ($N = 100$, $R = 2000$). Curves show internal Padé approximant [1, 1] and continuous-orthogonal-expansion approximants $n = 1$ and $n = 2$, obtained with weight (79). Computational errors in approximants, from Monte Carlo calculation of wave-vector integrals, are not shown in this or following figures, but have typical values $<0.4\%$.

by forming approximants to (53), regarded as a series in λ^2, substituting into (51), and solving the resulting integral equation by iteration. A form of (51) found to give stable convergence of the iteration process at all k for $u = 0$ is $G_k(0) = [G_k(0)/\Gamma_k(0)]^{1/2}$ (where we have taken $\nu = 0$, $\lambda = 1$). The results for $\Gamma_k(0)$ from the [0, 0] (direct-interaction approximation) and [1, 1] internal Padé approximants for spectrum shape E_1 are compared with the computer experiment in Fig. 11. The [1, 0] approximant was skipped because, as noted previously, it blows up in the high k limit.

A reduction in error in the [1, 1] approximation as compared with the direct-interaction approximation is apparent over the entire k range, with

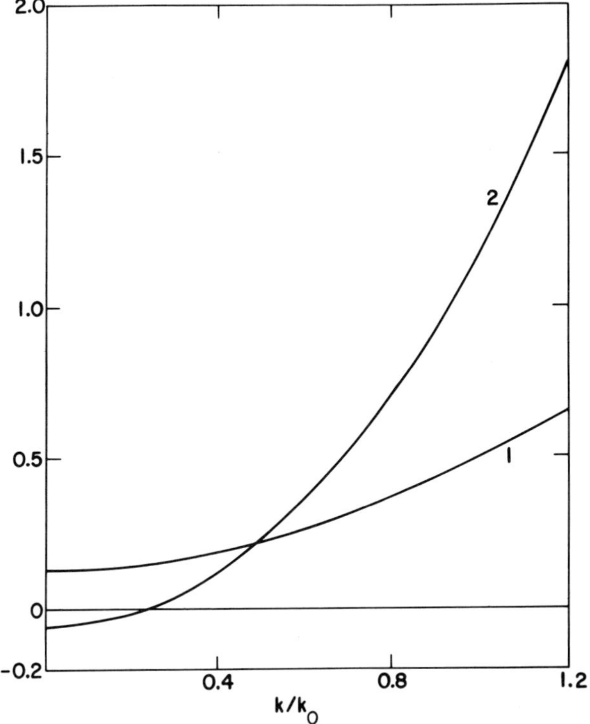

Fig. 12. Ratios $\bar{\Gamma}_k{}^6(0)/\Gamma_k{}^2(0)$, curve 2, and $-\bar{\Gamma}^4{}_k(0)/\bar{\Gamma}_k{}^2(0)$, curve 1, found from internal Padé approximant [1, 1], spectrum E_1.

the exception of an anomalous behavior at very small k/k_0. The latter phenomenon is associated with a reversal of sign of $\bar{\Gamma}_k{}^6(0)$ [defined as the total contribution in (53) $\propto \lambda^4$], and occurs also with spectrum E_2. Figure 12 shows the ratios $\bar{\Gamma}_k{}^4(0)/\bar{\Gamma}_k{}^2(0)$ and $\bar{\Gamma}_k{}^6(0)/\bar{\Gamma}_k{}^2(0)$ found from the [1, 1] solution. A similar behavior of these ratios is found if (53) is computed from the computer-experiment values of $G_k(0)$. The reversal of sign of $\bar{\Gamma}_k{}^6(0)$, which comes from the [6, 4] diagram, shows that, if $\Gamma_k(0)$ has an integral representation of the form

$$\Gamma_k(0) = \int_{-\infty}^{\infty} \mu_k(\alpha)(1 + i\alpha\lambda)^{-1}\,d\alpha = 2\int_0^{\infty} \mu_k(\alpha)(1 + \alpha^2\lambda^2)^{-1}\,d\alpha\,, \tag{77}$$

then $\mu_k(\alpha)$ cannot be positive for all α at low k, in contrast to the situation at high k.

The low k behavior recalls the danger with Padé approximants illustrated by the discussion of example (27). If the Stieltjes-transform density is indefinite, any given approximant may give a bad value, or blow up, even though the sequence of approximants tends to converge. The sign reversal of the λ^4 contribution at low k also has a computational disadvantage; it makes $\Gamma_k(0)$ sensitive to numerical error in computing $\bar{\Gamma}_k^6(0)$. The wave vector integrals in (53) were computed by a Monte Carlo method, and large samples were needed to obtain accuracy at low k.

Now we turn to the approximation of (53) by the continuous orthogonal expansion method. Again, we shall consider the high k limit first. Let us expand $\rho_k''(a)$ in (75) in Hermite functions, using as the c_n the coefficients of $\lambda G_k(u)$ in (56). As before, we choose the width q in (20) to make $b_2 = 0$ in (2). Now, this means $q = v_0 k$. Using the numerical values of the S_{2n} in (56), we find from (5) the following values of $2n!b_{2n}/v_0 k$ $(n = 0, \ldots, 7)$: 1, 0, 1, − 3, 17, − 140, 1437, − 17388. If the exact value $v_0 k G_k(0) = (\pi/2)^{1/2}$ is inserted in the approximants

$$[\Gamma_k(u)]_n = 2G_k(u) \sum_{m=0}^{n} b_{2m} \int_0^\infty w(a)\, \mathrm{He}_{2m}(a/v_0 k)/(1 + a^2[G_k(u)]^2)^{-1}\, da\,, \quad (78)$$

we find for $G_k(0)[\Gamma_k(0)]_n$ $(n = 0, \ldots, 7)$ the values 0.917667, 0.917667, 0.959064, 0.973863, 0.981717, 0.986719, 0.990105, 0.992462. If the approximants (78) are used in (51), and the resulting integral equations are solved by iteration ($\lambda = 1$, $\nu = 0$), the approximants $v_0 k[G_k(0)]_n$ thus found are $(n = 0, \ldots, 4)$ 1.33016, 1.33016, 1.28895, 1.27547, 1.26857.

These latter approximants converge toward the exact value 1.253314 substantially faster than the internal Padé approximants $[n, n]$ and $[n + 1, n]$. The error of $[G_k(0)]_4$, which uses five terms of (56), is + 1.2%, whereas the error of $[G_k(0)]_{3,3}$, which uses seven terms, is − 3.6%. There is a qualitative as well as quantitative improvement for $u = -ia$, $a \neq 0$. $[G_k(-ia)]_n$ at any n is continuous for all a and goes smoothly to zero as $a \to \infty$, in contrast to the pass-band behavior of the internal Padé approximants. (cf. Figs. 10 and 10a.)

The present approximants also differ in a significant way from the continuous-orthogonal-expansion approximations constructed in Sect. VI from the primitive perturbation expansion for $\Gamma_k(u)$. In the present case, $\Gamma_k(-ia)$ is determined for any a as an integral over $\rho_k''(a')$ to which all values of a' contribute at $\nu = 0$. In contrast, $\mathrm{Re}\,\Gamma_k(-ia) = \pi\rho_k'(a)$ at $\nu = 0$. Errors in approximating the density functions therefore have different effects on the error of $\Gamma_k(-ia)$ in the two cases.

The easiest way to extend the continuous orthogonal expansion for $\Gamma_k(0)$ to the full k range is to consider (53) as a series in λ^2, with the $\bar{\Gamma}_k^n(0)$ as coefficients, and work from the conjectured integral representation (77). In the high k limit, we took a $w(a)$ that made $b_2 = 0$. This procedure is suspect at low k because of the behavior shown in Fig. 12. The small values of $\bar{\Gamma}_k^4(0)/\bar{\Gamma}_k^2(0)$ and $\bar{\Gamma}_k^6(0)/\bar{\Gamma}_k^2(0)$ more likely are associated with negative regions in $\mu_k(\alpha)$ than with a narrow width. If we fix the width of w to make $b_2 = 0$, there is danger that w will fall off too fast at large α to give convergence. Instead, we adopt the safer, and equally simple, procedure of expanding $\mu_k(\alpha)$ about its high-k shape; that is, we expand about the hypothetical situation in which the dimensionless ratios $\bar{\Gamma}_n^{2n}(0)/\bar{\Gamma}_k^{2}(0)$ are independent of k. To implement this, we start with (78) at $n = 4$ as a sufficiently good approximation at high k [$\Gamma_k(0)$ is accurate to within 1.2%]. We take as the weight function

$$\begin{aligned} w(\alpha) &= K^{-1}(2\pi)^{-1/2}\exp(-\bar{\alpha}^2/2)[1 + He_4(\bar{\alpha})/4! \\ &\quad - 3He_6(\bar{\alpha})/6! + 17He_8(\bar{\alpha})/8!]\,, \\ \bar{\alpha} &= \alpha/K, \qquad K = v_0 k[G_k(0)]_4 = 1.2685745\,. \end{aligned} \tag{79}$$

Note that α in (77) is dimensionless and corresponds to $aG(0)$ in (75). Now we expand in the form

$$\mu_k(\alpha) = \sum_{n=0}^{\infty} b_{2n}P_{2n}(\alpha)w(\alpha)\,,$$

where the $P_n(\alpha)$ are the polynomials orthogonal on $w(\alpha)$. The first (even) three and their normalizations are

$$\begin{aligned} &P_0(\alpha) = 1\,, \qquad P_2(\alpha) = \bar{\alpha}^2 - 1\,, \qquad P_4(\alpha) = 3\bar{\alpha}^4 - 23\bar{\alpha}^4 + 11 \\ &\int_{-\infty}^{\infty} [P_0(\alpha)]^2 w(\alpha)\,d\alpha = 1\,, \qquad \int_{-\infty}^{\infty} [P_2(\alpha)]^2 w(\alpha)\,d\alpha = 3\,, \\ &\int_{-\infty}^{\infty} [P_4(\alpha)]^4 w(\alpha)\,d\alpha = 501\,. \end{aligned} \tag{81}$$

These polynomials and their normalizations, through $n = 4$, are independent of original choice of Gaussian $w(a)$ in (78); they are the same if we work from the truncation of (78) at $n = 4$ with $w(a)$ any weight normalized to one, and the He_{2n} replaced by the appropriate orthogonal polynomials on that weight, The relation of the coefficients in (81) to the denominators of the Padé approximants (76) should be noted.

Now we determine the b_{2n} in (80) by requiring that the expansion of (77) in powers of λ^2 reproduce (53). Thus,

$$\begin{aligned} &b_0 = \bar{\Gamma}_k^2(0)\,, \qquad -b_2 = \tfrac{1}{3}[\bar{\Gamma}_k^2(0) + \bar{\Gamma}_k^4(0)/K^2]\,, \\ &b_4 = \tfrac{1}{501}[11\bar{\Gamma}_k^2(0) + 23\bar{\Gamma}_k^4(0)/K^2 + 3\bar{\Gamma}_k^6(0)/K^2]\,, \ldots. \end{aligned} \tag{82}$$

Finally, the nth-order approximants $[G_k(0)]_n$ and $[\Gamma_k(0)]_n$ are found by iterative solution of the equation

$$G_k(0) = \left[2\sum_{m=0}^{n} b_{2n}(k)\int_0^{\infty} w(\alpha)P_{2n}(\alpha)(1 + \alpha^2)^{-1}\,d\alpha\right]^{-1}, \tag{83}$$

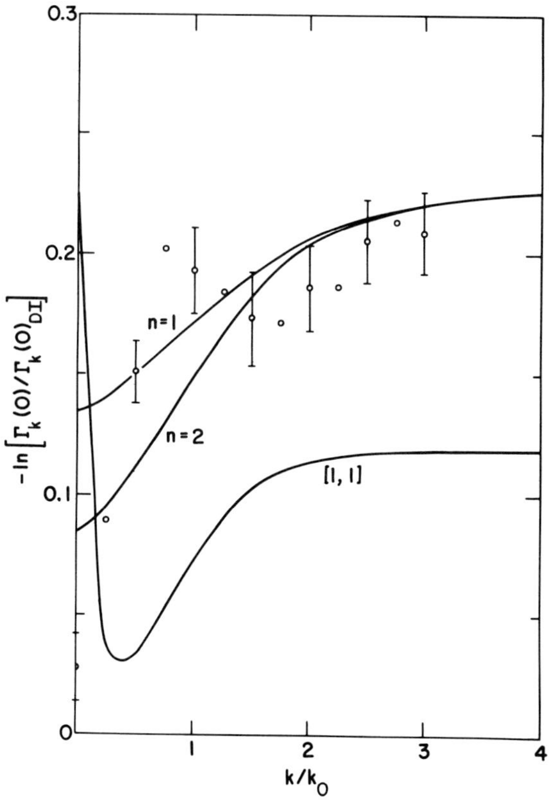

Fig. 13. Figure 11 repeated for spectrum E_2.

which is obtained by inserting the nth-order truncation of (77) into (51), with $\lambda = 1$, $\nu = 0$. In the high k limit, each of the approximants $n = 0$, 1, 2, 3, 4 gives back the original value $v_0 k G_k = K$, and each of these approximants gives the numerical values $b_0 = 1$, $b_2 = b_4 = b_6 = b_8 = 0$ when that value of $G_k(0)$ is inserted in (82). It should be noted that, once $w(\alpha)$ is chosen, the approximants to $\Gamma_k(0)$, considered as functionals of $G_k(0)$, are constructed by operations on the terms of (53) which are wholly linear, and independent of k. According to Sect. II, this process will converge to give the correct $G_k(0)$ if (77) is a healthy enough representation, a sufficient but not necessary condition being that $|\mu_k(\alpha)|^2/w(\alpha)$ is integrable over α and w gives complete orthogonal functions.

The approximants to $\Gamma_k(0)$ obtained by the procedure just described are shown in Figs. 11 and 13 for $n = 1$ and $n = 2$, for spectra E_1 and E_2. Note that the anomalous behavior shown by the Padé approximants at low k does not occur here. Moreover, the substantial difference in the faithfulness of approximation at low k for spectra E_1 and E_2 does not occur which was found in Sect. VI for the continuous orthogonal expansion of the primitive expansion.

The procedure just described does not give good approximants to $\Gamma_k(-ia)$ over the whole a range. The generalization of (77) to $u \neq 0$ is

$$\Gamma_k(u) = \int_{-\infty}^{\infty} \mu_k(\alpha, u)(1 + i\alpha\lambda)^{-1}\, d\alpha\ , \tag{84}$$

where, in general, $\mu_k(\alpha, u)$ must be complex for $u = -ia$. We could attempt to expand in the form

$$\mu_k(\alpha, -ia) = \sum_n b_{2n}(-ia, k) P_{2n}(\alpha) w(\alpha)\ , \tag{85}$$

but it seems clear by comparison with (75) that this will not converge well at large a. A crucial characteristic of (75) is the presence in the *denominator* of the complex quantity $G_k(u)$, which goes from pure real at $u = 0$ to asymptotically pure imaginary values as $iu \to \infty$.

It is of interest, therefore, to describe an alternative procedure, which offers more promise over the whole frequency range, and which may be regarded as a preliminary step toward treating (53) as a true functional power series. We start by noting that, when $\nu = 0$, λ is simply a scaling parameter whose dynamical role is trivial. Therefore, let us take $\lambda = 1$ at the start and seek a dimensionless expansion parameter with greater dynamical significance. In the high k limit, we have already used $v_0 k G_k(u)$ as such a variable [Eqs. (56) and (75)]. This quantity is unsuitable at low k, however, because there $\Gamma_k(u)$ is almost independent of $G_k(u)$ and is determined principally by the values of $G_{k'}(u)$ for $k' \sim k_0$. Instead, we take $\bar{\Gamma}_k^{2}(u)$ as our fundamental variable. This represents a known linear transformation of $G_k(u)$ (we assume that the inverse transformation exists) and has the advantage that $\bar{\Gamma}_k^{2}(u)$ behaves very much like $\Gamma_k(u)$ over the entire k range. In order to obtain a dimensionless variable, analogous to λ, we normalize $\bar{\Gamma}_k^{2}(u)$ by the direct-interaction value of $\bar{\Gamma}_k^{2}(0)$, obtained by solving the direct-interaction equation (73). The final choice of expansion variable is then

$$\begin{aligned} \Lambda_k(u) &= \bar{\Gamma}_k^{2}(u)/\gamma_k\ , \\ \gamma_k &= \bar{\Gamma}_k^{2}(0)]_{DI} = \int [\mathbf{k}\cdot\mathsf{P}(\mathbf{q})\cdot\mathbf{k}]\,U(q)[\mathrm{G}_{\mathbf{k}-\mathbf{q}}(0)]_{DI}\, d\mathbf{q}\ . \end{aligned} \tag{86}$$

We recall that the direct-interaction solution is already a fairly good uniform approximation to $\Gamma_k(0)$ over the whole k range. $\Lambda_k(u)$ is an implicit functional of $U(q)$ which, at $u = 0$, can be expected to vary only slightly with k over the entire range $0 \leq k \leq \infty$.

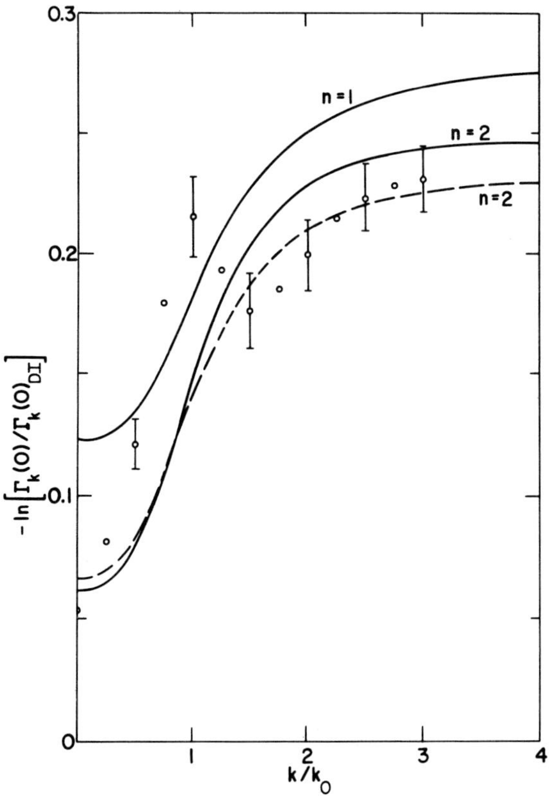

Fig. 14. Logarithmic deviation of continuous-orthogonal-expansion approximants $n = 1$ and $n = 2$ from direct-interaction values, spectrum E_1. Solid curves, Eq. (89). Dashed curve, Eq. (90). Points are computer experiment.

We now set $\lambda = 1$ in (53) and rewrite it instead as an analogous power series in $\Lambda_k(u)$.

$$\Gamma_k(u) = \sum_{n=1}^{\infty} \{\bar{\Gamma}_k^{2n}(u)/[\Lambda_k(u)]^{2n-2}\}[\Lambda_k(u)]^{2n-2}, \tag{87}$$

where the functions { } are regarded as coefficients. Equation (87) is contrived, but it is less artificial than the previous series in λ^2. $\Lambda_k(u)$ is a function whose behavior is intimately related to that of $\Gamma_k(u)$, which we want to find, and it does not seem unreasonable to express the relation between the two functions as a power series. In the high k limit, we find $\Lambda_k(u) = v_0 k G_k(u)$, $\{\bar{\Gamma}_k^{2n}(u)/[\Lambda_k(u)]^{2n-2}\} = (-1)^{n-1}S_{2n}$, and (87) gives back (56), the power series in $v_0 k G_k(u)$. Equstion (87) may be regarded as an extrapolation of (56) over the entire k range in which we attempt to keep the different wave numbers as independent as possible.

To form approximants to $\Gamma_k(u)$, we now treat $\Lambda_k(u)$ as we did λ, seeking an integral representation of the form

$$\Gamma_k(u) = \int_{-\infty}^{\infty} \mu_k{}'(\alpha, u)[1 + i\alpha\Lambda_k(u)]^{-1}\, d\alpha, \tag{88}$$

in analogy to (84). Here again $\mu_k{}'(\alpha, u)$ is complex in general, but now, because $\Lambda_k(u)$ appears in the denominator, there is much more hope of converging well to $\Gamma_k(-ia)$ at large a by expansion in real orthogonal functions. In the high k limit, (88) reduces to (75), so that $\mu_k{}'(\alpha, u)$ factors into a real function independent of u and a complex function independent of α.

We carry out calculations for $\Gamma_k(0)$ by again expanding about a hypothetical situation, in which now $\mu_k{}'(\alpha, 0)$ is independent of k. First, we shall do this in the form

$$[\Gamma_k(0)]_n = (2\pi)^{-1/2} \sum_{m=0}^{n} b_{2n}$$

$$\times \int_{-\infty}^{\infty} \exp(-\alpha^2/2) He_{2m}(\alpha)[1 + i\alpha\Lambda(0)]^{-1}\, d\alpha\,, \tag{89}$$

where $w(\alpha)$ is Gaussian and, in the high k limit, $b_2 = 0$. The b_{2n} are determined as explicitly linear functions of the quantities { } in (87), according to (5). At high k, this expansion reduces to (78).

The results obtained by using $[\Gamma_k(0)]_n$ in (51) ($\nu = 0$, $\lambda = 1$) and solving by iteration are shown in Figs. 14 and 15 for $n = 1$ and $n = 2$. It is reassuring that the results differ little from those obtained by (79)–(83), the procedure based on treating (53) as a series in λ^2. The present method can also be applied using the weight $w(\alpha)$ defined in (79), thereby getting higher accuracy at high k in the low orders. To do this, we replace (89) by

$$[\Gamma_k(0)]_n = \sum_{m=0}^{n} b_{2n} \int_{-\infty}^{\infty} w(\alpha) P_{2m}(\alpha)[1 + i\alpha\Lambda_k(0)]^{-1}\, d\alpha\,. \tag{90}$$

The results are shown in Figs. 14 and 15. We see that there is little difference from the Gaussian choice.

A more logically satisfying procedure than either of those described previously would be to regard (53) as a true functional power series in $\Lambda_k(u)$. The slow variation of $\Lambda_k(0)$ as a function of k suggests that it should be possible to represent $\Lambda_k(u)$ satisfactorily as the sum of some relatively low number r of orthogonal functions in k and u. The functional power series would then be approximated by an r-dimensional ordinary power series, and approximants could be constructed as suggested at the end of Sect. II.

The results obtained in the present section, both with Padé approximants and continuous orthogonal expansions, must be regarded as heuristic

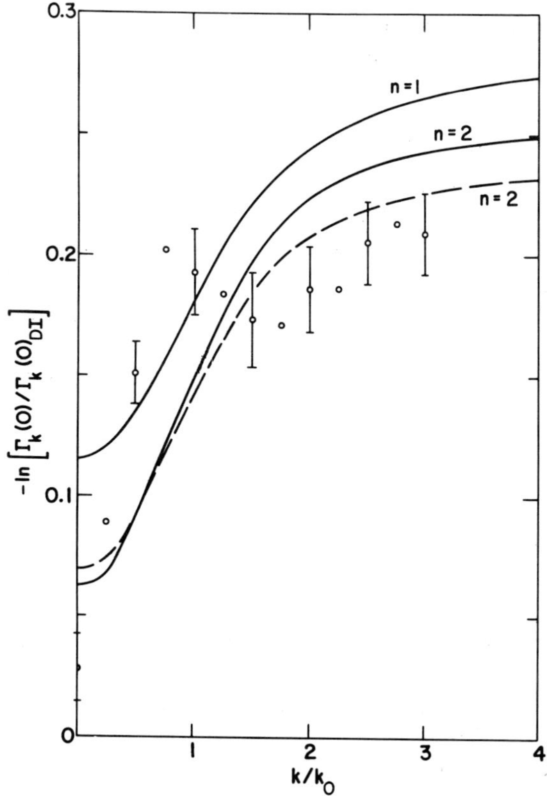

Fig. 15. Repetition of Fig. 14 for spectrum E_2.

until the conjectured integral representations on which they are based are justified, or convergence is proved in some alternative fashion. A simple counter example shows that this is not an empty question. Consider the random oscillator (57), but suppose that $\rho(a)$ is some function that vanishes at $a = 0$. Then, as ν goes from ∞ to 0,

$$G(0) = \int_0^\infty \langle q(t) \rangle \, dt$$

rises from zero to a maximum and falls back to zero. But then clearly $\Gamma(0) = [-\nu + 1/G(0)]/\lambda^2$ is a two-valued function of $G(0)$, since $G(0)$ takes the same value for two different values of ν. It follows that the irreducible expansion for $\Gamma(0)$ as a power series in $G(0)$ has two branches and that there is no single integral representation of the simple kind we have considered which covers both branches. Also, it is easy to see that if the methods of the present section were applied we would find only the branch that goes from $\nu = \infty$ to the value of ν which gives maximum $G(0)$. This example suggests that a necessary condition for the validity of our manipulations with the irreducible expansion is that the primitive expansion (50) have a single-valued inverse, giving $G_k{}^0(u)$ in terms of the $G_k(u)$. The physics of the turbulent diffusion problem suggests that this is so, but it is easy to construct examples with the same diagram structure where it is not.

In the high k limit, the integral representation (75) can be justified because the exact solutions are accessible. Orszag (1969) has deduced from the recursion relation for S_{2n} a differential equation for $\Gamma_k(u)$ as a function of $G_k(u)$:

$$d\Gamma/dG = -(\Gamma - G)/(G^2\Gamma) , \tag{91}$$

where we take $v_0 k = 1$ and omit argument and subscript. Study of (91) shows that $\rho_k''(a)$ in (75) has a singularity of logarithmic type at $a = 0$. Such a singularity is consistent with convergence of both Padé approximants and continuous orthogonal expansions, but it represents abehavior sufficiently different from the smooth density $\rho_k'(a)$ of the primitive expansion for $\Gamma_k(u)$ that caution is suggested in extrapolating the success of the approximants to wider applications.

Some kind of compromise between the primitive expansion, where the convergence theory is relatively complete, and the irreducible expansion, where the approximants seem heuristically to be superior, may be possible. We have already suggested, at the end of Sect. VI, using results from the direct-interaction approximation (a truncation of the irreducible expansion known to be self-consistent) to fix weight functions for applying the continuous orthogonal expansion method to the primitive expansion. It is also possible, in several ways, to express the exact $\Gamma_k(u)$ as an explicit expansion about the direct-interaction solution (Kraichnan, 1968, 1970a), a kind of half way stop between primitive and irreducible expansions.

ACKNOWLEDGMENTS

I am indebted to Dr. Steve Orszag for several fruitful discussions of this work and to Dr. J. R. Herring for collaboration in the numerical studies. This work was supported by the Office of Naval Research under Contract N 00014-67-C-0284.

REFERENCES

Baker, G. A., Jr. (1970). The Padé approximant method and some related generalizations (*in* this volume).

Kraichnan, R. H. (1961). *J. Math. Phys.* **2,** 124; **3,** 205.

Kraichnan, R. H. (1968). *Phys. Rev.* **174,** 240.

Kraichnan, R. H. (1970a). *J. Fluid. Mech.* **41,** 189.

Kraichnan, R. H. (1970b). *Phys. Fluids.* **13,** 22.

Masson, D. (1970). Hilbert space and the Padé approximant (*in* this volume).
Orszag, S. A. (1969). Unpublished.
Saffman, P. (1960), *J. Fluid Mech.* **8**, 273.
Shanks, D. (1955). *J. Math. and Phys.* **34**, 1.
Wall, H. S. (1948). "Analytic Theory of Continued Fractions." D. van Nostrand, Princeton, New Jersey.

CHAPTER 5 PADÉ APPROXIMANTS AND LINEAR INTEGRAL EQUATIONS*

J. S. R. Chisholm

University of Kent at Canterbury, England

I. INTRODUCTION

A variety of physical problems give rise to linear integral equations for functions of one or more variables. The theory of these equations forms an important branch of functional analysis, and is dealt with in a variety of standard textbooks; these books will be referred to through the names of their authors: Mikhlin, Hoheisel, and so on (Mikhlin, 1964; Hoheisel, 1968a; Pogorzelski, 1966; Tricomi, 1957; Reisz and Sz.-Nagy, 1955a).

The most powerful established method of solution of these equations is the Fredholm method, and for a certain class of integral equations the method can be shown to give approximations to the unique solution. A less powerful method of solution is an iterative one giving rise to the "Neumann series," also known as the perturbation series. This series can be regarded as a power series expansion in a parameter λ, and is usually convergent only when λ is sufficiently small. We can ask whether the series solution can be analytically continued to large values in λ by forming Padé approximants, thus extending the scope of the Neumann series. In this chapter, it is shown that the diagonal approximants $[n, n]$ in λ tend to the exact solution whenever the Fredholm method can be used; there is reason to believe that the Padé method will be more generally applicable than the Fredholm method, but this has not yet been proved.

In Sect. II, we discuss the general properties of classes of linear integral equations. In Sect. III, these properties are used to establish two

* This research was sponsored in part by the European Office of Aerospace Research, United States Air Force, Grant No. EOOAR-69-0046.

theorems on Padé approximants formed from the Neumann series; the second of these theorems is the main approximation theorem. In the final section, we study, as an example, the integral equation for two-particle scattering through a Yukawa potential; it is shown that this integral equation can be reduced to one of the class that can be solved either by the Fredholm method or by the Padé method.

II. PROPERTIES OF LINEAR INTEGRAL EQUATIONS

We shall be concerned with an equation for a function $\phi(x)$ of a single real variable, of the form

$$\phi(x) = f(x) + \int_a^b dy\, K(x, y)\phi(y) \,, \tag{1}$$

where the integration limits a and b may be infinite, and $f(x)$ and $K(x, y)$ are defined on the closed interval (a, b). We can regard

$$K \equiv \int_a^b dy\, K(x, y) \tag{2}$$

in (1) as an operator acting on the function ϕ. In operator form, the integral equation is

$$\phi = f + K\phi \,. \tag{3}$$

The function $K(x, y)$ is the *kernel* of the equation, and, in order to ensure a valid definition and a unique solution of (1), various restrictions can be imposed on ϕ, f, and K. It is usual to assume ϕ and f to be functions of finite norm $\|\ \|$ in the Hilbert space L^2 of functions defined on the interval (a, b); for instance,

$$\|\phi\|^2 = \int_a^b |\phi(x)|^2\, dx < \infty \,.$$

A sequence of functions $\{f_n\}$ is bounded if

$$\|f_n\| \leq B < \infty \qquad (n = 1, 2, 3, \ldots) \,.$$

We shall now discuss various assumptions that can be made about the operator K or kernel $K(x, y)$, and the consequences that follow from these assumptions.

A. Boundedness

The operator (2) is bounded (Hoheisel, 1968b; Reisz and Sz.-Nagy, 1955b) if there is a finite constant M such that

$$\|Kh\| \leq M\,\|h\| \tag{4}$$

for all functions h of finite norm. [This is a different condition from boundedness of the *function* $K(x, y)$, given by $|K(x, y)| \leq C < \infty$.]

The *norm* of a bounded operator K is the least value of M satisfying (5). An operator is unbounded if no finite M can be found such that (5) holds for all h.

B. Complete Continuity*

Suppose that T is a linear operator that transforms any function h in L^2 into another function Th of L^2; in particular, T could be of the form K in (2). Every linear operater of this type is continuous in the sense that it transforms a sequence $\{h_n\}$ converging in the mean to h into a sequence $\{Th_n\}$ converging to Th (Reisz and Sz.-Nagy, 1955b).

*Complete continuity** of a linear transformation is a stronger condition than continuity, and is of great importance in the theory of linear operators and integral equations. Suppose that $\{f_n\}$ is any bounded infinite sequence in L^2, so that $\{Tf_n\}$ is another infinite sequence in L^2. Then if, for all bounded $\{f_n\}$, $\{Tf_n\}$ contains a subsequence converging to some function h as $n \to \infty$, so that

$$||Tf_n - h|| \to 0 ,$$

then T is completely continuous (Hoheisel, 1968c; Reisz and Sz.-Nagy, 1955c). Equivalent definitions are given by Reisz and Nagy; they also give a necessary and sufficient condition for complete continuity (Reisz and Sz.-Nagy, 1955d).

C. Schmidt Kernels

A sufficient condition for $K(x, y)$ to be completely continuous is that it satisfies the *Schmidt condition* (Reisz and Sz.-Nagy, 1955d).

$$||K||^2 \equiv \int_a^b dx \int_a^b dy\, |K(x, y)|^2 < 1 . \tag{5}$$

Since αK is completely continuous if K is, where α is any complex constant, the condition (5) can be generalized to

$$||K||^2 \equiv \int_a^b dx \int_a^b dy\, |K(x, y)|^2 \leq B < \infty . \tag{6}$$

Kernels of this type are known as $\mathscr{L}^2$ kernels. It is often convenient to use the more general condition (6) and to generalize (1) to

$$\phi(x) = f(x) + \lambda \int_a^b dy\, K(x, y)\phi(y) , \tag{7}$$

where λ is any complex number.

* "Complete continuity" is often called "compactness."

One method of solving (7) is to substitute

$$\phi(x) = f(x) + \sum_{n=1}^{\infty} \lambda^n f_n(x) \tag{8}$$

and to equate coefficients of λ on each side of (7). This gives, for $n = 1, 2, 3, \ldots,$

$$f_{n+1}(x) = \int_a^b dy\, K(x, y) f_n(y) , \tag{9}$$

defining the terms in (8) iteratively; if K satisfies (6) and $f \in L^2$, these integrals are all finite (Reisz and Sz.-Nagy, 1955d). Putting $\lambda = 1$, the series solution of (1), known as the *Neumann series*, is

$$\begin{aligned} f(x) &= g(x) + \sum_{n=1}^{\infty} f_n(x) \\ &= g(x) + Kg(x) + K^2 g(x) + K^3 g(x) + \cdots, \end{aligned} \tag{10}$$

using (9) and the definition (2) of the operator K; powers of K denote repeated operations of K. The series (10) will be a solution of (1) if it is uniformly convergent in the interval (a, b). When K satisfies the Schmidt condition (5), this convergence is readily established, and the operator series,

$$(I - K)^{-1} \equiv I + K + K^2 + K^3 + \cdots, \tag{11}$$

where I is the identity operator, converges uniformly except on a set of measure zero (Tricomi, 1957b). Similarly, the series (8) is a solution of (7), and

$$(I - \lambda K)^{-1} \equiv I + \lambda K + \lambda^2 K^2 + \lambda^3 K^3 + \cdots \tag{12}$$

converges in the norm, if

$$|\lambda| < ||K||^{-1} . \tag{13}$$

The series (8) and (12) are not usually convergent for larger values of $|\lambda|$; it is, however, possible that formation of Padé approximants from the power series (8) in λ will give approximations to the solution when $|\lambda|$ does not satisfy (13). This problem, investigated by Chisholm (1963), is the main subject matter of Sect. III.

D. Kernels of Finite Rank

A kernel K_n is of finite rank n if it is expressible in the form

$$K_n(x, y) = \sum_{i=1}^{n} \alpha_i(x) \beta_i^*(y) , \tag{14}$$

where $\{\alpha_i\}$ and $\{\beta_i\}$ are sets of functions in L^2. In (14), it is assumed that n is the least possible number of "separable" terms in the sum (14); thus the sets of functions $\{\alpha_i\}$ and $\{\beta_i\}$ $(i = 1, 2, \ldots, n)$ are each linearly independent.

The scalar product of functions f and g in L^2 is

$$(f, g) = \int_a^b dy\, f(y) g^*(y) . \tag{15}$$

Substituting (14) in (7) gives

$$\phi(x) = f(x) + \lambda \sum_{j=1}^{n} (\phi, \beta_i) \alpha_i(x) . \tag{16}$$

The coefficients (ϕ, β_i) are complex numbers, so any solution of the equation must be of the form

$$\phi = f + \sum_{i=1}^{n} \xi_j \alpha_j , \tag{17}$$

where the coefficients ξ_j $(j = 1, 2, \ldots, n)$ have to be determined. Substituting (17) into (16) and equating coefficients of the (independent) functions ϕ_j, we find

$$\sum_{i=1}^{n} [\delta_{ij} - \lambda c_{ij}] \xi_i = \eta_j , \tag{18}$$

where

$$c_{ij} = (\alpha_i, \beta_j) , \qquad n_j = \lambda(f, \beta_j) \tag{19}$$

are known.

So the solution of the integral equation reduces to the solution of the system (18) of n linear equation for the parameters ξ_j $(j = 1, 2, \ldots n)$. The determinent of coefficients in (18) is denoted by

$$\Delta(\lambda) \equiv |\delta_{ij} - \lambda c_{ij}| , \tag{20}$$

and the minor associated with the (i, j) element by $\Delta_{ij}(\lambda)$. Then Δ is a polynomial in λ with λ^n as the highest possible power, whereas Δ_{ij} are polynomials with λ^{n-1} as highest possible power.

Provided λ does not satisfy

$$\Delta(\lambda) = 0 , \tag{21}$$

Eqs. (18) have the unique solution

$$\xi_i = \Delta^{-1} \sum_{j=1}^{n} \Delta_{ij} \eta_j , \tag{22}$$

and $f(x)$ is then uniquely determined by (17) as

$$\phi = f + \Delta^{-1} \sum_{i=1}^{n} \sum_{j=1}^{n} \Delta_{ij}(f, \beta_j)\alpha_i \,. \tag{23}$$

As a function of the parameter λ, this solution is a rational fraction, the highest possible power in both numerator and denominator being λ^n.

The homogeneous equation

$$f = \lambda \sum_{i=1}^{n} (f, \beta_i)\alpha_i \,, \tag{24}$$

obtained by putting $g(x) \equiv 0$, in general has the unique solution

$$f(x) \equiv 0 \,. \tag{25}$$

It has a nonzero solution only if $\Delta(\lambda) = 0$, that is, when λ is a pole of (22) or of the solution (23) of the nonhomogeneous equation.

III. INTEGRAL EQUATIONS AND PADÉ APPROXIMANTS

We can now establish the first theorem concerning Padé approximants and integral equations (Chisholm, 1963):

THEOREM 1. The exact solution of the integral equation (7), with a kernel (14) of finite rank n, is given by forming the Padé approximant $[n, n]$ from the first $(2n + 1)$ terms of the power series solution (8).

To establish this result, we simply note that the power series expansion of (23) in powers of λ must be identical with the power series (3), since they are both the exact solution for

$$|\lambda| < \|K_n\|^{-1} \,.$$

Since

$$\|K_n\|^2 = \int_a^b dx \int_a^b dy \left| \sum_{i=1}^{n} \alpha_i(x)\beta_i^*(y) \right|^2$$

$$\leq \int_a^b dx \sum_{i=1}^{n} |\alpha_i(x)|^2 \int_a^b dy \sum_{j=1}^{n} |\beta_j(y)|^2 < \infty \,,$$

using Cauchy's inequality, the radius of convergence of (8) in the λ plane is nonzero, the Schmidt condition (6) being satisfied. Since (23) is a rational fraction in λ, with highest possible power λ^n in both numerator and denominator, and since its expansion in λ agrees term by term with (8), it is the Padé approximant $[n, n]$ for the series (8).

The second theorem established by Chisholm (1963) deals with the solution by the Padé method of integral equations with completely continuous kernels K. We first note a basic property of completely continuous transformations: that they can be uniformly approximated by transformations of finite rank; thus, there is an infinite sequence of kernels $K_1, K_2, K_3 \ldots$ of rank 1, 2, 3, ... such that

$$\|(K - K_n)h\| < \varepsilon_n \|h\| , \tag{26}$$

with $\varepsilon_n \to 0$ as $n \to \infty$, for all $h \in L^2$.

The solution of (7) is written formally as

$$\phi = (I - \lambda K)^{-1} f ; \tag{27}$$

when K is replaced by K_n, the solution ϕ_n is given by Theorem 1:

$$\phi_n = (I - \lambda K_n)^{-1} f = [n, n]_n , \tag{28}$$

where $[n, n]_n$ is the Padé approximant in λ formed from the series

$$f + \sum_{r=1}^{\infty} (\lambda K_n)^r f . \tag{29}$$

Since K is completely continuous, the solution (27) is a meromorphic function of λ (Hoheisel, 1968d); also, eigenvalues of K_n in any compact region R of the λ plane approach eigenvalues of K as $n \to \infty$. Thus,

$$\lim_{n \to \infty} \phi_n = \phi$$

in R, except at the finite number of poles of ϕ. Using (28), we have

$$\lim_{n \to \infty} [n, n]_n = \phi \tag{30}$$

in R, except at the poles of ϕ; thus, the solution ϕ is the limit of a sequence of diagonal Padé approximonts. This result is not in itself very useful, since each approximant in the sequence is derived from a different Neumann series (29), with kernel K_n, not K. We need to establish a similar result for the sequence of approximants $[n, n]$ derived from the single series

$$f + \sum_{r=1}^{\infty} (\lambda K)^r f . \tag{31}$$

The limits of the coefficients in (29) are the coefficients in (31),

$$K_n{}^r f \to K^r f . \tag{32}$$

Thus the Padé approximants $[n, n]_n$ and $[n, n]$, given explicitely by Baker (1970), are approximately equal, except at the poles of $[n, n]_n$ and of $[n, n]$. Now the poles of $[n, n]_n$ tend to poles of ϕ as $n \to \infty$; however, unless the series (31) falls into certain special categories (for example, being series of

Stieltjes), we do not know where the poles of $[n, n]$ are. Thus, we have the second main result.

THEOREM 2. If $[n, n]$ are the diagonal Padé approximants formed from the Neumann series (31) of the integral equation (7) with a completely continuous kernel K, and if R is any compact region of the λ plane, then the solution ϕ of (7) is given by

$$\phi = \lim_{n\to\infty} [n, n]$$

in R, except at the finite number of poles of ϕ and at limit points of poles of $[n, n]$ as $n \to \infty$.

The main defect of this result is the lack of knowledge of the limit points of poles of $[n, n]$. This is one of the main unsolved problems in the theory of Padé approximants. Theorem 2 is essentially a theorem on the approximation of meromorphic functions. In practice, we know that Padé approximants can successfully approximate functions with branch cuts and essential singularities; it seems certain that the theorem is capable of generalization to wider classes of operators, bounded operators for example. A simple example of a bounded operator is the unit operator, with $K(x, y) = \delta(x - y)$. The solution (27) is then

$$\phi = (1 - \lambda)^{-1} f,$$

which is given by the Padé method from the Neumann series. It seems likely that the method can be applied also for some classes of unbounded operators.

IV. THE INTEGRAL EQUATION FOR POTENTIAL SCATTERING

The quantum-mechanical problem of two bodies interacting through a central potential can be expressed in terms of the Lippmann–Schwinger integral equation for the T matrix. The reduction of this integral equation to one with an L^2 kernel has been effected by Kowalski and Feldman (1961, 1963) and by Noyes (1965); another approach has been used by Brown *et al.* (1963). We shall follow the account of the first three authors' work given in Chapter 2 of the book by Watson *et al.* (1967).

The scattering matrix element from an initial state χ_a to a final state χ_b is

$$T_{ba} = [\chi_b, T(E_{a+})\chi_a], \tag{33}$$

where

$$E_+ = \lim_{\eta=0} (E + i\eta). \tag{34}$$

Matrix elements $\langle \mathbf{q}' |T(E)| \mathbf{q}\rangle$ satisfy the Lippman–Schwinger equation

$$\langle \mathbf{q}' |T(E)| \mathbf{q}\rangle = \langle \mathbf{q}' |V| \mathbf{q}\rangle + \int \frac{d^3\mathbf{q}''\langle \mathbf{q}' |V| \mathbf{q}''\rangle\langle \mathbf{q}'' |T(E)| \mathbf{q}\rangle}{E - E_{q''}}, \tag{35}$$

where

$$\begin{aligned} \langle \mathbf{q}' |V| \mathbf{q}\rangle &= (2\pi)^{-3}\int d^3r \exp[i\mathbf{r}\cdot(\mathbf{q}-\mathbf{q}')]\, V(r), \\ E_q &= \mathbf{q}^2/2m, \end{aligned} \tag{36}$$

and m is the reduced mass. The physical two-body scattering amplitude is given by

$$E_q = E_{q'}, \qquad E = E_{q''}. \tag{37}$$

Equation (35) involves matrix elements $\langle \mathbf{q}'' |T(E)| \mathbf{q}\rangle$ for which $E_{q''} \neq E$. In solving the three-particle problem, one needs to know two-particle matrix elements for which both conditions (37) are violated; both $\mathbf{q}$ and $\mathbf{q}''$ are "off the energy shell." For this reason, it is convenient to study Eq. (35) with arbitrary $\mathbf{q}$ and $\mathbf{q}'$.

For central potentials $V(r)$, we can decompose (35) into equations for each partial wave, writing

$$\langle \mathbf{q}' |V| \mathbf{q}\rangle = \sum_{lm} Y_l^m(\mathbf{q}/q)\, Y_l^{m*}(\mathbf{q}'/q')\, V_l(q', q) \tag{38}$$

and

$$\langle \mathbf{q}' |T(E)| \mathbf{q}\rangle = \sum_{lm} Y_l^m(\mathbf{q}/q)\, Y_l^{m*}(\mathbf{q}'/q')\, T_l(q', q, E). \tag{39}$$

Then the partial wave amplitudes T_l satisfy

$$T_l(q', q, E) = V_l(q', q) + \int_0^\infty dq''\, q''^2 \frac{V_l(q', q'')T_l(q'', q, E)}{E - E_{q''}}. \tag{40}$$

The kernel of this integral equation is

$$q''^2 V_l(q', q'')/(E - E_{q''}).$$

This is not continuous for E real and positive, because of the singularity when $q''^2 = 2mE$. *A fortiori*, it is not completely continuous. The singularity is eliminated by a standard technique (Watson *et al.*, 1967). We define q_E by

$$q_E^2 = 2mE, \qquad q_E > 0,$$

and the function

$$\tau(q', q_E) = V_l(q', q_E)/V_l(q_E, q_E). \tag{41}$$

Now put $q' = q_E$ in (40), multiply by $\tau(q', q_E)$, and subtract the result from (40) itself. This gives

$$T_l(q', q, E_+) = \tau(q', q_E) T_l(q_E, q, E_+) + V_l(q', q) - \tau(q', q_E) V_l(q_E, q) + \int_0^\infty dq'' \, \Lambda(q', q'', E_+) T_l(q'', q, E_+) , \tag{42}$$

with kernel

$$\Lambda(q', q'', E) = q''^2 [V_l(q', q'') - \tau(q', q_E) V_l(q_E, q'')](E - E_{q''})^{-1} . \tag{43}$$

Because of the choice of τ, the term in brackets vanishes when $E = E_{q''}$. If $V_l(q', q)$ is an analytic function, Λ has no singularity at $E = E_{q''}$. Equation (42) is therefore a nonsingular integral equation for $T_l(q', q, E_+)$.

We shall not discuss the details of the use of this equation, which are given by Watson *et al.* (1967). We shall be concerned only with showing that, for a certain class of potentials, Λ obeys the Schmidt condition. From this, it follows that the equation may be solved by using *either* the Fredholm method *or* the Padé method, justified by the theorems of Sect. III.

First we consider the Yukawa potential

$$V(r) = r^{-1} e^{-\mu r} ,$$

for which

$$\langle \mathbf{q}' \, | V | \, \mathbf{q} \rangle = \tfrac{1}{2} \pi^{-2} (|q - q'|^2 + \mu^2)^{-1} .$$

It follows (Noyes, 1965; Brown *et al.*, 1963) that (38) gives

$$V_l(q, q') = C(2qq')^{-1} Q_l \, [(q^2 + q'^2 + 2\mu^2)/2qq'] , \tag{44}$$

where C is a constant and Q_l is the Legendre function of the second kind.

In order to show that the kernel (43) is L^2, and hence completely continuous, we note that

$$Q_l(z) \propto z^{-1} \qquad \text{as} \qquad z \to \infty \tag{45}$$

and that

$$Q_l(z) \propto \log (1 - z) \qquad \text{as} \qquad z \to 1 . \tag{46}$$

In the condition (6), $x = q$, $y = q'$, and $(a, b) = (0, \infty)$. We substitute $K = \Lambda(q, q', E)$ in the integral (6), with Λ given by (43); in establishing the finiteness of the integral, we may omit the factor $q^2(E - E_q)^{-1}$, since there is no singularity when $E = E_q$, and $q^2/E_q \to 2m$ as $q \to \infty$. We therefore study the integral

$$\int_0^\infty dq \int_0^\infty dq' \, |V_l(q', q) - V_l(q', q_E) V_l(q_E, q)/V_l(q_E, q_E)|^2 . \tag{47}$$

One contribution to this integral is

$$\int_0^\infty dq \int_0^\infty dq' \, |V_l(q', q)|^2 \, . \tag{48}$$

If we write

$$q = K \cos \theta \, , \tag{49a}$$

$$q' = K \sin \theta \, , \tag{49b}$$

then (44) gives

$$V_l(q' \, q) = \frac{C}{K^2 \sin 2\theta} Q_l \left[\frac{K^2 + 2\mu^2}{K \sin 2\theta} \right] ,$$

and (48) becomes

$$\int_0^\infty dK \int_0^{(1/2)\pi} d\theta \, \frac{C^2}{K^3 \sin 2\theta} Q_l^2 \left[\frac{K^2 + 2\mu^2}{K^2 \sin 2\theta} \right] . \tag{50}$$

The factor K^{-3} ensures convergence except where the integrand diverges. When $K^2 \sin 2\theta \to 0$, (45) tells us that the integrand is not divergent if $\mu \neq 0$; if $\mu = 0$, the K integral diverges at $K = 0$; the long-range Coulomb potential does not therefore give an L^2 kernel.

When $\mu > 0$, the only other divergence arises when

$$\frac{K^2 + 2\mu^2}{K^2 \sin 2\theta} \to 1 \, ,$$

that is, when $\theta = \frac{1}{4}\pi$, $K \to \infty$. Writing $\theta = \frac{1}{4}\pi + \frac{1}{2}\lambda$, the divergence is in the region $K \geqslant R \gg \eta$, $|\lambda| \leqslant \delta$, where δ is a small positive constant. Using (46), the contribution of this region to (50) is bounded by a multiple of

$$\int_R^\infty K^{-3} \, dK \int_{-\delta}^{\delta} d\lambda \log^2 (1 - \sec \lambda) \, .$$

The λ integral behaves like

$$\int_{-\delta}^{\delta} d\lambda \, [\log (-\tfrac{1}{2}\lambda^2)]^2 \, ,$$

which is finite.

It is easy to check that the other integrals in (47) give rise to singularities that are no worse than those in (48). Thus, the kernel Λ is $\mathscr{L}^2$. It is also easy to check that a superposition of Yukawa potentials

$$V(r) = r^{-1} \int_\mu^\infty d\sigma(m) e^{-mr} \, , \tag{51}$$

where $d\sigma(m) > 0$ and

$$\int_\mu^\infty d\sigma(m) < \infty \, ,$$

gives rise in the same way to an $\mathscr{L}^2$ kernel in the equation for $T_l(q', q, E)$. These equations can therefore also be treated either by the Fredholm method or by the Padé method.

REFERENCES

Baker, G. A., Jr. (1970). The Padé approximant method and some related generalizations (*in* this volume), Eq. (II.2).

Brown, L., Fivel, D. I., Lee, B. W., and Sawyer, R. E. (1963). *Ann. Phys.* **23,** 2, 187.

Chisholm, J. S. R. (1963); *J. Math. Phys.* **4,** 12, 1506.

Hoheisel, G. (1968a). "Integral Equations." Ungar, New York.

Hoheisel, G. (1968b). "Integral Equations," p. 21. Ungar, New York.

Hoheisel, G. (1968c). "Integral Equations," p. 29. Ungar, New York.

Hoheisel, G. (1968d). "Integral Equations," p. 31. Ungar, New York.

Kowalski, K. L., and Feldman, D. (1961). *J. Math. Phys.* **2,** 499.

Kowalski, K. L., and Feldman, D. (1963). *J. Math. Phys.* **4,** 507.

Mikhlin, S. G. (1964). "Integral Equations," 2nd ed. Pergamon Press, Oxford.

Muskhelishvili, N. I. (1965). "Singular Integral Equations." Noordhoff, New York.

Noyes, H. P. (1965). *Phys. Rev. Letters* **15,** 538.

Pogorzelski, W. (1966). "Integral Equations and Their Applications," Vol. I. Pergamon Press, Oxford.

Reisz, F., and Sz.-Nagy, B. (1955a). "Functional Analysis." Ungar, New York.

Reisz, F., and Sz.-Nagy, B. (1955b). "Functional Analysis," p. 149. Ungar, New York.

Reisz, F., and Sz.-Nagy, B. (1955c). "Functional Analysis," pp. 203–208. Ungar, New York.

Reisz, F., and Sz.-Nagy, B. (1955d). "Functional Analysis," p. 224. Ungar, New York.

Tricomi, F. G. (1957a). "Integral Equations." Wiley (Interscience), New York.

Tricomi, F. G. (1957b). "Integral Equations," pp. 50–53. Wiley (Interscience), New York.

Watson, K. M., Nuttall, J., and Chisholm, J. S. R. (1967). "Topics in Several Particle Dynamics." Holden-Day, San Francisco.

CHAPTER 6 SERIES OF DERIVATIVES OF δ-FUNCTIONS*

J. S. R. Chisholm and A. K. Common[†]

University of Kent at Canterbury, England

I. INTRODUCTION

The meaning of a series of derivatives of δ functions,

$$g(k) = 2\pi \sum_{n=0}^{\infty} a_n(-1)^n \delta^{(n)}(k) , \tag{1}$$

has been discussed by Guttinger (1966). This series can be interpreted in two ways:

a. By integrating term-by-term with a suitable test function $\phi(k)$, giving

$$\begin{aligned} g(k)(\phi(k)) &= 2\pi \sum_{n=0}^{\infty} a_n(-1)^n \int_{-\infty}^{\infty} \phi(k)\delta^{(n)}(k)\, dk \\ &= 2\pi \sum_{n=0}^{\infty} a_n \phi^{(n)}(0) . \end{aligned} \tag{2}$$

The class of test functions $\phi(k)$ must be chosen so that the series (2) converges.

b. By considering (1) as the term-by-term Fourier transform of

$$f(x) = \sum_{n=0}^{\infty} a_n(-ix)^n , \tag{3}$$

* This research was sponsored in part by the European Office of Aerospace Research, United States Air Force, Grants Nos. AF EOAR 66-51 and EOOAR-69-0046.

† Present address: CERN, Geneva, Switzerland.

Then formally

$$g(k) = \int_{-\infty}^{\infty} dx\, e^{ikx} \sum_{n=0}^{\infty} a_n(-ix)^n \,, \tag{4}$$

this interpretation being invalid unless the series (3) converges for all real values of x. It is possible, however, that the series (3) may represent a function $f(x)$ defined in the whole x plane, apart from certain singularities, and that the integral

$$\int_{-\infty}^{\infty} dx\, e^{ikx} f(x) \,, \tag{5}$$

may exist and give a definition of $g(k)$ equivalent to (2). This happens, for instance, when (3) is a series of Stieltjes in the variable ix.

Suppose first that $f(x)$ is an entire function whose Taylor series (3) converges in the sense of convergence in the space Φ' of generalized functions corresponding to the test function space Φ on which $f(x)$ acts; Φ may be any of the test function spaces D, Z, $S_\beta{}^\alpha$ or S, defined for example by Guttinger (1966). The series (2) is then necessarily convergent in Ψ, the Fourier transform of the space Φ. When $f(x)$ is not entire, one can define the series (2) by

$$\begin{aligned} g(k)(\psi(k)) &= 2\pi \sum_{n=0}^{\infty} \frac{a_n n!}{2\pi i} \int_{\Gamma_0} \frac{\psi(k)\, dk}{k^{n+1}} \\ &= -i \int_{\Gamma_0} \left(\sum_{n=0}^{\infty} a_n n! \Big/ k^{n+1} \right) \psi(k)\, dk \end{aligned} \tag{6}$$

provided Γ_0 is a simple closed contour lying entirely within the region of convergence of the series

$$G(k) = \sum_{n=0}^{\infty} a_n n! \Big/ k^{n+1} \,, \tag{7}$$

where $\psi(k)$ is here assumed to belong to an entire analytic class of test functions. $G(k)$ is essentially the inverse Borel transform of $f(x)$ defined by (3). The singularities of $G(k)$ are restricted to the region $|k| \leq R^{-1}$ where

$$R = \lim\inf\, (|a_n|\, n!)^{-1/n} \,. \tag{8}$$

Provided R is nonzero, a finite contour Γ_0 can be found encircling all the singularities of $G(k)$, and (6) provides a definition of $g(k)$.

When $R = 0$, it is not possible to find a closed contour Γ_0 in the definition and some other method of interpreting (1) has to be found. Guttinger (1966, p. 519) suggests that one looks for a function $g_1(k)$ which is integrable along some finite or infinitely extended path Γ and which is a solution of the moment problem

$$2\pi\, n!\, a_n = \mu_n = \int_\Gamma g_1(k) k^n \, dk \qquad (n = 0, 1, 2, \ldots) . \tag{9}$$

Then

$$g(k)[\phi(k)] = 2\pi \sum_{n=0}^{\infty} a_n \phi^{(n)}(0) = \int_\Gamma g_1(k)\phi(k)\, dk . \tag{10}$$

The generalized function $g(k)$ defined by (1) is then equivalent to the function $g_1(k)$. The problem then is to solve the moment problem for $g_1(k)$ given the coefficients a_n. In this paper we discuss the solution of this problem and consider in particular the case when (3) is a series of Stieltjes. For this latter case, we show how an approximate form for $g(k)$ may be obtained from a knowledge of the first few coefficients a_n, using a generalization (Baker, 1967) of the Padé method [for a recent general review of the Padé method, see Baker (1966)] of approximating a power series.

II. AN EXAMPLE

We now describe the simple example which stimulated the investigation. Let the coefficients in (1) be given by

$$a_n = \alpha^{-n} , \tag{11}$$

where α is real and positive. Then (1) and (3) become

$$g(k) = \sum_{n=0}^{\infty} (-\alpha)^{-n} \delta^{(n)}(k) \tag{12}$$

and

$$f(x) = \sum_{n=0}^{\infty} (-ix/\alpha)^n . \tag{13}$$

The radius of convergence of the series (7) is then zero and we cannot find a finite contour Γ_0. Another way of expressing the difficulty is that (13) has a finite radius of convergence, but $f(x)$ has to be given over the range $(-\infty, \infty)$ of x in order to define the Fourier transform (4). We can, however, analytically continue $f(x)$ by forming a Padé approximant from (13), giving

$$f(x) = \alpha/(\alpha + ix) . \tag{14}$$

Then, remembering that we have chosen α to be positive, (4) becomes

$$g(k) = -i\alpha \int_{-\infty}^{\infty} dx \frac{e^{ikx}}{x - i\alpha} = 2\pi\alpha\theta(k)\, e^{-k\alpha} \qquad (k \neq 0) , \tag{15}$$

so that $g(k)$ is a function of unbounded support, as we expect. When

$k = 0$, the integral is undefined. If, however, we use the expression (15) for all k, we can easily see that (15) satisfies the requirements (Guttinger, 1966, p. 519) of being a generalized function equal to (1). For taking the contour Γ to be the whole real axis, the generalized moments of the function (15) are:

$$\mu_n = \int_\Gamma g(k) k^n \, dk = 2\pi\alpha \int_0^\infty dk \, k^n \, e^{-k\alpha} = 2\pi\alpha^{-n} \, n! = 2\pi \, n! \, a_n \, , \tag{16}$$

and therefore satisfy the condition (9).

Thus the two interpretations a and b of the series (1) give the same result in this example when the Padé method is used to continue the series (3) outside its circle of convergence. This suggests that there may be a more general relation between term-by-term interpretation of (1) as a generalized function, and forming Padé approximants of the Fourier-transformed series (3). In studying this relationship, we need to clear up the obscurity of the behavior of (15) at $k = 0$.

In Section III we study the form of $g(k)$ as a generalized function and discuss how the generalized moment problem (9) may be solved. In Section 4 we show that when $f(x)$ is a series of Stieltjes in ix, so that sequences of Padé approximants $[n, n + j]\,(ix)$ with $j \geq -1$ converge to an analytic function (Baker, 1966), the corresponding approximant to the generalized function is a Gammel–Baker (GB) approximant (Baker, 1967); further, we prove that when $j = -1$, sequences of these GB approximants converge to $g(k)$, providing a practical means of approximation.

III. EXPRESSIONS FOR $g(k)$

To study the series (1), we modify and extend the method of Guttinger (1966, p. 519). We start by noting that, for any given sequence of coefficients $\{a_n\}$, we can write*

$$a_n = \int_0^\infty u^n \, d\phi(u) \, , \tag{17}$$

where the measure $d\phi(u)$ is of bounded variation:

$$\int_0^\infty |d\phi(u)| \leq B < \infty \, . \tag{18}$$

The measure $d\phi(u)$ is by no means unique, in general. However, if the sequence $\{a_n\}$ obeys certain conditions (Baker, 1966), there are unique solutions of (17) and (18) when we make the extra requirement that $\phi(u)$ is a

* These results were first proved by R. P. Boas (unpublished). For a proof, see D. V. Widder (1946, p. 139).

bounded nondecreasing function of u. The series (3) is then said to be a series of Stieltjes.

Formally,

$$f(x) = \sum_{n=1}^{\infty} a_n(-ix)^n$$
$$= \int_0^{\infty} \frac{d\phi(u)}{1 + iux} . \tag{19}$$

Even if the series (3) is divergent, the integral (19) is well defined and convergent except when $-ix$ is real and positive. If (3) is a series of Stieltjes, then we know that sequences of Padé approximants $[n, n + j](ix)$, $j \geq -1$, to (3) converge to the function (19) in the cut x plane.

Assuming (17) and (19), we first show that the coefficients a_n, $(n \geq 1)$, can be expressed in the form

$$a_n = \frac{1}{(n-1)!} \int_0^{\infty} k^{n-1} g_0(k)\, dk , \tag{20}$$

where

$$g_0(k) = \int_{0+}^{\infty} e^{-k/u}\, d\phi(u) \tag{21}$$

for $k \geq 0$. To prove this, substitute (21) in the right-hand side of (20); with $m = n - 1$, this gives

$$\frac{1}{m!} \int_0^{\infty} k^m\, dk \int_{0+}^{\infty} e^{-k/u}\, d\phi(u) = \frac{1}{m!} \int_{0+}^{\infty} d\phi(u) \int_0^{\infty} k^m e^{-k/u}\, dk$$
$$= \int_{0+}^{\infty} u^{m+1}\, d\phi(u) = \int_{0+}^{\infty} u^n\, d\phi(u)$$
$$= \int_0^{\infty} u^n\, d\phi(u) = a_n .$$

In order to justify the interchange of integration above by Fubini's theorem*, we must assume that for any integer $r \geq 1$.

$$|d\phi(u)| < A_r u^{-r}\, du \qquad (A_r \text{ constant})$$

when u is sufficiently large. This condition follows from the convergence of the integrals (3.1) if $d\phi(u)$ is a positive measure. This assumption of positivity is necessary to justify the next step in the argument; further, the results of Section IV depend on the assumption that the series (3) is a series of Stieltjes, for which $d\phi(u)$ is positive. We therefore assume for the remainder of this paper that (3) is a series of Stieltjes.

* See for example, Widder (1946, pp. 25–26).

The next stage of the argument depends also on the choice of the test function space Φ. For a given series (1) or (3), we shall require absolute convergence of the series

$$\sum_{n=0}^{\infty} a_n \phi^{(n)}(0) .$$

Since $a_n > 0$ for a series of Stieltjes, we can for instance require that

$$|\phi^{(n)}(0)| \leq |A_n|/n^{1+\delta} a_n , \tag{22}$$

where $\delta > 0$ and

$$\limsup |A_n| = A < \infty .$$

We therefore define the space Φ to be the space Z when all functions $\phi(s) \in Z$ satisfy (22). Otherwise we define Φ to be the space of entire functions

$$\phi(z) = \sum_{n=0}^{\infty} c_n z^n ,$$

with norm $||\phi||$ defined by

$$||\phi|| = \sup_n |c_n|\, n!\, a_n\, n^{1+\delta} < \infty , \tag{23}$$

where $\delta > 0$ is fixed. This defines a Banach space satisfying the usual conditions for a test-function space (see for example, Friedman, 1963). Naturally, the more divergent the sequence $\{a_n\}$ is, the smaller is the Banach space.

We can now use Fubini's theorem to justify interchange of the order of integration in the following interpretation of (10) term-by-term as a generalized function:

$$\begin{aligned} g(k)[\phi(k)] &= 2\pi \sum_{n=0}^{\infty} a_n \phi^{(n)}(0) \\ &= 2\pi a_0 \phi(0) + 2\pi \sum_{m=0}^{\infty} \phi^{(m+1)}(0) \int_0^{\infty} (m!)^{-1} k^m g_0(k)\, dk \\ &= 2\pi a_0 \phi(0) + 2\pi \int_0^{\infty} dk\, g_0(k) \sum_{m=0}^{\infty} k^m \phi^{(m)}(0) \Big/ m! \\ &= 2\pi a_0 \phi(0) + 2\pi \int_0^{\infty} \phi'(k) g_0(k)\, dk . \end{aligned} \tag{24}$$

This defines $g(k)$ as a generalized function, since $g_0(k)$ is defined by (21) for $k \geq 0$. We can express (24) in several different ways by introducing the generalized function $\theta(k)$; then with the usual definition of $[g_0(k)\theta(k)]'$ as a generalized function, (24) becomes

$$g(k)[\phi(k)] = 2\pi a_0\phi(0) + 2\pi \int_{-\infty}^{\infty} \phi'(k) g_0(k)\theta(k)\, dk \tag{25}$$

$$= 2\pi a_0\phi(0) - 2\pi \int_{-\infty}^{\infty} \phi(k)[g_0(k)\theta(k)]'\, dk\,. \tag{26}$$

Thus $g(k)$ is the generalized function

$$g(k) = 2\pi a_0 \delta(k) - 2\pi[g_0(k)\theta(k)]'\,. \tag{27}$$

If g_0' tends to a finite limit as $k \to 0$, we may write

$$\begin{aligned} g(k) &= 2\pi a_0\delta(k) - 2\pi[g_0(k)\delta(k) + g_0'(k)\theta(k)] \\ &= 2\pi[a_0 - g_0(0)]\delta(k) - 2\pi g_0'(k)\theta(k)\,; \end{aligned} \tag{28}$$

the coefficient of $\delta(k)$ is not necessarily zero, since a_0 and $g_0(0)$, given by (17) and (21), may be different. If $g_0'(k)$ does not tend to a finite limit as $k \to 0$, we should use one of the expressions (24)–(27) as the definition of $g(k)$. For series of Stieltjes the measure $d\phi(u)$ is uniquely determined, so that $g_0(k)$ and $g(k)$ are uniquely defined by (21) and (24)–(27).

III. APPROXIMANTS TO $g(k)$

Although we have defined $g(k)$ as a generalized function by (27) and (21), we may not know the whole sequence of coefficients $\{a_n\}$ and hence the measure function $\phi(u)$. In this Section we consider the problem of obtaining an approximate form for $g(k)$. A useful practical method of computing a function from a finite number of terms of its series expansion is the Padé method (Baker, 1966). When the sequence of coefficients a_n are expressible in the form (17) with $\phi(u)$ bounded and nondecreasing, (3) is a series of Stieltjes in ix, and sequences $[n, n+j](ix)$ of diagonal Padé approximants converge to $f(x)$ provided that the sequence $\{a_n\}$ does not increase too rapidly, [with $\limsup a_n/(2n)! < \infty$ roughly] (Baker, 1966). This technique has been generalized recently by the discovery of Gammel-Baker approximants (Baker, 1967) and we will use this generalization to obtain approximants to the generalized function $g(k)$.

Since we are assuming that (3) is a series of Stieltjes, we can use the known convergence properties of the approximants $[n, n+j](ix)$; there are other conditions under which $[n, n+j](ix)$ have been proved to converge, and it is well known that convergence occurs in practice for many series not encompassed by known theorems. We therefore expect that the results of this Section will ultimately be capable of wide expansion.

We approximate $g(k)$ by finding an approximate form for $g_0(k)$ defined by (21). The first step is to note that for $k > 0$,

$$g_0(k) \equiv \int_{0+}^{\infty} e^{-k/u}\, d\phi(u)$$
$$= \frac{1}{2\pi i}\int_{0+}^{\infty} \frac{d\phi(u)}{u}\left[\int_C \frac{e^{-th}}{t(t-1/u)}\,dt\right] + \int_{0+}^{\infty} d\phi(u)\,, \tag{29}$$

where C is the t plane contour shown in Fig. 1 and consists of segments of the lines $\arg t = \pm\pi/4$ and of an arc of the circle $|t| = \delta$, where δ is a constant. Interchanging the order of integration, we get

$$g_0(k) = -\frac{1}{2\pi i}\int_C dt\,\frac{e^{-tk}}{t}\int_{0+}^{\infty}\frac{d\phi(u)}{1-ut} + \int_{0+}^{\infty} d\phi(u)$$
$$= -\frac{1}{2\pi i}\int_C dt\,\frac{e^{-tk}}{t}\int_{0+}^{\infty}\frac{d\phi(u)}{1-ut} + \int_{0}^{\infty} d\phi(u)\,, \tag{30}$$

since any $\delta(u)$ term in $d\phi(u)$ gives equal and opposite contributions to the two integrals on the right-hand side of (30).

The Padé approximant $[n, n+j](-t)$ to the function

$$\int_0^{\infty}\frac{d\phi(u)}{1-ut} = \sum_{n=0}^{\infty} a_n t^n \tag{31}$$

is of the form

$$\sum_{m=1}^{n}\frac{\alpha_{m,n}}{1-t\sigma_{m,n}} + \sum_{l=0}^{j}\beta_{l,n}(-t)^l\,; \tag{32}$$

when $j = -1$, the last summation is absent. For series of Stieltjes,

$$a_n > 0\,,\quad \alpha_{m,n} > 0\,,\quad \sigma_{m,n} > 0 \qquad (\text{all } m, n) \tag{33}$$

so that the poles of (32) lie on the positive real t axis. Also,

$$\sum_{m=1}^{n}\alpha_{m,n}(\sigma_{m,n})^l + \beta_{l,n} = a_l \qquad (0 \le l \le j) \tag{34}$$

and

$$\sum_{m=1}^{n}\alpha_{m,n}(\sigma_{m,n})^l = a_l \qquad (j < l \le 2n+j)\,. \tag{35}$$

From (34),

$$\sum_{m=1}^{n}\alpha_{m,n} + \beta_{0,n} = a_0 = \int_0^{\infty} d\phi(u) \tag{36}$$

and

$$\beta_{l,n} \le a_l = \int_0^{\infty} u^l\, d\phi(u) \qquad (0 \le l \le j)\,. \tag{37}$$

We now establish a key result.

THEOREM. When (31) is a series of Stieltjes in $(-t)$,

$$\lim_{n\to\infty}\left\{-\frac{1}{2\pi i}\int_C t^{-1}e^{-tk}[n, n+j](-t)\,dt\right\}, \tag{38}$$

exists for $k > 0$, where C is the contour of the Fig. 1.

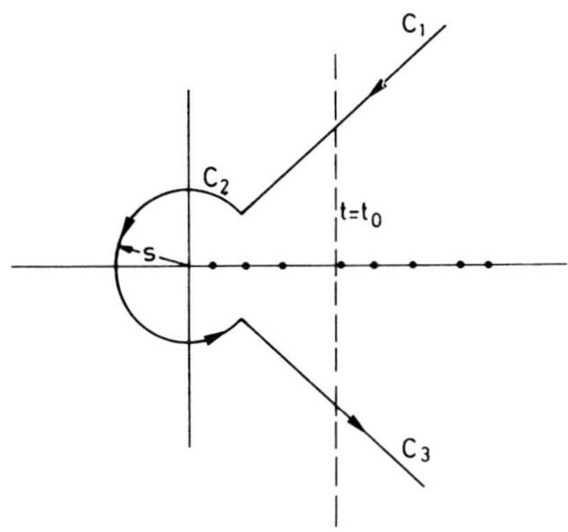

Fig. 1. The contour C.

To prove this result, divide the contour C into three parts C_1, C_2, and C_3 by its intersections with the line

$$\operatorname{Re} t = t_0 .$$

Then the part of (36) along C_1 (and likewise C_3) is in modulus equal to

$$\left|\frac{1}{2\pi}\int_{C_1} t^{-1}e^{-tk}\left[\sum_{m=1}^{n}\frac{\alpha_{m,n}}{1-t\sigma_{m,n}}+\sum_{l=0}^{j}\beta_{l,n}(-t)^l\right]dt\right|$$

$$\leq \frac{1}{2\pi}\int_{C_1}\left|t^{-1}e^{-tk}\left\{\left(\sum_{m=1}^{n}\alpha_{m,n}\right)\operatorname{Max}\left|\frac{1}{1-t\sigma_{m,n}}\right|+\sum_{l=0}^{j}\beta_{l,n}(-t)^l\right\}\right|dt .$$

Since $t = (1+i)\,\tau$ (with τ real) on C_1,

$$\operatorname*{Max}_{0\leq t<\infty}\left|\frac{1}{1-t\sigma_{m,n}}\right| = \begin{Bmatrix}\sqrt{2} & (\sigma_{m,n}\neq 0)\\ 1 & (\sigma_{m,n}=0)\end{Bmatrix}.$$

Using (34) and (35),

$$\left|\frac{1}{2\pi}\int_{C_1}\frac{e^{-tk}}{t}[n, n+j](-t)\,dt\right|$$

$$\leq \frac{1}{2\pi}\int_{C_1}\left|\frac{e^{-tk}}{t}\right|\left\{\sqrt{2}\,a_0+\sum_{l=0}^{j}a_l\,|t|^l\right\}dt .$$

The last integral is independent of n, and for $k > 0$, we can choose t_0 so that the integral is bounded in modulus by ε, for any $\varepsilon > 0$. In fact we choose t_0 so that

$$\int_{C_1} t^{-1} e^{-tk} \left[\int_0^\infty \frac{d\phi(u)}{1 - ut} \right] dt$$

is also bounded in modulus by ε and the same bounds hold for the corresponding integrals along C_3.

On C_2, $[n, n + j](-t)$ converges uniformly to

$$\int_0^\infty \frac{d\phi(u)}{1 - ut} .$$

Therefore given $\varepsilon > 0$ there exists n_0 such that, if L is the length of C_2,

$$\left| [n, n + j](-t) - \int_0^\infty \frac{d\phi(u)}{1 - ut} \right| < \frac{2\pi\varepsilon}{L} \left\{ \operatorname*{Max}_{(C_2)} \left| \frac{e^{-tk}}{t} \right| \right\}^{-1} , \qquad n \geq n_0 ;$$

the maximum is taken for any fixed $k > 0$. Therefore, for these values of n,

$$\left| \frac{1}{2\pi} \int_C \frac{e^{-tk}}{t} \left\{ [n, n + j](-t) - \int_0^\infty \frac{d\phi(u)}{1 - ut} \right\} dt \right| \leq 5\varepsilon .$$

Therefore the limit of (38) exists and equals

$$-\frac{1}{2\pi i} \int_C dt \frac{e^{-tk}}{t} \int_0^\infty \frac{d\phi(u)}{1 - ut} = g_0(k) - a_0 \tag{39}$$

by (30).

When $k > 0$ we can evaluate the integral in (38) by closing the contour C by a straight line parallel to the imaginary axis in the t plane and beyond all the poles of any particular approximant. The integral is then

$$\begin{aligned} -\frac{1}{2\pi i} \oint_C \frac{e^{-tk}}{t} & \left[\sum_{m=1}^{n} \frac{\alpha_{m,n}}{1 - t\sigma_{m,n}} + \sum_{l=0}^{j} \beta_{l,n} (-t)^l \right] dt \\ &= -\sum_{m=1}^{n} \alpha_{m,n} + \sum_{m=1}^{n} \alpha_{m,n} e^{-k/\sigma_{m,n}} - \beta_{0,n} \\ &= \sum_{m=1}^{n} \alpha_{m,n} e^{-k/\sigma_{m,n}} - a_0 , \end{aligned} \tag{40}$$

using Eq. (36). From (39), (40) and the theorem,

$$\begin{aligned} g_0(k) &= \int_{0+}^\infty e^{-k/u} \, d\phi(u) \\ &= \lim_{n \to \infty} \sum_{m=1}^{n} \alpha_{m,n} e^{-k/\sigma_{m,n}} \qquad (k > 0) . \end{aligned} \tag{41}$$

Now define

$$g_{0,n}(k) = \sum_{m=1}^{n} \alpha_{m,n} e^{-k/\sigma_{m,n}} \tag{42}$$

and, corresponding to (27),

$$g_n(k) = 2\pi a_0 \delta(k) - 2\pi[g_{0,n}(k)\theta(k)]' . \tag{43}$$

It is a simple corollary of the theorem that, as $n \to \infty$, $g_{0,n}(k) \to g_0(k)$ uniformly in any fixed interval (δ, ∞) with $\delta > 0$. We wish to show that $g_n(k) \to g(k)$ as a generalized function; using (25), (27) and (43), we require

$$\left| \int_0^\infty \phi'(k)[g_0(k) - g_{0,n}(k)]\, dk \right| \to 0 \tag{44}$$

as $n \to \infty$, for any test function $\phi(k)$. By dividing the range of integration into the ranges $(0, \delta)$ and (δ, ∞), it is not difficult to establish (44) provided that $|g_{0,n}(k)|$ is bounded in the range $(0, \infty)$, uniformly in n and k. From (42) and (36),

$$|g_{0,n}(k)| \leq \sum_{m=1}^{n} \alpha_{m,n} = a_0 - \beta_{0,n} .$$

Since $\beta_{0,n}$ is not necessarily positive, this does not in general provide a uniform bound. But if $j = -1$, the second summation in (32) is absent, and

$$|g_{0,n}(k)| \leq a_0 . \tag{45}$$

So when (32) is the $[n, n-1]$ approximant, (42) and (43) define an approximation to the generalized function $g(k)$, given by (27).

The sum in (41) has the form of a Gammel–Baker approximant to the corresponding integral. The convergence theorems for these approximants, established by Baker (1967), do not apply here, since $e^{-k/u}$ is not analytic in u at $u = 0$; we have had to modify Baker's method of proof of convergence.

This approximation theorem allows us to obtain an approximation to the generalized function $g(k)$ from a knowledge of the first few members of the sequence $\{a_n\}$. We proceed by forming an approximant $[n, n-1]$ from the series (3), and hence evaluating $\alpha_{m,n}$ and $\sigma_{m,n}$ in (32), remembering that the second summation is absent when $j = -1$. Substitution of these values in (42) gives the approximant to $g(k)$ through (24) or (27).

We can now return to the example of Section II. With $a_n \equiv \alpha^{-n}$ in (31), (32) becomes

$$1/[1 - (t/\alpha)]$$

consisting of only one term with $\alpha_{1,n} = 1$ and $\sigma_{1,n} = \alpha^{-1}$, for all n. Hence

(41) and (24) give

$$g_0(k) = e^{-\alpha k}$$

and

$$g(k)[\phi(k)] = 2\pi\phi(0) + 2\pi \int_0^\infty \phi'(k)\, e^{-\alpha k}\, dk$$

$$= 2\pi\alpha \int_0^\infty \phi(k) e^{-\alpha k}\, dk .$$

Thus Eq. (15) is seen to hold for all values of k. We also see that

$$g(k) = -2\pi g_0'(k)\theta(k) ,$$

since $a_0 - g_0(0) = 0$ in Eq. (28).

IV. CONCLUSIONS

We have been able to define the series (1) when the Fourier-transformed series (3) is a series of Stieltjes in ix; it is identified as the generalized function $g(k)$ given by (27) and (21), and has noncompact support on the real k axis. We have further shown that knowledge of a few of the coefficients $\{a_n\}$ allows us to calculate approximants to $g(k)$, using (41).

Physically, we are studying a particular type of multipole expansion in one dimension, and it is not generally realized that a multipole expansion can, in a strict mathematical sense, represent a spacially extended distribution of, say, charge. One important question can be asked: to what extent can an arbitrary charge distribution be represented by a multipole expansion at a single point? One can envisage reducing problems involving extended sources to simpler problems involving multipoles at the origin. The present work is limited to one dimension, and by the assumption that (3) is a series of Stieltjes. There seems to be no real reason why these restrictions should not ultimately be removed; but it has so far proved very difficult to establish any general theorems on Padé approximants. The condition that $\{a_n\}$ satisfy the determinate Stieltjes moment problem leads to the asymmetry in k of the generalized function $g(k)$; it may be possible to replace Stieltjes moments by Hamburger moments, thereby eliminating the asymmetry.

ACKNOWLEDGMENTS

This work was partly stimulated by a lecture series given in Canterbury by Professor W. Guttinger of Munich, under the auspices of the United States Air Force. The example discussed in Section II was worked out in collaboration with Professor Guttinger. We are also grateful for several discussions with Dr. R. Hughes Jones and Dr. J. Rennison, both of Canterbury. Finally we would like to thank Professor J. Leray for his comments which have led to improvements in this paper.

REFERENCES

Baker, G. A., Jr. (1966). "Advances in Theoretical Physics" (K. A. Brueckner, ed.), Vol. 1, pp. 1–58. Academic Press, New York.

Baker, G. A., Jr. (1967). *Phys. Rev.* **161,** 434.

Friedman, A. (1963). "Generalized Functions and Partial Differential Equations." Prentice-Hall, Englewood Cliffs, New Jersey.

Guttinger, W. (1966). *Fortschr. Physik* **14,** 517.

Widder, D. V. (1946). "The Laplace Transform," p. 139. Princeton Univ. Press, Princeton, New Jersey.

CHAPTER 7 HILBERT SPACE AND THE PADÉ APPROXIMANT*

D. Masson

University of Toronto, Canada

I. INTRODUCTION

The Padé approximant currently is being used to obtain approximate solutions to a wide variety of dynamical equations for various quantum-mechanical models. The models considered have encompassed both potential theory and quantum field theory, and the dynamical equations have included the Lippman–Schwinger equation (Tani, 1966; Chen, 1966; Masson, 1967b; Gammel and McDonald, 1966), Bethe–Salpeter equation (Nuttall, 1967; Nieland and Tjon, 1968), the N/D equation (Masson, 1967a; Common, 1967), and renormalized perturbation theory (Copley and Masson, 1967; Bessis and Pusterla, 1968; Copley *et al.*, 1968). In each case, one is interested in knowing whether or not the approximations converge and precisely what the limitations of the method are.

It is with this in mind that we examine here the general role that the Padé approximant plays as an approximation technique in Hilbert space. Since Hilbert space is the mathematical setting for all the models mentioned, the results that we obtain can be used both to justify present approximations and to suggest new ones. In Sect. II, we review those features of the moment problem which are essential to our discussion. In Sect. III, we apply these results to show how the Padé approximant may be used to calculate the matrix elements of the resolvent of a symmetric operator. In Sect. IV, we introduce a related but more general technique known as the method of moments and discuss the scope of this method for various classes of operators.

* Supported by the National Research Council of Canada.

An important feature of the approximation methods of Sects. III and IV is that in principle they can be used to perform calculations for unbounded operators. Such a feature is essential if one wants a method that is capable of performing quantum field theory calculations for strong interactions.

II. EXTENDED SERIES OF STIELTJES

The theory of Padé approximants for functions that are series of Stieltjes has been reviewed recently by Baker (1965). However, for such functions one is limited to a discussion of the Stieltjes moment problem {the moment problem over the interval $[0, \infty)$}. Since we shall be interested in the moment problem (Wall, 1948; Akhiezer, 1965) over the full interval $(-\infty, \infty)$, we shall have to consider a wider class of functions which we call extended series of Stieltjes. In this section, we review some of the basic facts concerning these latter functions and their related moment problem.

Consider the function

$$f(z) = \int_{-\infty}^{\infty} \frac{d\phi(x)}{1 - xz} \tag{1}$$

having the formal (not necessarily convergent) power series expansion

$$f(z) = \sum_{n=0}^{\infty} f_n z^n . \tag{2}$$

The coefficients f_n are assumed to be finite and are given by the moments

$$f_n = \int_{-\infty}^{\infty} x^n \, d\phi(x) . \tag{3}$$

Definition. The function $f(z)$ is an extended series of Stieltjes if and only if $\phi(x)$ is a bounded nondecreasing function with infinitely many points of increase in the interval $(-\infty, \infty)$.

This property of $f(z)$ may be characterized in terms of the determinants of the coefficients

$$D_f(m, n) = \det \begin{vmatrix} f_m & f_{m+1} & \cdots & f_{m+n} \\ f_{m+1} & f_{m+2} & \cdots & f_{m+n+1} \\ \vdots & \vdots & & \\ f_{m+n} & f_{m+n+1} & \cdots & f_{m+2n} \end{vmatrix} \tag{4}$$

due to the following theorem.

THEOREM 1. A necessary and sufficient condition for $f(z)$ to be an extended series of Stieltjes is

$$D_f(0, n) > 0 , \qquad n = 0, 1, 2, \ldots . \tag{5}$$

A proof of the theorem can be found in Wall (1948, Theorem 86.1). Let us also consider the functions

$$f_{2m}(z) = [f(z) - f_0 - f_1 z \cdots - f_{2m-1} z^{2m-1}]/z^{2m} \tag{6}$$

and

$$g_{2m}(z) = [-g(z) + g_0 + g_1 z \cdots + g_{2m-1} z^{2m-1}]/z^{2m} \tag{7}$$

for $m = 1, 2, \ldots$, where $g(z) = 1/f(z)$ has the power series expansion $\Sigma g_n z^n$.

THEOREM 2. If $f(z)$ is an extended series of Stieljes, then $f_{2m}(z)$ and $g_{2m}(z)$, $m = 1, 2, \ldots$, are also extended series of Stieltjes.

Proof. The fact that $f_{2m}(z)$ is an extended series of Stieltjes follows directly from the representation (1) and the definition (6). That is,

$$f_{2m}(z) = \int_{-\infty}^{\infty} \frac{x^{2m}\, d\phi(x)}{1 - xz} .$$

Taking

$$\psi(x) = \int_{-\infty}^{x} x^{2m}\, d\phi(x) ,$$

one has that

$$f_{2m}(z) = \int_{-\infty}^{\infty} \frac{d\psi(x)}{1 - xz} ,$$

where $\psi(x)$ is bounded and nondecreasing with infinitely many points of increase. To prove that $g_{2m}(z)$ is an extended series of Stieltjes, one need only prove it for $m = 1$ and then apply the first part of the theorem to $g_2(z)$. From the Hadamard relation (Baker, 1965)

$$D_g(2, n) = (-1)^{n+1}(g_0)^{2n+3} D_f(0, n + 1) ,$$

and since

$$D_{g_2}(0, n) = (-1)^{n+1} D_g(2, n) ,$$

we have that

$$D_{g_2}(0, n) = (g_0)^{2n+3} D_f(0, n + 1) > 0 .$$

Hence, from Theorem 1, $g_2(z)$ is an extended series of Stieltjes.

If one is given the coefficients f_n in the power series expansion (2) of an extended series of Stieltjes, the problem of constructing the function $\phi(x)$ [and hence from Eq. (1) obtaining $f(z)$] is known as the moment problem. The moment problem is said to be *determinate* if there is a *unique* bounded nondecreasing function $\phi(x)$ satisfying the moment conditions (3) and the supplementary conditions $\phi(-\infty) = 0$ and

$$\phi(x) = \lim_{\varepsilon\to 0} \tfrac{1}{2}[\phi(x+\varepsilon) + \phi(x-\varepsilon)] .$$

(The supplementary conditions are needed, since the moment conditions alone can only determine ϕ to within an arbitrary constant and at points where ϕ is continuous.) If the moment problem is not determinate, then there are *infinitely many* solutions, and the moment problem is said to be *indeterminate*.

Since $\phi(x)$ is given in terms of $f(z)$ by the Stieltjes inversion formula (Wall, 1948)

$$\phi(x) = -\lim_{\varepsilon\to 0+} \pi^{-1} \int_{-\infty}^{x} \operatorname{Im}\{(y+i\varepsilon)^{-1} f(1/(y+i\varepsilon))\}\, dy ,$$

there is a one-to-one correspondence between $\phi(x)$ and $f(z)$. The common extended series of Stieltjes' properties of $f(z)$, $f_{2m}(z)$, and $g_{2m}(z)$ thus imply the following theorem.

THEOREM 3. If $f(z)$ is an extended series of Stieltjes and the moment problem associated with $f_{2m}(z)$ or $g_{2m}(z)$ is determinate, then the moment problem associated with $f(z)$ is also determinate.

In the case where the moment problem is determinate, one can represent the function $f(z)$ for $\operatorname{Im} z \neq 0$ in terms of a converging continued fraction known as a real J-fraction (see Wall (1948), Chapter 17).

$$f(z) = \frac{f_0}{-a_0 z + 1 - [b_0^2 z^2/(-a_1 z + 1 - \cdots)]}, \qquad \operatorname{Im} z \neq 0 . \tag{8}$$

Moreover, this J-fraction converges uniformly with respect to z in any compact region in the upper or lower half-z plane (Wall, 1948, Chapter 5, Theorem 27.2).

Now the Nth approximant of the J-fraction is the $[N, N-1]$ Padé approximant of $f(z)$, which we shall denote by $[N, N-1]f(z)$. The uniqueness property of the Padé approximant (Baker, 1965) gives one the relations

$$\begin{aligned} &f_0 + f_1 z \cdots f_{2m-1} z^{2m-1} \\ &\qquad + z^{2m}[N, N-1] f_{2m}(z) = [N, N+2m-1] f(z) , \\ &g_0 + g_1 z \cdots g_{2m-1} z^{2m-1} \\ &\qquad - z^{2m}[N, N-1] g_{2m}(z) = ([N+2m-1, N] f(z))^{-1} . \end{aligned} \tag{9}$$

If one uses Theorem 3 and the relations (9) and one considers the uniformly convergent J-fractions for $f(z)$, $f_{2m}(z)$, and $g_{2m}(z)$, then one has the following result.

THEOREM 4. If $f(z)$ is an extended series of Stieltjes and the moment problems associated with $f_{2(m+1)}(z)$ and $g_{2m}(z)$ are determinate, then, for fixed $j = 0, \pm 1, \pm 2, \ldots \pm m$, the sequence $[N, N + 2j + 1] f(z)$ converges to $f(z)$ for $\operatorname{Im} z \neq 0$. The convergence is uniform with respect to z in any compact region in the upper or lower half-z plane.

An important feature of the Padé approximant is the location of its poles and zeros. If $f(z)$ is an extended series of Stieltjes and one expands $[N, N - 1] f(z)$ in partial fractions, one has

$$[N, N-1] f(z) = \sum_{i=1}^{N} \frac{\mu_i}{1 - z/z_i},$$

where the μ_i are positive and the z_i are real. This can be written in terms of a bounded nondecreasing step function $\phi_N(x)$ having points of increase at $x = 1/z_i$:

$$[N, N-1] f(z) = \int_{-\infty}^{\infty} \frac{d\phi_N(x)}{1 - zx},$$

where

$$\phi_N(x) = \sum_{i=1}^{N} \mu_i \theta(x - 1/z_i)$$

and $\theta(x) = 1$ for $x \geq 0$ and $\theta(x) = 0$ for $x < 0$.

When the moment problem is determinate, one has that the solution $\phi(x)$ is given, at its points of continuity, by

$$\phi(x) = \lim_{N \to \infty} \phi_N(x).$$

It should also be noted that (1) if $\phi(x)$ is constant for $-\infty < x < a$ and/or $b < x < \infty$, the same is true of $\phi_N(x)$; (2) if $\phi(x)$ is constant for $c < x < d$, then $\phi_N(x)$ has at most one point of increase in this interval. For the details of these properties, we refer the reader to Akhiezer (1965, Chapters I and II).

These properties may be used to extend the region of uniform convergence of the J-fraction to include certain portions of the real z axis where $f(z)$ is analytic. For example, if $f(z)$ is an extended series of Stieltjes which is analytic for $a^{-1} < z < b^{-1}$ with $a < 0$ and $0 < b$ [that is, $\phi(x)$ is constant for $x < a$ and $x > b$], then it can be shown that the J-fraction converges uniformly for z in any compact region that is a positive distance from the real segments $(-\infty, a^{-1})$ and (b^{-1}, ∞).

We have outlined some of the main features of the moment problem for the determinate case. It is, of course, important to know whether a given moment problem is determinate or indeterminate. A useful criterion in terms of the coefficients of the power series (2) is given by:

Carleman's Theorem. If $f(z)$ is an extended series of Stieltjes and if the series $\sum (1/f_{2n})^{1/2n}$ is divergent, then the moment problem is determinate. (For a proof, see Wall (1948), Theorem 88.1.)

III. SYMMETRIC OPERATORS AND THE PADÉ APPROXIMANT

There is a close connection between the theory of Padé approximants and the theory of symmetric operators in a Hilbert space. Here we show how the Padé approximant may be used to calculate matrix elements of the resolvent of a symmetric operator. We shall first review some of the basic concepts of Hilbert space theory.

A set of vectors $\mathscr{S}$ is *dense* (in $\mathscr{H}$) if for any vector h in the Hilbert space $\mathscr{H}$ there is a sequence of vectors h_n in $\mathscr{S}$ such that $h_n \to h$. By $h_n \to h$, one means that $\lim_{n\to\infty} \|h - h_n\| = 0$, where the *norm* of a vector f is $\|f\| = (\langle f | f \rangle)^{1/2}$. Let $\mathscr{D}(A)$ be the set of vectors in the *domain* of an operator A. If $\mathscr{D}(A)$ is dense, then A is said to be *densely defined.* If A is densely defined and $\langle f | Ag \rangle = \langle Af | g \rangle$ for all f and g in $\mathscr{D}(A)$, A is called *symmetric.* Let the operator A be densely defined, and consider the set $\mathscr{D}(A^+)$ of all vectors g such that for a fixed g and all f in $\mathscr{D}(A)$ a g' exists with $\langle Af | g \rangle = \langle f | g' \rangle$. This defines a linear mapping $g' = A^+ g$ [the g' is unique because $\mathscr{D}(A)$ is dense]. The operator A^+ is called the *adjoint* of A. If $\mathscr{D}(A) = \mathscr{D}(A^+)$ and $Af = A^+ f$ for all f in $\mathscr{D}(A)$, then $A = A^+$ and A is said to be *self-adjoint.* Note that in order for an operator to be self-adjoint it is necessary but not sufficient that it be symmetric. The operators of basic interest in quantum mechanics are normally assumed to be self-adjoint. For such operators, an important property that is essential to our discussion is given by the *spectral theorem* (Riesz and Sz.-Nagy, 1955).

Let A be a self-adjoint operator. The spectral theorem gives one a *unique* integral representation of the operator. As an example, consider the case when A has only a point spectrum. That is, A has eigenvalues a_n and corresponding eigenvectors ϕ_n (for convenience, we take the ϕ_n to be nondegenerate and to have unit norm), where the vectors ϕ_n span the space $\mathscr{H}$. For any vector f in $\mathscr{D}(A)$ and any vector g in $\mathscr{H}$, one has

$$\langle g | Af \rangle = \int_{-\infty}^{\infty} \lambda d \langle g | E(\lambda) f \rangle , \tag{10}$$

where

$$E(\lambda) = \sum_n \theta(\lambda - a_n) \, |\phi_n\rangle \langle \phi_n | .$$

One writes Eq. (10) formally as

$$A = \int_{-\infty}^{\infty} \lambda \, dE(\lambda) \,. \tag{10'}$$

The operator $E(\lambda)$ is called an orthogonal resolution of the identity. It is a family of projection operators depending on the real parameter λ, which satisfies the conditions

$$\begin{aligned} E(\lambda)E(\mu) &= E(\lambda) \qquad \text{for} \quad \lambda \le \mu \,, \\ E(\lambda)E(\mu) &= E(\mu) \qquad \text{for} \quad \lambda > \mu \,, \\ \lim_{\lambda \to -\infty} E(\lambda) &= 0 \,, \\ \lim_{\lambda \to \infty} E(\lambda) &= I \,, \\ \lim_{\varepsilon \to 0+} E(\lambda + \varepsilon) &= E(\lambda) \,. \end{aligned} \tag{11}$$

(The limits are in the sense of strong convergence. For the definition of strong convergence, see Sect. IV.)

As a consequence, one has that $E_f(\lambda) = \langle f \mid E(\lambda) f \rangle$ is a bounded nondecreasing function of λ, is right continuous, and has $E_f(-\infty) = 0$ and $E_f(\infty) = \langle f \mid f \rangle$.

Equations (10) and (11) and hence the properties of $E_f(\lambda)$ also hold when the spectrum is discrete and/or continuous and degenerate. That is, any self-adjoint operator determines one and only one family of projection operators $E(\lambda)$ satisfying Eqs. (10′) and (11).

Equation (10′) can be generalized to a function of A by means of the formula

$$f(A) = \int_{-\infty}^{\infty} f(\lambda) \, dE(\lambda) \,. \tag{10''}$$

The particular function $R(zA) = (1 - zA)^{-1}$ which is trivially related to the resolvent of A thus has the representation

$$R(zA) = \int_{-\infty}^{\infty} \frac{dE(\lambda)}{1 - z\lambda} \,. \tag{12}$$

If the vector f is in the domain of A^n for all n and one takes diagonal matrix elements of $R(zA)$, one has that the function of the complex variable z given by

$$R_f(z) = \langle f \mid R(zA) f \rangle = \int_{-\infty}^{\infty} \frac{dE_f(\lambda)}{1 - z\lambda} \tag{13}$$

is either an extended series of Stieltjes or a rational fraction [in case $E_f(\lambda)$ has an infinite or finite number of points of increase, respectively]. In the former case, in order to ensure that the resulting moment problem is determinate, it is convenient to introduce the notion of an *analytic vector* (Nelson, 1959).

DEFINITION. A vector f is said to be analytic (with respect to an operator A) if and only if

$$\Sigma \frac{||A^n f||\, s^n}{n!} < \infty \qquad \text{for some} \quad s > 0\,. \tag{14}$$

THEOREM 5. If A is a self-adjoint operator and f is an analytic vector, then, for fixed $j = 0, \pm 1, \pm 2, \ldots$, the sequence $[N, N+2j+1]R_f(z)$ converges to $R_f(z)$ for $\operatorname{Im} z \neq 0$. The convergence is uniform with respect to z in any compact region in the upper or lower half-z plane.

Proof. If $R_f(z)$ is a rational fraction, then, since $[N, N+2j+1]R_f(z)$ is a unique rational fraction (unique if one agrees to cancel common factors in the numerator and denominator) having the same power series development as $R_f(z)$ up to and including the term $z^{2(N+j)+1}$, one has for sufficiently large N that $[N, N+2j+1]R_f(z) \equiv R_f(z)$. If $R_f(z)$ is not a rational fraction, it is an extended series of Stieltjes. Since f is an analytic vector, the power series (14) converges, and consequently the inequality

$$\frac{||A^n f||}{n!}\, s^n < c$$

holds for some $c > 0$. That is,

$$\left(\frac{1}{||A^n f||}\right)^{1/n} > \left(\frac{s^n}{cn!}\right)^{1/n}.$$

Since $\lim_{n\to\infty} (s^n/cn!)^{1/n} = s/en$ and the $2n$th coefficient in the power-series expansion of $R_f(z)$ is $||A^n f||^2$, we have that

$$\Sigma \left(\frac{1}{||A^n f||^2}\right)^{1/2n} = \infty\,.$$

The present theorem then follows from Carleman's theorem and Theorem 4.

The condition that A be self-adjoint can actually be weakened. We must, however, introduce the notion of the *closure* of an operator.

The operator A is *closed* if it has the property that, whenever the sequences f_n and Af_n converge to f and g, respectively, then f is in $\mathscr{D}(A)$ and $g = Af$. If operators A and B satisfy the conditions that $\mathscr{D}(A) \subset \mathscr{D}(B)$ and $Af = Bf$ for any f in $\mathscr{D}(A)$, then B is an *extension* of A, and one writes $A \subseteq B$. Now any symmetric operator A has a closed symmetric extension, and there exists a unique minimal, closed, symmetric extension $\bar{A}$ (minimal in the sense that any closed extension of A is an extension of $\bar{A}$) called the *closure* of A.

The following theorem tells us that, for a self-adjoint operator, the set of all analytic vectors is dense.

Nelson's Theorem (1959). Let A be a symmetric operator. Then $\bar{A}$ is self-adjoint if and only if $\bar{A}$ has a dense set of analytic vectors.

One can now use Nelson's theorem to replace in Theorem 5 the condition that A be self-adjoint by the weaker condition that A be symmetric. Consider the linear manifold $\mathscr{L}_f$ that is formed from all the linear combinations of a finite number of the vectors $f, f_1, \ldots, f_m, \ldots,$ where $f_m = A^m f$. By including all the limit points of $\mathscr{L}_f$, one obtains a closed linear manifold $\mathscr{H}_f$ that is a Hilbert space.

Consider the operator $A' \subseteq A$, where $\mathscr{D}(A') = \mathscr{L}_f$ and $A'x = Ax$ for any x in $\mathscr{L}_f$. One has that $\bar{A}'$ is a closed symmetric operator in $\mathscr{H}_f$. Now one can easily show that f_m and any linear combination of a finite number of the f_m are analytic vectors with respect to both A and A'. [For example, $\|A^n f_m\| = \|A^{n+m} f\| < c(n+m)!/s^{n+m}$, and hence $\Sigma \|A^n f_m\| \, t^n/n!$ converges for $0 < t < s$.] Thus any vector in $\mathscr{L}_f$ is analytic with respect to A'. Since $\mathscr{L}_f$ is dense in $\mathscr{H}_f$, one has from Nelson's theorem that $\bar{A}'$ is self-adjoint in $\mathscr{H}_f$. Theorem 5 can then be applied to $\bar{A}'$. Finally, one will note that, for any z such that $R(zA)$, $R(z\bar{A}')$, and $R(zA')$ all exist and any x in $\mathscr{D}[R(zA')]$, one has

$$\langle x \,|\, R(zA)x \rangle = \langle x \,|\, R(z\bar{A}')x \rangle = \langle x \,|\, R(zA')x \rangle$$

[since $R(z\bar{A}') \supseteq R(zA')$ and $R(zA) \supseteq R(zA')$]. In particular, the equality is true for $\operatorname{Im} z \neq 0$ and $x = f$.

Although the proof of convergence of the sequence of Padé approximants was restricted to the diagonal matrix elements of a resolvent, it should be noted that a nondiagonal matrix element can always be written as a sum of diagonal matrix elements by means of the identity

$$\begin{aligned} \langle f \,|\, Rg \rangle = \tfrac{1}{2}\{ & \langle f+g \,|\, R(f+g) \rangle - i \langle f+ig \,|\, R(f+ig) \rangle \\ & + (i-1)\langle f \,|\, Rf \rangle + (i-1)\langle g \,|\, Rg \rangle \} \,. \end{aligned} \tag{15}$$

Hence one can use the Padé approximant to calculate $\langle f \,|\, Rg \rangle$ by applying it to the individual terms on the right-hand side of Eq. (15).

The Padé approximant may also be used to investigate the spectrum of an operator. The matrix elements of the resolvent of an operator A have singularities at the eigenvalues of A. The poles of the Padé approximant to $R_f(z)$ are thus giving one an approximation to the eigenvalues of A. For a symmetric A, the poles of $[N, N-1]R_f(z)$ give an approximation that is in fact identical to a variational method. By choosing a trial vector $\psi = f + \alpha_1 Af + \cdots \alpha_{N-1} A^{N-1} f$ and varying the parameters $\alpha_1, \ldots, \alpha_{N-1}$, one can obtain a set of approximate eigenvalues $\{\lambda_i\}$, where each λ_i is an extremum for the expression $\langle \psi \,|\, A\psi \rangle / \langle \psi \,|\, \psi \rangle$. It can be shown that the set of poles $\{z_i\}$ of $[N, N-1]R_f(z)$ and the set $\{1/\lambda_i\}$ are equal [see Vorobyev (1965), p. 62].

We have seen that the Padé approximant provides us with a technique of constructing a sequence of approximations that converge to the matrix elements of the resolvent of a symmetric operator.

There is an even deeper connection between the theory of operators and the results that we have outlined previously. This connection is not of direct interest to our present discussion, and we refer the reader to the books by Akhiezer (1965) and Stone (1932). It does, however, lead one to a closely related but more general approximation than we have presented here. This more general approach is called the *method of moments* (Vorobyev, 1965). The virtues of the method are (1) one is not restricted to self-adjoint operators, (2) one deals with a sequence of approximations to the operator itself, and (3) one can handle much more general operator-valued functions than $R(zA)$. The method is presented in the following section.

IV. METHOD OF MOMENTS

Let A be a closed linear operator in a Hilbert space $\mathscr{H}$. (If A is not closed but has closed extensions, we could deal with $\bar{A}$ the closure of A.) Consider a vector f in the domain of A^n for arbitrary n. From the sequence of vectors $f_0, f_1, \ldots, f_n \ldots$, where $f_n = A^n f$, we can construct an orthonormal set of vectors $\{e_n\}$ using the Schmidt orthogonalization procedure. If one takes all linear combinations of a finite number of the vectors f_n or e_n, one has a linear manifold $\mathscr{L}_f$. By including all the limit points of $\mathscr{L}_f$, one has a complete space $\mathscr{H}_f$ that is a separable Hilbert space and a subspace of $\mathscr{H}$. (That is, if g_n is a sequence of vectors in $\mathscr{L}_f$ with $g_n \to g$ in $\mathscr{H}$, then g is also in $\mathscr{H}_f$.)

We shall be concentrating our attention not on A but on the operator $\bar{A}'$, which is the closure of A' where $A' \subseteq \bar{A}' \subseteq A$ and $\mathscr{D}(A') = \mathscr{L}_f$. With this understood, we shall use the symbol B instead of $\bar{A}'$. Since $\mathscr{L}_f$ is dense in $\mathscr{H}_f$ and $\mathscr{L}_f \subset \mathscr{D}(A')$, we have that B is a closed, densely defined linear operator in $\mathscr{H}_f$.

If the vectors $f_0, f_1, \ldots, f_n$ are linearly independent, one has

$$e_n = P_n(B)f, \tag{16}$$

where

$$P_n(B) = \alpha_n(B^n + \alpha_{n,n-1}B^{n-1} + \cdots \alpha_{n,0}). \tag{17}$$

The coefficients $\alpha_{n,i}$ are uniquely determined from the linear equations

$$\sum_{k=0}^{n-1} c_{ik}\alpha_{n,k} + c_{in} = 0, \qquad i = 0, 1, \ldots, n-1, \tag{18}$$

where

$$c_{ik} = \langle f_i | f_k \rangle . \tag{19}$$

The normalization factor α_n is chosen so that $\langle e_n | e_n \rangle = 1$ and can be taken to be real and positive.

The orthogonalization procedure would terminate at $n = m$, the largest value of n such that $f_0, f_1, \ldots, f_n$ are linearly independent. If m exists, the space $\mathscr{H}_f$ has finite dimension $m + 1$. If m does not exist, $\mathscr{H}_f$ is an infinite dimensional Hilbert space having $\{e_n\}$ as an orthonormal basis. With respect to this basis, B has the matrix representation

$$M = \begin{bmatrix} a_{00} & a_{01} & \cdots \\ a_{10} & a_{11} & \cdots \\ 0 & a_{21} & \cdots \\ 0 & 0 & \cdots \\ \vdots & & \end{bmatrix}, \tag{20}$$

where $a_{ij} = \langle e_i | Be_j \rangle$.

In the case when $\mathscr{H}_f$ has finite dimension $m + 1$, the polynomials $P_n(B)$ satisfy the recursion relation

$$BP_n(B) = a_{n+1n}P_{n+1}(B) + a_{nn}P_n(B) + \cdots a_{0n}P_0(B) \tag{21}$$

for $n < m$. If for $n = m$ one defines the polynomial $Q_{m+1}(B)$ by means of the equation

$$BP_m(B) = \alpha_m Q_{m+1}(B) + a_{mm}P_m(B) + \cdots a_{0m}P_0(B) , \tag{22}$$

then the eigenvalues and eigenvectors of B can be determined in terms of $Q_{m+1}(B)$. That is, $Q_{m+1}(B) = (B - \lambda_1)(B - \lambda_2) \cdots (B - \lambda_{m+1})$, where $(B - \lambda_i)^{-1}Q_{m+1}(B)f$ is an eigenvector of B having the eigenvalue λ_i. Note that the eigenvalues are nondegenerate, since otherwise this would contradict the assumption that $f_0, f_1, \ldots, f_m$ were linearly independent.

In the case when $\mathscr{H}_f$ has infinite dimension, the recursion relation (21) is valid for all n. Since M is an infinite-dimensional matrix, the eigenvalue problem is nontrivial. However, one would suspect that with increasing n the roots of the polynomials $P_n(B)$ would give increasingly better approximations to the eigenvalues of B.

In either the finite- or infinite-dimensional case, one can consider the following sequence of approximations to B. Define the matrix B_N by setting to zero all but the first $N + 1$ rows and columns of M. In terms of the operator $E_N = \sum_{n=0}^{N} |e_n\rangle\langle e_n|$, which projects onto the subspace $\mathscr{H}_N$ spanned by the vectors $e_0, e_1, \ldots, e_N$, one has that

$$B_N = E_N B E_N . \tag{23}$$

Note that, for any polynomial $Q_n(B)$ of degree $n \leq N$, one has $Q_n(B_N)f = Q_n(B)f$. In particular, one has $e_n = P_n(B_N)f$ for $n \leq N$. From Eq. (21), one can then show that $P_{N+1}(B_N)f = 0$. It follows that, if λ_i is a root of the polynomial $P_{N+1}(\lambda)$, then $(B - \lambda_i)^{-1}P_{N+1}(B)f$ is an eigenvector of B_N having eigenvalue λ_i, where $\lambda_i \neq \lambda_j$ for $i \neq j$.

We wish to consider in what sense the sequence of operators B_N approximates B. That is, does the sequence B_N converge to B? Do the eigenvalues and eigenvectors of B_N converge to those of B? Do operator valued functions $f(B_N)$ converge to $f(B)$?, etc. The case when $\mathscr{H}_f$ has finite dimension is less intersting, since $B_m \equiv B$. Henceforth we shall assume that $\mathscr{H}_f$ is infinite dimensional. In order to discuss the foregoing convergence questions, we review the standard definitions of *weak*, *strong*, and *uniform* convergence.

Consider an operator A in $\mathscr{H}$, a sequence of operators A_N, and a set of vectors $\mathscr{S} \subset \mathscr{D}(A)$ such that for any vector g in $\mathscr{S}$ and all N greater than some M one has g in $\mathscr{D}(A_N)$:

1. $A_N \underset{w}{\rightarrow} A$ in $\mathscr{S}$ if, for any g in $\mathscr{S}$ and any h in $\mathscr{H}$, one has $\lim_{N\to\infty} \langle h | A_N g \rangle = \langle h | Ag \rangle$.
2. $A_N \underset{s}{\rightarrow} A$ in $\mathscr{S}$ if, for any g in $\mathscr{S}$, one has $\lim_{N\to\infty} \|A_N g - Ag\| = 0$ (or, equivalently, $A_N g \rightarrow Ag$).
3. $A_N \underset{u}{\rightarrow} A$ in $\mathscr{S}$ if $\lim_{N\to\infty} \|A_N - A\|_{\mathscr{S}} = 0$.

The expression $\|A\|_{\mathscr{S}}$ denotes the smallest number a such that $\|Ag\| \leq a\,\|g\|$ for all g in $\mathscr{S}$. If $\|A\|_{\mathscr{S}} < \infty$, the operator is *bounded in* $\mathscr{S}$. If $\|A\|_{\mathscr{H}} < \infty$, the operator is said to be *bounded* and to have *norm* $\|A\| = \|A\|_{\mathscr{H}}$.

One should note that uniform convergence implies strong convergence, which in turn implies weak convergence.

We shall now apply these definitions of convergence to the sequence of operators B_N, but first we shall derive the following lemma.

LEMMA 1. If T and T_N are closed, linear, and uniformly bounded and $T_N \underset{s}{\rightarrow} T$ in $\mathscr{S}$, where $\mathscr{S}$ is dense in $\mathscr{H}_f$, then $T_N \underset{s}{\rightarrow} T$ in $\mathscr{H}_f$.

Proof. For any g in $\mathscr{H}_f$, there exists a sequence g_n in $\mathscr{S}$ such that $g_n \rightarrow g$. The lemma follows from the inequality

$$\begin{aligned} \|T_N g - Tg\| &\leq \|T_N(g - g_n)\| + \|T(g - g_n)\| + \|(T_N - T)g_n\| \\ &\leq 2M\,\|g - g_n\| + \|(T_N - T)g_n\|\,, \end{aligned}$$

where $\|T\|, \|T_N\| \leq M$. Both $\|g - g_n\|$ and $\|(T_N - T)g_n\|$ can be made arbitrarily small by choosing n and N sufficiently large.

THEOREM 6. $B_N \underset{s}{\rightarrow} B$ in $\mathscr{L}_f$.

Proof. Any vector g in $\mathscr{L}_f$ is of the form $Q(B)f$, where $Q(B)$ is a polynomial of some degree n. For $N \geq n + 1$, one has

$$BQ(B)f - B_N Q(B)f = 0 .$$

COROLLARY. If B is bounded, then $B_N \underset{s}{\rightarrow} B$ in $\mathscr{H}_f$.

Proof. From Theorem 6, one has that $B_N \underset{s}{\rightarrow} B$ in $\mathscr{L}_f$. The corollary follows by noting that $\mathscr{L}_f$ is dense in $\mathscr{H}_f$ and that $||B_N|| = ||E_N B E_N|| \leq |E_N||\,||B||\,||E_N|| = ||B||$ and then applying Lemma 1.

An operator B is completely continuous if there exist denumerable sets of vectors ϕ_i and ψ_i such that the sequence of operators

$$\sum_{i=0}^{N} |\psi_i\rangle\langle\phi_i| \underset{u}{\rightarrow} B .$$

For the special class of completely continuous operators, one can replace strong convergence by the stronger statement of uniform convergence (Vorobyev, 1965).

THEOREM 7. A necessary and sufficient condition for B to be completely continuous is $B_N \underset{u}{\rightarrow} B$ in $\mathscr{H}_f$.

The sufficiency part of the theorem is trivial, since it follows from our definition of complete continuity. For the necessity, one must establish that one can choose $\phi_i = e_i$ and $\psi_i = \sum_{j=0}^{i+1} a_{ji} e_j$. For the proof of this, we refer the reader to Vorobyev (1965, Theorem IV).

In many applications, one is interested in solving a linear vector equation of the form

$$y = g + zBy . \tag{24}$$

When a unique solution to this equation exists, it is given by

$$y = R(zB)g , \tag{25}$$

where $R(zB) = (1 - zB)^{-1}$. One can obtain approximate solutions to Eq. (24) by solving the corresponding equation with B replaced by B_N:

$$y_N = g + zB_N y_N . \tag{26}$$

The solution to this latter equation may be written explicitly in terms of a ratio of polynomials in z. For $z \neq 1/\lambda_i$, where λ_i is an eigenvalue of B_N, one has that [for the derivation of an equivalent form of Eq. (27), see

Vorobyev (1965)]

$$R(zB_N) = \frac{\det\begin{vmatrix} c_{00}, & c_{01}, & \ldots, & c_{0N+1} \\ c_{10}, & c_{11}, & \ldots, & c_{1N+1} \\ \vdots & & & \vdots \\ c_{N0}, & c_{N1}, & \ldots, & c_{N,N+1} \\ z^{N+1}(I - E_N), & z^N, & z^{N-1}(zB_N + 1), \ldots, & z^N B_N^N + z^{N-1}B_N^{N-1} \cdots 1 \end{vmatrix}}{\det\begin{vmatrix} c_{00}, & c_{01}, & \ldots, & c_{0N+1} \\ c_{10}, & c_{11}, & \ldots, & c_{1N+1} \\ \vdots & & & \vdots \\ c_{N0}, & c_{N1}, & \ldots, & c_{NN+1} \\ z^{N+1}, & z^N, & \ldots, & 1 \end{vmatrix}} \tag{27}$$

One should note that the denominator of $R(zB_N)$ is proportional to the polynomial $z^{N+1}P_{N+1}(1/z)$, which has zeros at the reciprocal eigenvalues of B_N.

Under suitable conditions, one can show that $R(zB_N) \underset{s}{\rightarrow} R(zB)$ in $\mathscr{H}_f$ and consequently that $y_N \rightarrow y$ for arbitrary g. In order to formulate these conditions, it is useful to introduce the concept of the numerical range of an operator. If B is a bounded operator, then $W(B)$, the numerical range of B, is the set of all points b such that $\langle g | Bg \rangle = b\langle g | g \rangle$ for some $g \neq 0$ in $\mathscr{H}_f$.

THEOREM 8. If B is bounded and $1/z \neq 0$ is not in $W(B)$, then $R(zB_N) \underset{s}{\rightarrow} R(zB)$ in $\mathscr{H}_f$. The convergence is uniform with respect to z for $1/z$ in a compact region C that does not contain the origin and that is a positive distance from $W(B)$.

Proof. If $1/z$ is not in $W(B)$, then it is a positive distance d from $W(B)$, and one has a bound on $R(zB)$, namely, $\|R(zB)\| \leq (|z|\, d)^{-1}$, (see Stone, 1932 Theorem 4.20). If one takes only vectors g from $\mathscr{H}_N$, the subspace onto which E_N projects, one obtains a subset of the numerical range $W_N(B) \subset W(B)$. In this case, since $(g, Bg) = (g, E_N B E_N g)$, one has that $W(B_N) = W_N(B) \cup \{0\}$. Hence, $\|R(zB_N)\| \leq (|z|\, \delta)^{-1}$, where $\delta = \min\,[|z|^{-1}, d]$. Consider the difference

$$R(zB_N) - R(zB) = zR(zB_N)(B_N - B)R(zB)\,.$$

For any vector g in $\mathscr{H}_f$, one has $y = R(zB)g \in \mathscr{H}_f$. Now

$$\|[R(zB_N) - R(zB)]g\| \leq |z|\, \|R(zB_N)\|\, \|(B_N - B)y\|\,.$$

Therefore, $f_N(z) = ||[R(zB_N) - R(zB)]g|| \leq \delta^{-1} ||(B - B_N)y||$, which can be made arbitrarily small, since $B_N \underset{s}{\rightarrow} B$ in $\mathscr{H}_f$.

In order to show that the convergence is uniform for $1/z$ in C, we consider

$$|f_N(z) - f_N(z')| = |\,||[R(zB_N) - R(zB)]g|| - ||[R(z'B_N) - R(z'B)]g||\,| \\ \leq ||[R(zB_N) - R(z'B_N)]g - [R(zB) - R(z'B)]g|| .$$

Using the identity $R(zT) - R(z'T) = (z - z')R(zT)TR(z'T)$ for $T = B$ or B_N, this becomes

$$|f_N(z) - f_N(z')| \leq |z - z'|\,[||R(zB_N)B_NR(z'B_N)g|| + ||R(zB)BR(z'B)g||] \\ \leq \frac{2\,|z - z'|\,||g||\,||B||}{|zz'|\,D^2},$$

with $D = \min\,[|z_0|^{-1}, d']$, where $1/|z_0|$ is the largest value of $1/|z|$ for all $1/|z|$ in C and d' is the distance from C to $W(B)$. Since C is compact, there are, for any $\eta > 0$ and all $1/z$ in C, a finite number of points $1/z_k$ in C such that $|1/z - 1/z_k| < \eta$ for at least one k. Since

$$f_N(z) \leq |f_N(z) - f_N(z_k)| + f_N(z_k) \leq \frac{2\,||B||\,||g||\,|z^{-1} - (z_k)^{-1}|}{D^2} + f_N(z_k) ,$$

we may choose an $\eta < \varepsilon D^2/(2\,||B||\,||g||)$ and an $M(z_k, \varepsilon)$ such that $f_N(z_k) < \varepsilon$ for all $N > M(z_k, \varepsilon)$. For all $1/z$ in C and all $N > M > \max\,[M(z_k, \varepsilon)]$, one then has $f_N(z) \leq 2\varepsilon$.

One will note that, since

$$||[R(zB_N) - R(zB)]g|| \leq ||(B - B_N)R(zB)g||/\delta$$

and

$$||(B - B_N)R(zB)g|| \leq ||B - B_N||\,||R(zB)||\,||g|| ,$$

one has that for B completely continuous both $B_N \underset{u}{\rightarrow} B$ and $R(zB_N) \underset{u}{\rightarrow} R(zB)$ in $\mathscr{H}_f$. For completely continuous B, one may also weaken the restriction on z. If z is not a reciprocal eigenvalue of B, then $R(zB)$ exists and is bounded. Consider the identity $R(zB_N) = [1 - zR(zB)(B_N - B)]^{-1}R(zB)$. Since $||zR(zB)(B_N - B)|| \leq |z|\,||R(zB)||\,||B_N - B||$, one can find an M such that, for all $N > M$, $||zR(zB)||\,||(B_N - B)|| < 1$. For such N, one then has $||R(zB_N)|| \leq (1 - |z|\,||R(zB)||\,||B_N - B||)^{-1}\,||R(zB)||$. Let R and r be the maximum values of $||R(zB)||$ and $|z|$ for all z in a compact region that does not contain a reciprocal eigenvalue of B, and choose an M' such that, for all $N > M'$, $rR\,||B_N - B|| < 1$. One then has

$$||[R(zB_N) - R(zB)]g|| \leq r(1 - rR\,||B_N - B||)^{-1}R\,||B_N - B||\,||g|| .$$

We have thus proved the theorem.

THEOREM 9. If B is completely continuous and z is finite and not equal to a reciprocal eigenvalue of B, then $R(zB_N) \underset{u}{\rightarrow} R(zB)$ in $\mathscr{H}_f$. The convergence is uniform with respect to z for z in a compact region that does not contain a reciprocal eigenvalue of B.

This theorem implies that, for a completely continuous B, the nonzero eigenvalues and corresponding eigenvectors of B_N converge to eigenvalues and eigenvectors of B. For a precise statement on the rate of convergence, we refer the reader to Vorobyev (1965), Chapter II, page 37.

In the particular case of a symmetric operator, one can make much stronger statements for bounded and unbounded operators. Also, there is then a direct connection between the method of moments and the Padé approximant. First let us indicate this connection.

If B is symmetric, then the matrix in Eq. (20) is tridiagonal. That is, $a_{nm} = 0$ for $|n - m| > 1$, and the recursion relation (21) reduces to three terms. Since the constants $c_{ik} = \langle B^i f \mid B^k f \rangle$, one has $c_{ik} = c_{0k+i}$ and real. Consequently, the α_{nm} in Eq. (2) are real, together with a_{nn} and $a_{nn-1} = a_{n-1n}$. Since B is symmetric, $B_N = E_N B E_N$ is also symmetric, and hence the roots of the polynomial $P_{N+1}(z)$ are all real. The function $\langle f \mid R(zB) \mid f \rangle = R_f(z)$ has a power-series development $R_f(z) = \Sigma c_n z^n$, where we have defined $c_{0n} = c_n$. Using the formula (11.1) in Baker (1965) for the $[N + 1, N]$ Padé approximant and comparing this with Eq. (27), we see that

$$[N + 1, N]\, R_f(z) = \langle f \mid R(zB_N) f \rangle . \tag{28}$$

An equivalent observation is that, in the continued fraction expansion of $R_f(z)$,

$$R_f(z) = \frac{c_0}{-a_0 z + 1 - [b_0{}^2 z^2/(-a_1 z + 1 - \cdots)]},$$

one has $a_n = a_{nn}$ and $b_n = a_{nn+1}$.

If B is symmetric and bounded, then it is self-adjoint. Since the numerical range of a symmetric operator is confined to the real axis, one knows that $W(B)$ is contained in a finite interval $[a, b]$, where a and b are the lower and upper bounds of B and $\|B\| = \max(|a|, |b|)$. One can modify Theorem 8 to state

THEOREM 10. If B is self-adjoint and bounded, having lower and upper bounds a and b, then $R(zB_N) \underset{s}{\rightarrow} R(zB)$ in $\mathscr{H}_f$ for $1/z$ not in $[a, b]$. The convergence is uniform with respect to z for $1/z$ in a compact region that does not contain points in the interval $[a, b]$ or the point 0.

This theorem can be compared to Theorem 5 of Sect. III. Here we make no mention of the fact that f is an analytic vector. However, B being bounded in $\mathscr{H}_f$ implies that any vector in $\mathscr{H}_f$ is an analytic vector.

The spectrum of the self-adjoint bounded operator B (consisting of the eigenvalue and continuous spectrum) is contained in the interval $[a, b]$. There may, however, be gaps in the spectrum so that $R(zB)$ is defined and bounded for $1/z$ in $[c, d]$, a subinterval of $[a, b]$. It can be shown that B_N can have at most one nonzero eigenvalue in $[c, d]$. There is, however, nothing to guarantee that, for sufficiently large N, B_N will have no eigenvalue in $[c, d]$ (unless the spectrum to the left of c or to the right of d consists entirely of eigenvalues). Consequently, the eigenvalues of the sequence of operators B_N may have an accumulation point $\lambda \in [c, d]$, and the strong limit of $R(zB_N)$ may not exist for $z = 1/\lambda$. In Vorobyev (1965) Chapter III, it is shown how one may overcome this difficulty. If, for some N, B_N has an eigenvalue in $[c, d]$, one considers B'_N a modified B_N with $B'_N = B_N - \lambda_N |g_N\rangle\langle g_N|$, where λ_N is the offending eigenvalue and g_N the corresponding eigenvector. One can then shown that $R(zB'_N) \underset{s}{\to} R(zB)$ in $\mathscr{H}_f$ for any finite z such that $1/z$ is not in the spectrum of B.

It is remarkable that Theorem 10 can be generalized to the case when B is unbounded and self-adjoint. Before proving this, we need the following lemma.

LEMMA 2. If B is self-adjoint, then, for $\operatorname{Im} z \neq 0$, the set of vectors $S = (1 - zB)\,\mathscr{L}_f$ is dense in $\mathscr{H}_f$.

Proof. Since $\operatorname{Im} z \neq 0$, one has that $(1 - zB)^{-1}$ exists and has domain $\mathscr{H}_f$. Thus, any vector h in $\mathscr{H}_f$ can be written as $h = (1 - zB)g$, where $g = (1 - zB)^{-1}h$. Since B is the closure of an operator with domain $\mathscr{L}_f$, one has that there exists a sequence $g_n \to g$ such that $h_n = (1 - zB)g_n \to h$, where g_n is in $\mathscr{L}_f$.

THEOREM 11. If B is self-adjoint (not necessarily bounded), then $R(zB_N) \underset{s}{\to} R(zB)$ in $\mathscr{H}_f$ for $\operatorname{Im} z \neq 0$. The convergence is uniform with respect to z for z in a compact region C that is a positive distance d from the real axis.

Proof. Let $R(zB)g$ be in $\mathscr{L}_f$, and consider

$$\begin{aligned} f_N(z) &= \|[R(zB_N) - R(zB)]g\| = |z|\,\|R(zB_N)(B_N - B)R(zB)g\| \\ &\leq |z|\,\|R(zB_N)\|\,\|(B_N - B)R(zB)g\|\,. \end{aligned}$$

Since $\|R(zB_N)\| \leq |z|/|\operatorname{Im} z|$ and $B_N \underset{s}{\to} B$ in $\mathscr{L}_f$, we have that $R(zB_N) \underset{s}{\to} R(zB)$ in $S = (1 - zB)\,\mathscr{L}_f$. From Lemma 2, we have that S is dense in $\mathscr{H}_f$. Applying Lemma 1 one has that $R(zB_N) \underset{s}{\to} R(zB)$ in $\mathscr{H}_f$. To show that the convergence is uniform for z in C, we may follow a procedure analogous to the proof of Theorem 10. It is thus sufficient to show that, for z and z' in C, one has $|f_N(z) - f_N(z')| \leq |z - z'|\,\beta$, where β is constant.

As before, we have

$$|f_N(z) - f_N(z')| \\ \leq |z - z'| \, (||R(zB_N) B_N R(z' B_N) g|| + ||R(zB) BR(z' B) g||) \, .$$

Now $||R(zT)|| \leq |z_0|/d$, where $|z_0|$ is the largest value of $|z|$ in C and $T = B$ or B_N. Also,

$$||TR(zT)|| = |z|^{-1} \, ||zTR(zT) - R(zT) + R(zT)|| \\ \leq d^{-1}[1 + (|z_0|/d)] \, .$$

One can thus choose $\beta = 2 \, ||g|| \, |z_0| \, [1 + (|z_0|/d)]/d$.

If one considers the spectral decompositions

$$R(zB) = \int \frac{dE(\lambda)}{1 - z\lambda}$$

and

$$R(zB_N) = \int \frac{dE_N(\lambda)}{1 - z\lambda}$$

[note that $E_N(\lambda) \neq E_N E(\lambda) E_N$], then, using the fact that $R(zB_N) \underset{s}{\rightarrow} R(zB)$, one can prove the following remarkable theorem.

THEOREM 12. If B is self-adjoint, then $E_N(\lambda) \underset{s}{\rightarrow} E(\lambda)$ in $\mathscr{H}_f$ at all λ that are not eigenvalues of B.

Proof. For any g in $\mathscr{H}_f$, one has that $\phi_N(\lambda) = \langle g \,|\, E_N(\lambda) g \rangle$ is a sequence of bounded nondecreasing functions of λ. Consequently, there exists a nondecreasing function $\phi(\lambda)$ and a subsequence $\phi_{N_j}(\lambda)$ such that $\phi_{N_j}(\lambda) \rightarrow \phi(\lambda)$ at all points of continuity of $\phi(\lambda)$ (see Wall, 1948, Theorem 6.42). Suppose that $\phi_N(\lambda)$ itself does not converge. One then has at least two subsequences $\phi_{N_j}(\lambda)$ and $\phi_{N_k}(\lambda)$ having different limits $\phi_1(\lambda)$ and $\phi_2(\lambda)$. This contradicts the fact that $\langle g \,|\, R(zB_N) g \rangle = \int d\phi_N(\lambda)/(1 - z\lambda)$ converges to a unique analytic function $\langle g \,|\, R(zB) g \rangle = \int d\langle g \,|\, E(\lambda) g \rangle/(1 - z\lambda)$ for $\operatorname{Im} z \neq 0$. Hence, $\phi_N(\lambda)$ converges to $\langle g \,|\, E(\lambda) g \rangle$ for λ not an eigenvalue of B [note that $\langle g \,|\, E(\lambda) g \rangle$ may be discontinuous at the eigenvalues of B]. Since the diagonal matrix elements converge and any nondiagonal matrix element can be written as a sum of diagonal terms, we have that $E_N(\lambda) \underset{w}{\rightarrow} E(\lambda)$ in $\mathscr{H}_f$. Using the fact that $E_N(\lambda)$ and $E(\lambda)$ are projection operators, we may deduce from this that $E_N(\lambda) \underset{s}{\rightarrow} E(\lambda)$ in $\mathscr{H}_f$. That is, consider, for any g in $\mathscr{H}_f$,

$$||[E(\lambda) - E_N(\lambda)] g||^2 \\ = \langle g \,|\, E(\lambda) h \rangle - \langle g \,|\, E_N(\lambda) h \rangle + \langle g \,|\, E_N(\lambda) h \rangle - \langle h \,|\, E_N(\lambda) g \rangle \, ,$$

where $h = E(\lambda) g$. If we add and subtract the term $\langle g \,|\, E(\lambda) g \rangle = \langle g \,|\, E(\lambda) h \rangle$, we have

$$\|[E(\lambda) - E_N(\lambda)]g\|^2 = \langle h|[E(\lambda) - E_N(\lambda)]g\rangle + \langle g|[E_N(\lambda) - E(\lambda)]g\rangle - \langle g|[E_N(\lambda) - E(\lambda)]h\rangle,$$

where each term can be made arbitrarily small.

With our knowledge of the convergence of $E_N(\lambda)$, we can turn our attention to the problem of the convergence of operator-valued functions of B_N.

THEOREM 13. If B is self-adjoint and $f(\lambda)$ is a continuous polynomial bounded function, then $f(B)$ is a closed densely defined operator and $f(B_N) \xrightarrow[s]{} f(B)$ in $\mathscr{L}_f$.

Proof. The operator $f(B)$ is defined by the formal expression $f(B) = \int f(\lambda)\, dE(\lambda)$ and has a domain $\mathscr{D}[f(B)]$ consisting of all vectors g such that $\int |f(\lambda)|^2\, d\langle g|E(\lambda)g\rangle < \infty$ (see Riesz and Sz.-Nagy, 1955, Chapter IX). Now $|f(\lambda)| < P(\lambda)$ for some polynomial $P(\lambda)$. Since $\int \lambda^n\, d\langle g|E(\lambda)g\rangle < \infty$ for any n and all g in $\mathscr{L}_f$, one has that $\mathscr{L}_f \subset \mathscr{D}[f(B)]$. Applying the same definition to $f(B_N)$, one has that $f(B_N) = \int f(\lambda)\, dE_N(\lambda)$ is a closed bounded operator in $\mathscr{H}_f$. If for any g in $\mathscr{L}_f$ and any h in $\mathscr{H}_f$ one considers the expression

$$\langle h|[f(B_N) - f(B)]g\rangle = \int_{-\infty}^{\infty} f(\lambda)\, d\langle h|[E_N(\lambda) - E(\lambda)]g\rangle,$$

then one can show that this vanishes in the limit $N \to \infty$. This is a result of a general theorem on sequences of Stieltjes integrals which states that, if (1) $\phi_N(\lambda)$ and $\phi(\lambda)$ are of bounded variation on $[a, b]$, (2) $\int_a^b |d\phi_N(\lambda)| < M$, (3) $f(\lambda)$ is continuous for $a \le \lambda \le b$, and (4) $\phi_N(\lambda) \to \phi(\lambda)$ at $\lambda = a$ and b and a dense set of points in $[a, b]$ then

$$\lim_{N\to\infty} \int_a^b f(\lambda)\, d\phi_N(\lambda) = \int_a^b f(\lambda)\, d\phi(\lambda)$$

(see Wall (1948), Theorem 64.1). In our case, we have

$$\phi_N(\lambda) = \langle h|E_N(\lambda)g\rangle$$

and

$$\phi(\lambda) = \langle h|E(\lambda)g\rangle,$$

both of bounded variation, and $\int_{-\infty}^{\infty} |d\phi_N(\lambda)| \le \|g\|\,\|h\|$. Since $E_N(\lambda) \xrightarrow[s]{} E(\lambda)$ in $\mathscr{H}_f$ for λ not an eigenvalue of B and the eigenvalues are countable, one has $\phi_N(\lambda) \to \phi(\lambda)$ for a dense set of points λ. Choosing $-b, b \ne$ an eigenvalue of B, we may apply the theorem stated to show that

$$\lim_{N\to\infty} \int_{-b}^{b} f(\lambda)\, d\phi_N(\lambda) = \int_{-b}^{b} f(\lambda)\, d\phi(\lambda).$$

Therefore,

$$\left|\lim_{N\to\infty}\int_{-\infty}^{\infty} f(\lambda)\,d\phi_N(\lambda) - \int_{-b}^{b} f(\lambda)\,d\phi(\lambda)\right|$$

$$= \left|\lim_{N\to\infty}\left[\int_{-\infty}^{-b} + \int_{b}^{\infty}\right] f(\lambda)\,d\phi_N(\lambda)\right|$$

$$\leq \|h\| \lim_{N\to\infty}\left[\int_{-\infty}^{-b} + \int_{b}^{\infty}\right] P(\lambda)\,d\langle g\,|\,E_N(\lambda)g\rangle$$

$$\leq \frac{\|h\|}{b^2}\lim_{N\to\infty}\int_{-\infty}^{\infty} \lambda^2 P(\lambda)\,d\langle g\,|\,E_N(\lambda)g\rangle$$

$$\leq \frac{\|h\|}{b^2}\lim_{N\to\infty}\langle g\,|\,B_N^2 P(B_N)g\rangle$$

$$= \frac{\|h\|}{b^2}\langle g\,|\,B^2 P(B)g\rangle\,.$$

For the last equality, we have used the fact that g is in $\mathscr{L}_f$. Taking the limit $b\to\infty$, we have the desired result. One can thus conclude that $f(B_N) \underset{w}{\to} f(B)$ in $\mathscr{L}_f$. This can be extended to a statement of strong convergence by considering

$$\|[f(B_N) - f(B)]g\|^2 = \langle g\,|\,[|f(B_N)|^2 - |f(B)|^2]g\rangle$$
$$+ 2\,\mathrm{Re}\,\{\langle f(B)g\,|\,[f(B) - f(B_N)]g\rangle\}\,.$$

The second term goes to zero, since $f(B_N) \underset{w}{\to} f(B)$ in $\mathscr{L}_f$. The first term goes to zero, since the function $|f(\lambda)|^2$ is also continuous and polynomial bounded, and hence $|f(B_N)|^2 \underset{w}{\to} |f(B)|^2$ in $\mathscr{L}_f$. In particular, if $f(\lambda)$ is bounded by a constant, it defines a bounded operator $f(B)$, and one can use Lemma 1 to state that $f(B_N) \underset{s}{\to} f(B)$ in $\mathscr{H}_f$.

V. DISCUSSION

Throughout the paper, we have avoided discussing the indeterminate moment problem. In Sect. II we stated results only for the determinate case, whereas in Sect. III we restricted ourselves to analytic vectors. Also, when discussing the resolvent of an unbounded operator in Sect. IV, we assumed that the operator B was self-adjoint. The indeterminate moment problem is associated with the case where B is symmetric but not self-adjoint. There are then infinitely many self-adjoint extensions of B, each one corresponding to a different solution to the moment problem. The details of this correspondence are contained in Akhiezer (1965).

The method of moments can be viewed as a generalized Padé approximant method for operators. Instead of working with the coefficients in the power-series expansion of a function, one uses the operator coefficients of the power-series expansion of $(1 - zB)^{-1}$ and constructs an approximation to $(1 - zB)^{-1}$ which is a ratio of polynomials in z [Eq. (27)] of a very similar form to the $[N, N-1]$ Padé approximant. In the limit of large N, the Padé approximant reconstructs, from the coefficients in the power series, the extended series of Stieltjes $f(z)$ and its associated weight function $\phi(x)$. In the method of moments, one reconstructs $(1 - zB)^{-1}$, and the analogy is complete when B is self-adjoint, since one also obtains its resolution of the identity $E(\lambda)$. Finally, Theorem 13 has its analogy in the recent work of Baker (1967) which discusses generalized approximants for functions that are not series of Stieltjes.

The method of moments is a special perturbation technique, and many of the theorems of Sect. IV are related to more general theorems in the theory of linear operators (Kato, 1966). The method of moments has the advantage of prescribing a particular type of perturbation which lends itself readily to practical calculations.

For applications of the method to quantum mechanics, it is natural to consider the Hamiltonian operator H. One could use the method of moments to approximate $(E + i\varepsilon - H)^{-1}$ and to construct an approximate resolution of the identity for H. In potential scattering, this would enable one to calculate the scattering amplitude from its asymptotic expansion for large energy E.

One could also use the method to sum the Born series. If $H = H_0 + \lambda V$, where V is the potential, then the scattering amplitude demands a knowledge of

$$[1 - \lambda V^{1/2}(E + i\varepsilon - H_0)^{-1} V^{1/2}]^{-1} .$$

Although $V^{1/2}(E + i\varepsilon - H_0)^{-1} V^{1/2}$ is not self-adjoint, it is completely continuous for a large class of potentials.

ACKNOWLEDGMENT

The author wishes to thank Professor E. Prugovečki for many valuable discussions and his critical reading of the manuscript.

REFERENCES

Akhiezer, N. I. (1965). "The Classical Moment Problem." Oliver and Boyd, London.

Baker, G. A., Jr. (1965). "Advances in Theoretical Physis" (K. A. Brueckner, ed.), Vol. 1, pp. 1–58. Academic Press, New York.

Baker, G. A., Jr. (1967). *Phys. Rev.* **161,** 434.

Bessis, D., and Pusterla, M. (1968). *Nuovo Cimento* **54A,** 243.

Chen, A. (1966). *Nuovo Cimento* **52A,** 474.

Common, A. K. (1967). *J. Math. Phys.* **8,** 1669.

Copley, L. A., and Masson, D. (1967). *Phys. Rev.* **164,** 2059.
Copley, L. A., Elias, D. K., and Masson, D. (1968). *Rhys. Rev.* **173,** 1552.
Gammel, J. L., and McDonald, F. A. (1966). *Phys. Rev.* **142,** 1254.
Kato, T. (1966). "Perturbation Theory for Linear Operators." Springer-Verlag, New York.
Masson, D. (1967a). *J. Math. Phys.* **8,** 512.
Masson, D. (1967b). *J. Math. Phys.* **8,** 2308.
Nelson, E. (1959). *Ann. Math.* **70,** 573.
Nieland, H. M., and Tjon, J. A. (1968). *Phys. Letters* **27,** 309.
Nuttall, J. (1967). *Phys. Rev.* **157,** 1312.
Riesz, F., and Sz.-Nagy, B. (1955). "Functional Analysis." Ungar, New York.
Stone, M. (1932). "Linear Transformations in Hilbert Space." American Mathematical Society, New York, Vol. 15.
Tani, S. (1965). *Phys. Rev.* **139,** B1011.
Tani, S. (1966). *Ann. Phys.* (*N. Y.*) **37,** 451.
Vorobyev, Yu. V. (1965). "Method of moments in Applied Mathematics." Gordon and Breach, New York.
Wall, H. S. (1948). "Analytic Theory of Continued Fractions." Van Nostrand, Princeton, New Jersey.

CHAPTER 8 THE CONNECTION OF PADÉ APPROXIMANTS WITH STATIONARY VARIATIONAL PRINCIPLES AND THE CONVERGENCE OF CERTAIN PADÉ APPROXIMANTS*

J. Nuttall

Texas A & M University

I. INTRODUCTION

The aim of this chapter is to describe the connection between Padé approximants and approximations derived from stationary variational principles. We go on to discuss some possible implications about the convergence of Padé approximants and conclude by proving a new result about the convergence of Padé approximants to a certain class of entire functions, which goes some way toward verifying our speculations.

It is shown that a sequence of approximants running parallel to the diagonal of the Padé table may be obtained by restricting the trial function in a variational principle to a particular sequence of finite-demensional subspaces, each one contained in the next. The essence of this result is apparent in the compact formula quoted by Baker, Eq. (15).

Cini and Fubini (1953, 1954) first proposed the approximation scheme that uses the type of trial function (11) in the Schwinger variational principle. However, it was not until recently that Nuttall (1966, 1967) and independently Bessis and Pusterla (1968), showed the equivalence of this scheme to the Padé approximant. It is perhaps interesting to point out that from the variational viewpoint Cini and Fubini (1954) suggested the

* Supported in part by the Air Force Office of Scientific Research, Office of Aerospace Research, U.S. Air Force, under Grant No. 918-67.

use of what amounts to the operator Padé approximant to sum the renormalized perturbation expansion in field theory, a topic discussed elsewhere in this volume (Gammel, Kubis and Menzel, Chapter 13).

Probably the main interest of the results lies in their possible bearing on the question of the convergence of Padé approximants that are not related to series of Stieltjes. So far, no new convergence theorems have been established from the viewpoint of variational principles, for little attention seems to have been paid to the convergence of stationary principles. We quote a theorem proved recently which may give some indication of the type of convergence to be expected in the case of Padé approximants. However, in the last section we prove a new result that shows that Padé approximants to a certain class of entire functions converge in measure (the Padé approximants for large N are nearly correct almost everywhere), and we expect that this result can be generalized.

The author does not believe that the Padé method is likely to be the most efficient way of providing numerical solutions of integral equations of the Lippmann–Schwinger type, including the Bethe–Salpeter equation. It is probably better to choose simple trial functions for use in the variational principle rather than the complicated functions (7) needed to obtain the Padé approximant. Thus, for example, the calculations of Schwartz and Zemach (1966) and others (Levine *et al.*, 1967) are probably a cheaper way to obtain a solution of the Bethe–Salpeter equation of given accuracy than the Padé approach (see Nuttall, 1966). However, it is in cases where a consistent formulation in terms of integral equations is difficult or impossible to achieve, such as quantum field theory and the Heisenberg model, that the Padé method comes into its own.

II. DERIVATION OF PADÉ APPROXIMANTS FROM STATIONARY VARIATIONAL PRINCIPLES

We illustrate the connection between Padé approximants and stationary variational principles by considering a particular case, the S-wave scattering of a single particle in nonrelativistic potential theory. Generalizations will be mentioned later.

The Hamiltonian will be taken (with $\hbar = 1$) as

$$H = -\frac{1}{2m}\frac{1}{r^2}\frac{d}{dr}\left(r^2\frac{d}{dr}\right) + \lambda v(r) = T + \lambda v\,, \qquad (1)$$

and a scalar product of two real functions $\alpha(r)$, $\beta(r)$ is to be defined by

$$\langle\alpha|\beta\rangle = \int_0^\infty dr\, r^2\alpha(r)\beta(r)\,. \qquad (2)$$

To describe scattering at a given energy $E = k^2/2m$, we need a wave function $\psi(r)$, bounded at $r = 0$, which satisfies

$$(E - H)|\psi> = 0$$

and has the asymptotic form

$$\psi(r) \underset{r\to\infty}{\sim} (kr)^{-1}(\sin kr + \tan\delta \cos kr) , \tag{3}$$

where δ is the phase shift. The function $|\psi>$ may also be described as the solution of the Lippmann–Schwinger integral equation

$$|\psi> = |0> + \lambda G_0 v|\psi> , \tag{4}$$

where the Green's function G_0 may be written as

$$< r|G_0|r'> = -2mkj_0(kr_<)n_0(kr_>) \tag{5}$$

and the state $|0>$ represents the function $(\sin kr)/kr$.

An expansion for $|\psi>$ in powers of the potential strength λ is obtained by iterating (4) and is

$$|\psi> = \sum_{i=0}^{\infty} \lambda^i |i> , \tag{6}$$

where the state $|i>$ is defined as

$$|i> = (G_0 v)^i|0> . \tag{7}$$

Inserting (6) into the relation between $\tan\delta$ and $|\psi>$, which is

$$t \equiv -\frac{\tan\delta}{2mk\lambda} = <0|v|\psi> , \tag{8}$$

we produce the Born series

$$t = \sum_{i=0}^{\infty} a_i \lambda^i , \tag{9}$$

where

$$a_i = <0|v|i> .$$

For large enough values of $|\lambda|$, these series will diverge.

A powerful method of obtaining numerical solutions to problems in scattering theory has proved to be the application of stationary variational principles, the two most important being those of Schwinger and Kohn (see, for example, Moiseiwitsch, 1966).

The Schwinger principle states that, for small variations of $|\phi>$ and $|\phi'>$ about $|\psi>$, the following quantity I is stationary, and at $|\phi> = |\phi'> = |\psi>$ takes on the value shown in brackets.

$$I = [t] = <0|v|\phi> + <\phi'|v|0> - <\phi'|\{v - \lambda v G_0 v\}|\phi> . \tag{10}$$

It leads to no loss of generality to take $|\phi> = |\phi'>$ in (10).

In practice, we take an approximation to $|\psi>$, $|\phi> = |\psi(N)>$, which depends on N parameters and calculate the expression I. The parameters are then varied until I is stationary, thus determining the approximate $|\psi(N)>$. It is simplest to assume that the dependence on the parameters is linear, and a natural choice for $|\psi(N)>$, bearing in mind (6), first suggested by Cini and Fubini (1953, 1954), is

$$|\psi(N)> = \sum_{i=0}^{N-1} c_i |i> , \tag{11}$$

where the real coefficients c_i are to be varied. In this case, we find that I is given by

$$\begin{aligned} I &= -\left\{\sum_{i,j=0}^{N-1} c_i c_j [a_{i+j} - \lambda a_{i+j+1}] - 2\sum_{i=0}^{N-1} c_i a_i\right\} \\ &= -\{\boldsymbol{c}^T M \boldsymbol{c} - 2\boldsymbol{c}^T \boldsymbol{a}\} \end{aligned} \tag{12}$$

in a matrix notation with

$$M_{ij} = a_{i+j} - \lambda a_{i+j+1} . \tag{13}$$

The form (12) is stationary when $\boldsymbol{c}$ is given by

$$\boldsymbol{c} = M^{-1}\boldsymbol{a} , \tag{14}$$

and here I, which leads to an estimate for tan δ accurate to second order, is

$$I^{(N)} = t^{(N)} = \boldsymbol{a}^T M^{-1} \boldsymbol{a} . \tag{15}$$

Comparison with (15) and (16) of Baker's article shows that $I^{(N)}$ is just the $[N, N-1]$ Padé approximant to t.

Other Padé approximants may be derived from the same approach. Suppose that, instead of varying all the coefficients c_i, we fix the first M of them to be equal to the values they take on in the Born series, i.e.,

$$c_i = \lambda^i , \qquad i = 0, 1, \ldots, M-1 .$$

In this case, I given by (12) becomes

$$\begin{aligned} I = -\Bigg\{&\sum_{i,j=0}^{M-1} \lambda^{i+j} M_{ij} + 2\sum_{i=0}^{M-1}\sum_{j=M}^{N-1} \lambda^i c_j M_{ij} \\ &+ \sum_{i,j=M}^{N-1} c_i c_j M_{ij} - 2\sum_{i=0}^{M-1} \lambda^i a_i - 2\sum_{i=M}^{N-1} c_i a_i\Bigg\} \\ = &\sum_{i=0}^{M-1} a_i \lambda^i - \sum_{i,j=M}^{N-1} c_i c_j M_{ij} + 2\lambda^M \sum_{i=M}^{N-1} c_i a_{i+M} . \end{aligned} \tag{16}$$

If the remaining c_i are varied, I has a stationary value given by $I^{(N,M)}$,

where

$$I^{(N,M)} = \sum_{i=0}^{M-1} a_i \lambda^i + \lambda^M \sum_{i,j=M}^{N-1} a_{i+M} a_{j+M} (M^{-1})_{ij} \tag{17}$$

and the inverse of the matrix M is worked out with i, j restricted by $M \leq i, j \leq N-1$. Again we see from Baker's Eq. (19) that $I^{(N,M)}$ is a Padé approximant to t, and this time it is

$$I^{(N,M)} = [N-M, M+N-1].$$

Thus from the Schwinger principle we obtain Padé approximants whose numerator and denominator differ in degree by $-1, 1, 3, 5, \ldots$.

Padé approximants that fill in the gaps in this sequence are found when we use the same trial function $|\psi(N)>$ in the Kohn principle. The Kohn principle states that, for functions $|\phi>$ having the asymptotic form (3), the following expression J is stationary for variations of $|\phi>$ about $|\psi>$, and the value of J at $|\phi> = |\psi>$ is t:

$$J = -\frac{\tan \delta_{tr}}{2mk\lambda} - \lambda^{-1} <\phi|(E-H)\phi>. \tag{18}$$

Here $\tan \delta_{tr}$ is obtained from the asymptotic form of $|\phi>$.

The function $|\psi(N)>$ is an allowed trial function for the Kohn method if we choose $c_0 = 1$, for (5) shows that all but the first term of $|\psi(N)>$ approach a multiple of $(\cos kr)/kr$ for large large r, and in fact

$$-\frac{\tan \delta_{tr}}{2mk\lambda} = \lambda^{-1} \sum_{i=1}^{N-1} c_i a_{i-1}.$$

Thus we have

$$\begin{aligned} J &= \lambda^{-1} \Big\{ \sum_{i=1}^{N-1} c_i a_{i-1} - \sum_{i=0}^{N-1} \sum_{j=1}^{N-1} c_i c_j a_{i+j-1} \\ &\quad + \lambda \sum_{i=0}^{N-1} \sum_{j=0}^{N-1} c_i c_j a_{i+j} \Big\} \\ &= a_0 - \lambda^{-1} \Big\{ \sum_{i,j=1}^{N-1} c_i c_j [a_{i+j-1} - \lambda a_{i+j}] - 2\lambda \sum_{i=1}^{N-1} c_i a_i \Big\} \end{aligned} \tag{19}$$

An analysis similar to before shows that the stationary value of J, $J^{(N)}$ is

$$J^{(N)} = [N-1, N-1].$$

If, in addition to choosing $c_0 = 1$, we fix $c_i = \lambda^i$, $i \leq M-1$, then we obtain another approximate value $J^{(N,M)}$ that can be shown to be

$$J^{(N,M)} = [N-M, N+M-2].$$

Thus in this fashion we find Padé approximants whose numerator and denominator differ in degree by 0, 2, 4, 6, ... To obtain Padé approximants for which the degree of the denominator is at least two more than that of the numerator, it would appear that other variational principles would be needed.

It is clear that, since Baker's Eq. (19) also holds for matrix Padé approximants, they also may be derived from variational principles exactly as in the foregoing.

In some cases in scattering theory, where inelastic processes are allowed, it is sometimes not clear what would be a reasonable Padé approximant to use. The variational approach, which may be adapted to these cases, should provide a useful way of obtaining Padé-type approximations if suitable trial functions are used.

It might appear that the variational derivation of Padé approximants is useful only in scattering theory, but in fact the arguments are more general. In many power series, it is possible to write the coefficient of the i th power as

$$a_i = < \chi' | K^i \chi > ,$$

where K is an operator and $| \chi >$ and $| \chi' >$ are functions in a suitable space. Even if the series diverges, it may be that $(1 - \lambda K)^{-1}$ exists and serves to define the sum of the series by

$$f(\lambda) = < \chi' | (1 - \lambda K)^{-1} | \chi > = \sum_{i=0}^{\infty} a_i \lambda^i . \tag{20}$$

A variational principle of the Schwinger type gives $f(\lambda)$ as the stationary value of the following functional for variations of $|\phi>$, $|\phi'>$ about functions $| \psi >$, $| \psi' >$.

$$[f(\lambda] = < \chi' | \phi > + < \phi' | \chi > - < \phi' | [1 - \lambda K] | \phi > . \tag{21}$$

Here $| \psi >$, $| \psi' >$ satisfy

$$\begin{aligned} | \psi > &= | \chi > + \lambda K | \psi > , \\ < \psi' | &= < \chi' | + \lambda < \psi' | K. \end{aligned} \tag{22}$$

Let us define the space S_N spanned by

$$| \chi >, K | \chi >, \ldots, K^{N-1} | \chi >$$

and S_N' spanned by

$$< \chi' |, < \chi' | K, \ldots, < \chi' | K^{N-1} ,$$

with P_N, P_N' the orthogonal projection operators onto these spaces. Then an approximation for $f(\lambda)$ is obtained from (21) by restricting $| \phi >$,

$|\phi'\rangle$ to vary in S_N, S_N', respectively, and finding the resulting stationary value. An analysis similar to the foregoing shows that it is again the $[N, N-1]$ Padé approximant that is obtained.

To connect this approach with our discussion of the Schwinger principle in scattering theory, we identify

$$|\chi\rangle = v^{1/2}|0\rangle, \qquad \langle\chi'| = \langle 0|v^{1/2}, \qquad K = v^{1/2}G_0v^{1/2}.$$

For reasonable potential, K is an L^2 operator and $|\chi\rangle$, $|\chi'\rangle$ are normalizable functions so that the whole analysis can be carried out in Hilbert space, which was not the case with our first discussion of the Schwinger principle.

In the special case that v is a potential of one sign only, K is a self-adjoint operator and $|\chi\rangle = |\chi'\rangle$. In this situation, we have an extended series of Stieltjes and convergence of the Padé approximants. A neat way of proving convergence is to use the bounds that can be derived from the effective potential method (Nuttall, 1967). It is found that, in the case of a positive potential,

$$|\delta^{(N)}| \leq |\delta|,$$

where $\delta^{(N)}$ is obtained from the equation

$$\tan\delta^{(N)} = -2mk\lambda I^{(N)}$$

This result holds for any real value of λ, and provides an extension of the work of Common (1968).

III. CONVERGENCE OF PADÉ APPROXIMANTS

Very little is known about the convergence of sequences of Padé approximants in a rigorous sense except in the case of series of Stieltjes or closely related series. The connection with stationary variational principles lends support to the belief that Padé approximants converge in some sense, but so far has not led to any significant mathematical advances. Unfortunately, there seems to be little known about the convergence of approximations derived from stationary principles in contrast to the case of minimum principles, where there is a solid body of theory.

Let us return to the principle (21) for K an L^2 operator and $|\chi\rangle$, $|\chi'\rangle$ in the Hilbert space $\mathscr{H}$ of square integrable functions. We may define the linear subspace S as the closure of all points in any S_N, whatever N, and similarly S'. It is possible that S, S' are not the whole Hilbert space. In this case, it might be objected that, since $|\phi\rangle$, $|\phi'\rangle$ are never allowed to vary over the entire space, it is possible that they may never be able to represent accurately $|\psi\rangle$, $|\psi'\rangle$, and so may never lead to a sequence that converges to $f(\lambda)$. However, it is not difficult to see that in fact $|\phi> \in S$, $|\phi'> \in S'$ even if the Born series diverges, so that there is no need to vary $|\phi>$ and $|\phi'>$ over the whole of $\mathscr{H}$.

In the limit of large N, if such makes any sense, the Padé method tells us to find $|\phi>$ by solving

$$P'|\phi> = P'|\chi> + \lambda P'K|\phi> , \tag{23}$$

where P' projects onto S'. This equation may have several solutions, one of which is the $|\phi>$ obtained from (4). However, another solution would give the same stationary value $f(\lambda)$. These heuristic arguments indicate that the Padé method tries to solve the problem in as streamlined a fashion as possible.

There has recently been some work on the convergence of the Kohn method[11] which, although not applicable to Padé approximants, is perhaps worth mentioning as an example of the type of convergence that might be expected from a stationary variational principle. In a typical application of the Kohn method, a trial function $\phi^{(N)}(r)$ is chosen of the form

$$\begin{aligned}\phi^{(N)}(r) = (kr)^{-1}[\sin kr - 2mk\lambda t^{(N)} \cos kr(1 - e^{-\alpha r})] \\ + \sum_{i=0}^{N-1} C_i\chi_i(r) ,\end{aligned} \tag{24}$$

where $t^{(N)}$ is the trial value of t and the $\chi_i(r)$ are taken from a complete set of normalized functions. The parameters $t^{(N)}$, C_i are determined by making J stationary with respect to their variation. This leads to an analysis similar to that of the previous section, and the resulting expression for $t^{(N)}$ involves the inverse of a matrix.

Suppose we regard λ as fixed, real. Then it is possible that, for some real values of the energy E, the matrix will be singular (Schwartz, 1961), leading to an infinite estimate $t^{(N)}$. Within a given range R $(a \leq E \leq b)$, the number of such singularities will in general increase with N. However, it has been shown (Nuttall, 1969), for reasonable potentials and provided that the functions χ_i form a core, that there is convergence in the following sense.

Given any δ, $\delta_1 > 0$, however small, an N_0 can be found such that, for all $N > N_0$,

$$|t^{(N)}(E) - t(E)| < \delta_1$$

for all $E \in R$ except for a set of measure less than δ.

Thus, for most values of E, most terms of the sequence $t^{(N)}(E)$ converge to the correct result $t(E)$. Whereas the number of singularities increases, the strength of each, and indeed their total strength approaches zero.

This type of convergence is called convergence in measure by mathematicians (Halmos, 1950).

The unwanted singularities in the Kohn method are reminiscent of the spurious poles and zeroes that are sometimes found in numerical evaluations of Padé approximants (Baker, 1965). If these spurious poles are present, however large the value of N, it seems likely that Padé approximants will converge only in some sense similar to the foregoing. Instead of averaging over E, we need to fix E and vary the potential strength λ in the preceding discussion, and the extension of the analysis that applied to the straightforward Kohn method to the case of Padé approximants does not appear to be trivial.

IV. THE CONVERGENCE OF PADÉ APPROXIMANTS TO CONVERGENT POWER SERIES

With the remarks of the previous section in mind, we now investigate whether the class of functions for which diagonal (or parallel to the diagonal) Padé approximants are proved to converge can be enlarged if the definition of convergence is relaxed. We shall show that indeed the $[N, N-1]$ Padé approximants to a certain class of entire functions converge in measure within any given bounded domain of the complex plane. Undoubtedly it will be possible to use similar techniques to extend the result to a larger class of functions.

We say that a sequence of functions $f_n(\lambda)$ of the complex variable λ converges in measure (Halmos, 1950) to $g(\lambda)$ within a closed bounded region D if, given δ, $\bar{\delta}$, however small, it is possible to find n_0 such that, for each $n > n_0$, $|f_n(\lambda) - g(\lambda)| < \delta$ for all $\lambda \in d_n$, where the region $d_n \subset D$ has area $D - \bar{\delta}$.

Suppose a function $g(\lambda)$ is given by the power series

$$g(\lambda) = \sum_{i=0}^{\infty} a_i \lambda^i , \tag{25}$$

which we assume to be convergent everywhere. A useful starting point is Eq. (20) of Baker's article which reads, with $M = N - 1$, $n = N$,

$$[N, N-1] = \sum_{i=0}^{2N-1} a_i \lambda^i + \lambda^{2N} \mathbf{w}^T \mathbf{V}^{-1} \mathbf{w} , \tag{26}$$

where $\mathbf{V}$ is given by Eq. (16) and the vector $\mathbf{w}$ by

$$\mathbf{w} = (a_N, a_{N+1}, \ldots, a_{2N-1}) . \tag{27}$$

As $N \to \infty$, the first term on the right of (26) converges to $g(\lambda)$ for all λ, so that our concern is with the second term, which we call the correction Γ:

$$\Gamma = \lambda^{2N} \mathbf{w}^T \mathbf{V}^{-1} \mathbf{w} , \tag{28}$$

The $[N, N-1]$ Padé approximant has N poles at $\lambda = \lambda_1, \lambda_2, \ldots, \lambda_N$, and the difficulty in proving convergence is the determination of the values of λ_i. We shall bypass this problem and prove a result that holds for any set of λ_i. After a little manipulation, it is possible to rewrite Γ as

$$\Gamma = -\lambda^{2N} P(\lambda) R^{-1}(\lambda) , \tag{29}$$

where

$$R(\lambda) = \prod_{i=1}^{N} (\lambda - \lambda_i) = \sum_{i=0}^{N} r_i \lambda^i \tag{30}$$

and

$$P(\lambda) = \sum_{i=0}^{N-1} p_i \lambda^i , \tag{31}$$

with

$$p_i = \sum_{j=i+1}^{N} r_j a_{2N-j+1} . \tag{32}$$

We are interested in the convergence of $[N, N-1]$ inside the circle $|\lambda| < 1$, and the worst situation arises when $|\lambda_i| < 2$ for all i, so that Γ may have N poles within the circle $|\lambda| = 1$. In this case, bearing in mind that r_j is a homogeneous polynomial in the λ_i, we can bound $P(\lambda)$ within $|\lambda| \leq 1$, $|\lambda_i| < 2$:

$$|P(\lambda)| \leq 4^N aN , \tag{33}$$

where a is the maximum value of $|a_i|$, $i = N, \ldots, 2N-1$.

Now for a particular choice λ_i let us define A to be that area of the complex plane for which

$$|R(\lambda)| < x^N , \tag{34}$$

and suppose that $A_{\max}$ is the maximum value of A for given N, x but varying λ_i. If $|\lambda - \lambda_i| \geq x$ for all λ_i (i.e., λ is anywhere within a region formed by excluding a circle of radius x about each λ_i), then $|R(\lambda)| \geq x^N$. Thus it is clear that

$$A_{\max} \leq N\pi x^2 . \tag{35}$$

We now choose $x(N)$ in such a way that $A_{\max} \to 0$ as $N \to \infty$. Thus a possible choice would be $x = N^{-1/2-\varepsilon}$, $\varepsilon > 0$.

Using (33) and (34) in (29) with the foregoing choice of x leads to the result that, for all λ within $|\lambda| < 1$ except those included within an area $A_{\max}$ (which $\to 0$),

$$|\Gamma| < 4^N N^{N(1/2+\varepsilon)+1} a . \tag{36}$$

We have shown that (36) holds for all $|\lambda_i| \leq 2$, but it is not difficult to see that this restriction may be lifted without changing (36). For suppose that $|\lambda_i| \geq 2$ if and only if $i \geq M$, $1 \leq M < N$. We rearrange (25) to read

$$\Gamma = -\lambda^{2N} p(\lambda) r^{-1}(\lambda) , \tag{38}$$

where

$$p(\lambda) = \rho(\lambda) \Big/ \prod_{i=M}^{N} (\lambda - \lambda_i) , \tag{38}$$

$$r(\lambda) = \prod_{i=1}^{M-1} (\lambda - \lambda_i) . \tag{39}$$

Since the power of a given λ_i in $\rho(\lambda)$ is never higher than the first, $p(\lambda)$ is bounded for large λ_i, and in fact $p(\lambda)$ obeys (33). Moreover, if $|\lambda_i| \leq 1$, $i \geq M$, then $|r(\lambda)| \geq |R(\lambda)|$ for all λ within $|\lambda| \leq 1$, so that the area for which $|r(\lambda)| \leq x^N$ is certainly no greater than $A_{\max}$. Thus again we get (36) for all λ in the unit circle save for those in an area $\leq A_{\max}$.

So for those $g(\lambda)$ for which $a \to 0$ fast enough to make the bound of (36) approach zero, we have convergence in measure of the Padé approximants $[N, N-1]$ as $N \to \infty$.

For example, for any function $g(\lambda)$ whose coefficients are bounded by $(N!)^{-1}$ for large N, we can see that $|[N, N-1] - g(\lambda)|$ can be made as small as we please for all but a vanishing region of any given bounded area of the complex plane. This class of fuctions includes many that are not series of Stieltjes. It should perhaps be stressed that, although the result is extremely weak, it does not require any assumptions about the behavior of the Padé approximants, an advantage over the theorems quoted by Baker.

In our discussion, we have made no attempt to obtain the strongest result that might be achieved by this line of argument. We have been unable to determine $A_{\max}$, but a plausible hypothesis is that $A_{\max} = \pi x^2$ (obtained when all λ are equal). If this were correct, it would probably mean that the result applied to all or nearly all entire functions. We might speculate further that a similar theorem will hold for a substantial class of functions that are the ratio of two entire functions (i.e. meromorphic).

Note added in proof: All the speculations in the last paragraph have now been proved. $A_{\max}$ is indeed πx^2, and the author has shown that $[N, N+J]$ Padé approximants to any meromorphic function converge in measure [*J. Math. Anal. Appl.* (to be published)].

REFERENCES

Baker, G. A., Jr. (1965). "Advances in Theoretical Physics" (K. A. Brueckner, ed.), Vol. 1, pp. 1–58. Academic Press, New York.

Bessis, D., and Pusterla, M. (1968). *Nuovo Cimento* **54**, 243.

Cini, M., and Fubini, S. (1953) *Nuovo Cimento* **10,** 1695.
Cini, M., and Fubini, S. (1954). *Nuovo Cimento* **11,** 142.
Common, A. K. (1968). *J. Math. Phys.* **9,** 32.
Halmos, P. R. (1950). "Measure Theory" Van Nostrand, Princeton, New Jersey.
Levine, M. J., Wright, J., and Tjon, J. A. (1967). *Phys. Rev.* **154,** 1433.
Moiseiwitsch, B. L. (1966). "Variational Principles." Wiley (Interscience), New York.
Nuttall, J. (1966). *Phys. Letters* **23,** 492.
Nuttall, J. (1967). *Phys. Rev.* **157,** 1312.
Nuttall, J. (1969). *Ann. Phys.* (N. Y.) **52,** 428.
Schwartz, C. (1961). *Ann. Phys.* (N. Y.) **16,** 36.
Schwartz, C., and Zemach, C. (1966). *Phys. Rev.* **141,** 1454.

CHAPTER 9 APPROXIMATE N/D SOLUTIONS USING PADÉ APPROXIMANTS*

D. Masson

University of Toronto, Canada

I. INTRODUCTION

The analytic properties of the partial-wave scattering amplitude can be used to perform dynamical calculations for relativistic models of scattering. One obtains a partial-wave integral equation that can be solved using the N/D method of Chew and Mandelstam (1960).

The N and D equations take the form of a Fredholm integral equation, which can be solved by standard techniques provided that the kernel is in the Schmidt class. Here we demonstrate an alternative to the N/D method (Masson, 1967). It involves constructing a formal power-series solution by means of iteration and then applying the Padé approximant to the power series. The result is an approximate amplitude that satisfies exact two-body unitarity. When the modified kernel of the integral equation for N is in the Schmidt class, we are able to show that diagonal sequences of Padé approximants converge to the N/D solution. The essential feature is that for a range of energy the scattering amplitude is related to a series of Stieltjes in the parameter that characterizes the strength of its left-hand singularity.

II. METHOD

To demonstrate the method, we consider the simple model of elastic s-wave scattering of two spinless particles of equal mass μ. The scattering

* Supported by the National Research Council of Canada.

amplitude $T(\lambda, s)$ is assumed to be a real analytic function of the square of the total center-of-mass energy s, cut along the real axis, which satisfies the dispersion relation

$$T(\lambda, s) = \lambda B(s) + \sum_n \frac{R_n(\lambda)}{s - s_n(\lambda)} + \frac{1}{\pi} \int_{s_R}^{\infty} \frac{\rho(s') \, |T(\lambda, s')|^2 \, ds'}{s' - s}, \tag{1}$$

where $s_R = 4\mu^2$, $\rho(s) = [(s - s_R)/s]^{1/2}$, and

$$B(s) = \frac{1}{\pi} \int_{-\infty}^{s_L} \frac{\sigma(s') \, ds'}{s' - s}, \tag{2}$$

with $s_L < s_R$.

In Eq. (1), we have made an explicit separation of the left-hand singularities into a given "force" term $\lambda B(s)$ and bound state terms $R_n(\lambda)/\{s - s_n(\lambda)\}$, with $s_n(\lambda) \leq s_R$. The force term is real for $s > s_L$, and λ is a real parameter that characterizes the strength of the force. The physical amplitude is given for $s > s_R$ by

$$T_+(\lambda, s) = e^{i\delta(s)} \sin \delta(s)/\rho(s),$$

where

$$T_+(\lambda, s) = \lim_{\varepsilon \to 0+} T(\lambda, s + i\varepsilon).$$

If we assume that the bound state terms are not present for a sufficiently weak force term, then for $|\lambda|$ less than some critical value λ_c one has

$$T(\lambda, s) = \lambda B(s) + \frac{1}{\pi} \int_{s_R}^{\infty} \frac{\rho(s') \, |T(\lambda, s')|^2 \, ds'}{s' - s}, \tag{3}$$

which we may treat as a nonlinear integral equation for $T(\lambda, s)$.

One can obtain a formal solution to Eq. (3) by iteration:

$$T(\lambda, s) = \sum_{i=1}^{\infty} t_i(s) \lambda^i, \tag{4}$$

with

$$t_1 = B(s),$$

$$t_2 = \frac{1}{\pi} \int_{s_R}^{\infty} \frac{\rho(s') B^2(s') \, ds'}{s' - s},$$

$$\vdots,$$

where, in order to have $t_i(s)$ well defined, it is sufficient to have

$$\lim_{s \to \infty} |B(s)| \ln s < \infty.$$

The power series (4) is, of course, only a useful representation of $T(\lambda, s)$ inside its circle of convergence and must diverge when there are bound states present. In order analytically to continue (4) outside its domain of convergence, one can form its $[N, M]$ Padé approximant (Baker, 1965)

$$[T(\lambda, s)]_{N,M} = P_{N,M}(\lambda, s)/Q_{N,M}(\lambda, s)\,, \tag{5}$$

where $P_{N,M}$ and $Q_{N,M}$ are polynomials in λ of degree M and N, respectively. Since one is now allowing for singularities in λ due to the vanishing of the denominator $Q_{N,M}$, the expression (5) can be a good approximation outside the circle of convergence of (4). The bound state energies $s_n(\lambda)$ in Eq. (1) are thus being approximated by zeros of $Q_{N,M}(\lambda, s)$.

The two features of the approximants (5) which we wish to stress are as follows:

(1) The $[N, M]$ approximant satisfies exact unitarity for $N \geq M \geq 1$.
(2) In the limit as $N, M \to \infty$, the $[N, M]$ approximant converges for $s_L < s < s_R$.

These statements are proved in the next section.

III. PROOF OF UNITARITY AND CONVERGENCE

One of the useful features of the N/D method is that the unitarity requirement

$$\operatorname{Im} T_+(\lambda, s) = \rho(s)\,|T(\lambda, s)|^2\,, \qquad s > s_R\,, \tag{6}$$

is automatically satisfied even for approximate N and D. This feature is true also of the Padé approximant method.

THEOREM 1. If λ is real, then $[T(\lambda, s)]_{N,M}$ satisfies exact unitarity for $N \geq M \geq 1$.

Proof. The unitarity requirement can be expressed as $\operatorname{Im}(\lambda/T_+(\lambda, s)) = -\lambda\rho(s)$ for $s > s_R$. Thus, $f(\lambda) = \lambda/T_+(\lambda, s) + i\lambda\rho(s)$ has a power series expansion with real coefficients, and $\operatorname{Im}[f(\lambda)]_{M-1,N} = 0$. Now, if $N \geq M$, one has $[f(\lambda)]_{M-1,N} = \lambda/[T_+(\lambda, s)]_{N,M} + i\lambda\rho(s)$, since both sides are ratios of polynomials of degree N in the numerator and $M - 1$ in the denominator and have the same power-series expansions up to and including the term λ^{N+M-1} (the uniqueness property of the Padé approximant). Therefore, $\operatorname{Im}(\lambda/[T_+(\lambda, s)]_{N,M}) = -\lambda\rho(s)$ for $N \geq M \geq 1$.

In order to prove the convergence of the sequence of Padé approximants, we use a modified version of the N/D method due to Ball (1965). A solution to Eq. (3) can be written as $T = N/D$ with

$$N(s) = \lambda B(s) + \lambda \int_{s_R}^{\infty} ds'\, K(s, s'; s_0) N(s') \tag{7}$$

and

$$D(s) = 1 - \frac{(s - s_0)}{\pi} \int_{s_R}^{\infty} \frac{N(s')\rho(s')\,ds'}{(s' - s)(s' - s_0)}, \tag{8}$$

where

$$K(s, s'; s_0) = \frac{1}{\pi} \frac{B(s')(s' - s_0) - (s - s_0)B(s)}{s' - s} \frac{\rho(s')}{s' - s_0}, \tag{9}$$

and s_0 is an arbitrary point less than s_R. As it stands, the kernel K is not symmetric, however, one can consider the modified kernel

$$k(s, s'; s_0) = K(s, s'; s_0) \left(\frac{\rho(s)}{\rho(s')} \frac{(s' - s_0)}{(s - s_0)} \right)^{1/2}, \tag{10}$$

which satisfies

$$k^*(s, s'; s) = k(s', s; s_0) = k(s, s'; s_0) \tag{11}$$

for $s, s' > s_0$.

If

$$\int_{s_R}^{\infty} \int_{s_R}^{\infty} |k(s, s'; s_0)|^2 \, ds\, ds' < \infty, \tag{12}$$

then $k(s, s'; s_0)$ is in the Schmidt class and is a completely continuous, symmetric kernel (Riesz and Sz.-Nagy, 1955). It has a countable set of real eigenvalues $\lambda_i(s_0)$ and corresponding orthonormal eigenfunctions $\phi_i(s; s_0)$ with (Smirnov, 1964)

$$\phi_i(s; s_0) = \lambda_i(s_0) \int_{s_R}^{\infty} k(s, s'; s_0)\phi_i(s'; s_0)\, ds', \tag{13}$$

where for convenience we have taken the eigenvalues to be nondegenerate.

For the moment, we shall consider the eigenvalues to be finite in number. One then has

$$k(s, s'; s_0) = \sum_i \phi_i(s; s_0)\phi_i(s'; s_0)/\lambda_i(s_0) \tag{14}$$

or, in terms of the original kernel,

$$K(s, s'; s_0) = \sum_i v_i(s; s_0)u_i(s'; s_0)/\lambda_i(s_0), \tag{15}$$

where

$$u_i(s; s_0) = \phi_i(s; s_0) \left(\frac{\rho(s)}{s - s_0} \right)^{1/2}, \tag{16}$$

$$v_i(s; s_0) = \phi_i(s; s_0) \left(\frac{\rho(s)}{s - s_0} \right)^{-1/2}, \tag{17}$$

and

$$v_i(s; s_0) = \lambda_i(s_0) \int_{s_R}^{\infty} K(s, s'; s_0) v_i(s'; s_0) \, ds' . \tag{18}$$

For $\lambda \neq \lambda_i(s_0)$, one has a unique solution to Eq. (7):

$$N(s) = \lambda B(s) + \lambda^2 \sum_i \frac{B_i(s_0) v_i(s; s_0) / \lambda_i(s_0)}{1 - \lambda/\lambda_i(s_0)} , \tag{19}$$

where

$$B_i(s_0) = \int_{s_R}^{\infty} ds' \rho(s') B(s') v_i(s'; s_0) / (s' - s_0) . \tag{20}$$

One should also note the relations

$$K(s, s'; s) = \pi^{-1} B(s') \rho(s') / (s' - s) \tag{21}$$

and

$$B_i(s_0) = \pi v_i(s_0; s_0) / \lambda_i(s_0) , \tag{22}$$

which follow from Eq. (9), (18), and (20). From Eqs. (15), (16), (17), (21), and (22), one has

$$B(s) = \sum_i v_i(s; s_0) B_i(s_0) . \tag{23}$$

Using this in expression (19), one has

$$N(s) = \lambda \sum_i \frac{B_i(s_0) v_i(s; s_0)}{1 - \lambda/\lambda_i(s_0)} . \tag{24}$$

Although Eq. (24) was derived for $s_0 < s_R$ and s in $[s_R, \infty)$, one can continue this expression analytically in the variable s. In particular, one can consider $s_L < s < s_R$ and $s_0 = s$. Using Eq. (22) and the fact that $D(s_0) = 1$, one then gets

$$T(\lambda, s) = \frac{\lambda}{\pi} \sum_i \frac{B_i^2(s) \lambda_i(s)}{1 - \lambda/\lambda_i(s)} , \tag{25}$$

where $\lambda_i(s)$ is real for $s < s_R$ and $B_i(s)$ is real for $s_L < s < s_R$.

When there are an infinity of eigenvalues, one must be careful about the convergence of the various infinite series that appear in Eqs. (14), (15), (19), (23), (24), and (25). In general, the equality in Eqs. (14) and (15) is in the sense of convergence in the mean rather than pointwise. That is,

$$\lim_{n\to\infty} \int_{s_R}^{\infty} \int_{s_R}^{\infty} \left(k(s, s'; s_0) - \sum_{i=1}^{n} \frac{\phi_i(s; s_0) \phi_i(s'; s_0)}{\lambda_i(s_0)} \right)^2 ds \, ds' = 0 , \tag{14'}$$

where we have taken $|\lambda_i| \leq |\lambda_{i+1}|$. If

$$\int_{s_R}^{\infty} \frac{B^2(s')\rho(s')}{s' - s_0} < \infty , \tag{26}$$

then the summation in Eq. (19) is absolutely and uniformly convergent (Smirnov, 1964) for s in $[s_R, \infty)$ and also convergent for $s = s_0$. In addition, one has Eq. (23) in the sense that

$$\lim_{n\to\infty} \int_{s_R}^{\infty} \left(B(s') - \sum_{i=1}^{n} v_i(s'; s_0) B_i(s_0) \right)^2 \frac{\rho(s')}{s' - s_0} \, ds' = 0 . \tag{23'}$$

If Eq. (23) is not pointwise convergent, we can only justify the once subtracted forms of Eqs. (24) and (25). For example, the once subtracted form of Eq. (25),

$$T(\lambda, s) = \lambda B(s) + \frac{\lambda^2}{\pi} \sum_i \frac{B_i^2(s)}{1 - \lambda/\lambda_i(s)} , \tag{27}$$

is valid provided only that conditions (12) and (26) are satisfied. We shall, however, assume that Eq. (25) is also valid. This requires the pointwise convergence of Eq. (23), at least for $s = s_0$.

In proving the convergence of the sequence of Padé approximants, we consider the following three mutually exclusive cases:

(a) There are a finite number of eigenvalues.

(b) There are an infinite number of eigenvalues, all of which have the same sign.

(c) There are an infinite number of eigenvalues, not all of the same sign.

THEOREM 2a. If the kernel K (or k) is of finite rank r, then $[T(\lambda, s)]_{N,M} = N/D$ for $N, M \geq r$.

Proof. If K is of finite rank (Smirnov, 1964) r, then there are r eigenvalues λ_i, and it follows from Eq. (25) that N/D is a ratio of polynomials in λ of degree r. From the uniqueness of the Padé approximant, it follows that $[T(\lambda, s)]_{N,M} \equiv N/D$ for $N, M \geq r$. One should note that K (or k) has finite rank r if and only if $B(s)$ consists of r poles.

THEOREM 2b. If the kernel k is not of finite rank but is semidefinite, (Smirnov, 1964) then

$$\lim_{N\to\infty} [T(\lambda, s)]_{N,N+j} = N/D$$

for $s_L < s < s_R$, $|\lambda| < \infty$ and $\lambda \neq \lambda_i(s)$.

Proof. The kernel is positive (negative) semidefinite if

$$\int_{s_R}^{\infty}\int_{s_R}^{\infty} k(s, s'; s_0) f(s) f(s')\, ds\, ds' \geq 0 \qquad (\leq 0)$$

for arbitrary, real, continuous $f(s)$. From Eqs. (2), (9), and (10), this is seen to be equivalent to the condition

$$\int_{-\infty}^{s_L} dt\, \sigma(t) g^2(t)(t - s_0) \geq 0 \qquad (\leq 0)\,,$$

where

$$g(t) = \int_{s_R}^{\infty} \left(\frac{\rho(s)}{s - s_0}\right)^{1/2} \frac{f(s)}{t - s}\, ds\,.$$

Hence, if $s_0 > s_L$, the kernel is semidefinite if and only if $\sigma(t)$ does not change sign in the interval $(-\infty, s_L)$. This is indeed the situation for many bootstrap models (Ball, 1965). Now, if the kernel is semidefinite, then all the eigenvalues $\lambda_i(s)$ $(s < s_R)$ have the same sign. Without loss of generality, we can assume them to be positive. From Eq. (25), one then has that $T(\lambda, s)/\lambda$ is a series of Stieltjes in λ. That is, if one defines

$$\phi(x, s) = \sum_i \theta(x - 1/\lambda_i(s)) B_i^2(s) \lambda_i(s)/\pi\,,$$

where $\theta(x) = 1$ for $x \geq 1$ and $\theta(x) = 0$ for $x < 1$, then Eq. (25) assumes the form

$$T(\lambda, s)/\lambda = \int_0^{\infty} \frac{d\phi(x, s)}{1 - x\lambda}\,,$$

where this is a Stieltjes integral and $\phi(x, s)$ is a bounded nondecreasing function of x having finite moments

$$t_n(s) = \int_0^{\infty} x^n\, d\phi(x) = \sum_i B_i^2(s)(\lambda_i(s))^{1-n}/\pi\,.$$

From the standard theorems on Padé approximants applied to functions that are series of Stieltjes (Baker, 1965), it follows that, for $|\lambda| < \infty$ and $\lambda \neq \lambda_i(s)$, $[T(\lambda, s)]_{N,N+j}$ converges to $T(\lambda, s)$ as $N \to \infty$.

THEOREM 2c. If the kernel k is not of finite rank and is not semidefinite, then one still has

$$\lim_{N\to\infty} [T(\lambda, s)]_{N,N+j} = N/D \tag{28}$$

for $j = \pm 1, \pm 3, \ldots$, $s_L < s < s_R$, $|\lambda| < \infty$ and $\lambda \neq \lambda_i(s)$.

Proof. From Eq. (27), one has that $f(\lambda, s) = [T(\lambda, s) - \lambda B(s)]/\lambda^2$ is an extended series of Stieltjes. That is,

$$f(\lambda, s) = \int_{-\infty}^{\infty} \frac{d\psi(x)}{1 - x\lambda},$$

where

$$\psi(x) = \sum_i \theta(x - 1/\lambda_i(s)) B_i^2(s)/\pi$$

is a bounded nondecreasing function of x having finite moments. It follows that, for $|\lambda| < \infty$ and $\lambda \neq \lambda_i(s)$, one has (Wall, 1948)

$$\lim_{N\to\infty} [f(\lambda, s)]_{N,N+j} = f(\lambda, s)$$

for $j = \pm 1, \pm 3, \ldots.$ Since

$$[T(\lambda, s)]_{N,N+j} = \lambda B(s) + \lambda^2 [f(\lambda, s)]_{N,N+j-2}$$

for $j \geq 1$, one has that

$$\lim_{N\to\infty} [T(\lambda, s)]_{N,N+j} = T(\lambda, s)$$

for $j = 1, 3, \ldots.$

One also has that the function $g(\lambda, s)$ defined by

$$T(\lambda, s) = \lambda B(s)/(1 - \lambda g(\lambda, s))$$

is an extended series of Stieltjes in λ. This can be seen by using the Hadamard relation (Baker, 1965)

$$D_{1/P}(1, n) = (-1)^{n+1} D_P(1, n)/P_0^{2n+2}, \tag{29}$$

where $P(\lambda)$ is any power series

$$P(\lambda) = P_0 + P_1\lambda + P_2\lambda^2 + \cdots,$$

and

$$D_P(m, n) = \det \begin{vmatrix} P_m & P_{m+1} & \cdots & P_{m+n} \\ P_{m+1} & P_{m+2} & \cdots & P_{m+n+1} \\ & \vdots & & \\ P_{m+n} & & \cdots & P_{m+2n} \end{vmatrix}.$$

Applying Eq. (29) to obtain a relation between the determinants of the coefficients of the functions $g(\lambda, s)$ and $f(\lambda, s)$, one gets

$$D_g(0, n) = D_f(0, n)/(B(s))^{2n+2}.$$

Now a necessary and sufficient condition that a function $f(\lambda)$ be an extended series of Stieltjes is (Wall, 1948)

$$D_f(0, n) > 0, \qquad n = 0, 1, 2, \ldots.$$

Thus it follows that $g(\lambda, s)$ is an extended series of Stieltjes. Since

$$[T(\lambda, s)]_{N,N+j} = \lambda B(s)/(1 - \lambda[g(\lambda, s)]_{N+j-1,N-1})$$

for $j \leq 1$ and since

$$\lim_{N\to\infty} [g(\lambda, s)]_{N+j-1,N-1} = g(\lambda, s)$$

for $j = \pm 1, \pm 3, \ldots$ and for values of λ where $g(\lambda, s)$ is not singular, we also have Eq. (28) for $j = -1, -3, \ldots$.

IV. DISCUSSION

We have shown that, in a dynamical model that is based on analyticity and unitarity, the Padé approximants provide one with a convergent sequence of approximate scattering amplitudes satisfying exact two-body unitarity.

Since the $[N, N]$ Padé approximant would be exact if the left-hand cut consisted of N poles, it is closely related to an approximation in which the kernel (9) is replaced by a sum of N separable kernels. The [1, 1] approximant

$$[T(\lambda, s)]_{1,1} = \lambda B(s)\left(1 - \frac{\lambda}{\pi} B^{-1}(s) \int_{s_R}^{\infty} \frac{B^2(s')\rho(s')\, ds'}{(s' - s)}\right)^{-1} \tag{30}$$

has a particularly simple form and has been derived by means of an different approximation by Shaw (1964) and Vasavada (1968). The approximation (30) has also been found to be numerically superior to the standard determinantal approximation (Reinfelds and Smith, 1966).

The interest in the use of Padé approximants is, however, not just in providing better or alternative methods of calculation for models that are already solvable. There is the hope that one may be able to apply the method to perturbation expansions in quantum field theory or relativistic S-matrix theory. Although the results illustrated here can be generalized to include higher angular momentum and many-channel scattering, a more difficult and outstanding task is to extend the usefulness of the Padé approximant method to problems that involve many-body unitarity (Gammel and McDonald, 1966).

REFERENCES

Baker, G. A., Jr. (1965). "Advances in Theoretical Physics" (K. A. Brueckner, ed.), Vol. 1, pp. 1–58. Academic Press, New York.

Ball, J. S. (1965), *Phys. Rev.* **137,** B1573.

Chew, G. F., and Mandelstam, S. (1960). *Phys. Rev.* **119,** 467.

Gammel, J. L., and McDonald, F. A. (1966). *Phys. Rev.* **142,** 1245.
Masson, D. (1967). *J. Math. Phys.* **8,** 512.
Reinfelds, J., and Smith, J. (1966). *Phys. Rev.* **146,** 1091.
Riesz, F., and Sz.-Nagy, B. (1955). "Functional Analysis." Ungar, New York.
Shaw, G. L. (1964). *Phys. Rev. Letters* **12,** 345.
Smirnov, V. I. (1964). "A Course of High Mathematics," Vol. 4. Pergamon Press, Oxford.
Vasavada, K. V. (1968). *Phys. Rev.* **165,** 1830.
Wall, H. S. (1948). "Analytic Theory of Continued Fractions," Chapters XVII and XX. Van Nostrand, Princeton, New Jersey.

CHAPTER 10 THE SOLUTION OF THE N/D EQUATIONS USING THE PADÉ APPROXIMANT METHOD

*A. K. Common**

University of Kent at Canterbury, England

I. INTRODUCTION

We give here an account of our work on the application of the Padé approximant method to the solution of N/D equations, contained in a previous article (Common, 1967), and of later work on the same subject by Sweig (1968). In the last section, we shall extend this work to cover a class of scattering amplitudes not considered previously.

The N/D equations were introduced by Chew and Mandelstam (1960) to give a construction for the partial wave scattering amplitudes which automatically satisfies the unitarity condition. The proof of the existence of the N/D decomposition has been established for a large number of amplitudes and has been applied to the calculation of the low-energy parameters for $\pi-\pi$ and $\pi-N$ scattering, as well as providing a framework for the "bootstrap" procedure.

The usual problem one has to solve is to find the partial wave amplitude, given its unphysical singularities and inelasticity. With the given information, one can construct integral equations for N and D. In general, solutions for N or D cannot be obtained in closed form, and so various means of approximate solution have been suggested. For instance, it has been suggested by Martin (1964) that the unphysical singularities be replaced by poles, and then the kernel of the integral equation for N becomes degenerate so that the equation can be solved. An alternative suggestion was made by Pagels (1965) that the spectral integral of the kinematical

* Present address: CERN, Geneva, Switzerland.

factor over the physical region be replaced by a number of pole terms. The kernel of the integral equation for D becomes degenerate, and so it can be solved as in the previous case.

The use of Padé approximants to obtain such pole approximations was suggested by Baker (1965) and Bander (1964). The Padé approximants give sequences of approximate solutions for N or D, and the interesting mathematical problem is whether any such sequences converge to the exact solutions. The answer is that one can find such sequences for a large class of scattering amplitudes, and so the Padé approximant method is very useful for the study of the N/D equations. The foregoing class of amplitudes is shown here for the first time to include those whose imaginary part, on the unphysical cut, goes to a constant at infinity and in particular to include the physically important "ρ bootstrap" amplitude.

II. THE INTEGRAL EQUATIONS FOR N AND D AND THEIR APPROXIMATE SOLUTION

In this section, we show how the integral equations for N or D are obtained and give the corresponding approximate solutions. We consider the elastic scattering of two equal scalar particles, so that the singularities of the partial wave amplitude $f_l(s)$ corresponding to angular momentum l are on the real axis, where s is the four-momentum squared.

The unitarity condition is

$$\rho_l(s)\,|f_l(s)|^2 = \operatorname{Im} f_l(s) \leq 1\,, \qquad \infty > s \geq 4\,, \tag{1}$$

where the kinematical factor $\rho_l(s) = [(s-4)/s]^{1/2}$. It can be proved, using rather weak assumptions about the asymptotic behavior of $f_l(s)$ in the complex direction (Atkinson, 1966), that (1) implies

$$\lim_{s\to-\infty} |\operatorname{Im} f_l(s)| \leq 1\,. \tag{2}$$

We consider two classes of amplitudes satisfying (2) defined by the following conditions:

(a) $|\operatorname{Im} f_l(s)| < C\,|s|^{-\mu}$ for some $\mu > 0$, when $-\infty < s \leq -s_1$;
(b) $\lim_{s\to-\infty} |\operatorname{Im} f_l(s)| = \lambda \leq 1$, where $-s_1$ is the beginning of the lefthand cut.

In the remainder of this section and in the next section, we consider amplitudes in class (a), and in the final section deal with amplitudes in class (b).

We assume for simplicity that there are no bound states of the two particles in the scattering channel and that there are no CDD poles. The amplitude may then be written as

$$f_l(s) = N_l(s)/D_l(s)\,, \tag{3}$$

where $N_l(s)$ has the left-hand cut and $D_l(s)$ the right-hand cut.

If the scattering is completely elastic, $N_l(s)$ and $D_l(s)$ satisfy the relations

$$D_l(s) = 1 - \frac{(s - s_0)}{\pi} \int_4^\infty \frac{\rho_l(s') N_l(s')\, ds'}{(s' - s_0)(s' - s)}\,, \tag{4}$$

$$N_l(s) = \frac{1}{\pi} \int_{-\infty}^{-s_1} \frac{\operatorname{Im} f_l(s') D_l(s')\, ds'}{(s - s)}\,, \tag{5}$$

where $D_l(s)$ has been normalized to unity at the subtraction point $s = s_0$. If $l > 0$, there are extra conditions on N_l and D_l to ensure that the amplitude $f_l(s)$ has the correct behaviour as $s \to 4$. To avoid this, one can, instead of (3), make the decomposition

$$f_l(s) = s^l N_l^0(s)/D_l^0(s)\,. \tag{6}$$

We shall not consider such forms for $f_l(s)$, but note in passing that our results can be extended to this case.

Returning to (5) and (6) and eliminating $N_l(s)$ or $D_l(s)$ in turn, we get

$$\frac{D(s)}{s} = \frac{1}{s} + \frac{1}{\pi} \int_{-\infty}^{0} \frac{x[F(x) - F(s)] \operatorname{Im} f(x)}{(x - s)} \left[\frac{D(x)}{x}\right] dx\,, \tag{7}$$

and

$$N(s) = B(s) + \frac{1}{\pi} \int_4^\infty \frac{\rho(x) N(x)}{x(x - s)} [xB(x) - sB(s)]\, dx\,, \tag{8}$$

where

$$F(s) = \frac{s}{\pi} \int_4^\infty \frac{\rho(x)\, dx}{x^2(x - s)}\,, \tag{9}$$

$$B(s) = \frac{1}{\pi} \int_{-\infty}^{0} \frac{\operatorname{Im} f(x)\, dx}{(x - s)}\,. \tag{10}$$

(In the preceding equations and for the remainder of this work, we take $s_0 = 0 = s_1$ and drop the suffix l.)

When the imaginary part of the amplitude is given on the left-hand cut, (7) and (8) are integral equations for $D(s)/s$ and $N(s)$, respectively, and, if they can be solved, the amplitude can be obtained in the whole cut plane.

In general, the integral equation cannot be solved in closed form, and approximate methods have to be used. Pagels (1965) suggested that (7) be solved by approximating its kernel with a degenerate kernel. He made the approximation

$$F(s) = \sum_{r=1}^{N} \frac{c_r}{(s - a_r)}, \tag{11}$$

c_r and a_r being real constants. The singularities of the approximant (11) lie on the right-hand cut if $a_r \geq 4$ for $r = 1, \ldots, N$, as do those of the exact function $F(s)$.

The kernel of (7) is then approximated by

$$\begin{aligned} k_N(s, x) &= \frac{x}{\pi(x - s)} \sum_{r=1}^{N} c_r \left[\frac{1}{(x - a_r)} - \frac{1}{(s - a_r)}\right] \operatorname{Im} f(x) \\ &= -\frac{x}{\pi} \sum_{r=1}^{N} \frac{c_r}{(x - a_r)(s - a_r)} \operatorname{Im} f(x) \end{aligned} \tag{12}$$

and so is degenerate. If $D_A(s)$ is the corresponding approximate solution for $D(s)$, then from (7) for s on the left-hand cut

$$D_A(s) = 1 - s \sum_{r=1}^{N} \left[\frac{c_r}{s - a_r}\right] N_A(a_r), \tag{13}$$

where $N_A(s)$ is the corresponding approximate solution for $N(s)$. Substituting for $D_A(s)$ into (5),

$$N_A(s) = B(s) - \sum_{r=1}^{N} \frac{c_r[sB(s) - a_r B(a_r)] N_A(a_r)}{(s - a_r)}. \tag{14}$$

Finally, substituting in (4),

$$D_A(s) = 1 + s \sum_{r=1}^{N} \frac{c_r N_A(a_r)}{(s - a_r)} + s\left[F(s) - \sum_{r=1}^{N} \frac{c_r}{(s - a_r)}\right]. \tag{15}$$

The values of $N_A(s)$ for $s = a_r$ $(r = 1, 2, \ldots, N)$ are obtained by setting $s = a_1, s = a_2, \ldots, s = a_N$ in turn in (14) and then solving the resulting N linear equations. Therefore, an approximate solution can be obtained for the scattering amplitude given its imaginary part on the left-hand cut.

In an exactly analogous way, (8) may be solved by making the approximation

$$B(s) = \sum_{r=1}^{N} \frac{c'_r}{(s - a'_r)} \tag{16}$$

for s on the right-hand cut, where the poles of the approximant lie on the left-hand cut. This method was suggested by Martin (1964) and gives the following approximate solutions for N and D:

$$D_A(s) = 1 + s \sum_{r=1}^{N} \frac{c_r' D_A(a_r')}{(s - a_r')} [F(a_r') - F(s)] , \tag{17}$$

$$N_A(s) = B(s) D_A(s) - \sum_{r=1}^{N} \frac{c_r'}{(s - a_r')} [D_A(s) - D_A(a_r')] . \tag{18}$$

The constants $D_A(a_r')$ are obtained from (17) in the same way as the $N_A(a_r)$ were obtained from (14).

III. CONVERGENT SEQUENCES OF APPROXIMATE SOLUTIONS TO THE N/D EQUATIONS

We shall show in this section that the Padé approximant method gives convergent sequences of approximations to $F(s)$ and $B(s)$ of the form (11) and (16), respectively, and that, for amplitudes in class (a) (see Sect. II), the corresponding approximant solutions for N and D converge to the exact solutions.

First of all, we consider Eq. (7) and its approximate solutions (13)–(15), and obtain the approximation (11). We write

$$F(s) = \frac{s}{\pi} \int_4^\infty \frac{\rho(x)\, dx}{x^2(x - s)} = -\frac{1}{\pi} \int_4^\infty \frac{\rho(x)\, dx}{x^2} + \frac{1}{\pi} \int_4^\infty \frac{\rho(x)\, dx}{(x - s) x} . \tag{19}$$

Making the substitutions

$$s = 4\omega/(1 + \omega) , \qquad x = 4/(1 - y) , \tag{20}$$

and dropping the constant term that does not alter the integral equation (7), we find that

$$F(s) = (1 + \omega) G(\omega) ,$$

where

$$G(\omega) = \int_0^1 \frac{\rho[4/(1 - y)]\, dy}{1 + \omega y} . \tag{21}$$

The function $G(\omega)$ is a series of Stieltjes in ω, since $\rho > 0$, with unit radius of convergence. We have the well-known result that the $[N; N + j]$ Padé approximants to $G(\omega)$ converge to the function as $N \to \infty$ for ω in the whole complex plane cut from -1 to $-\infty$. In particular, we have convergence for $-1 < \omega \leq 0$, which corresponds to $-\infty < s \leq 0$. The approximants for $j = -1$ may be written in the form

$$[N, N - 1] = \sum_{r=1}^{N} \frac{\delta_r}{\alpha_r + \omega} , \tag{22}$$

where $\delta_r > 0$ and $\alpha_r > 1$, since the poles of the approximant have to lie on the cut of the function $G(\omega)$ and the residues are positive. The corresponding approximant for $F(s)$ is

$$F(s) = (1 + \omega) \sum_{r=1}^{N} \frac{\delta_r}{(\alpha_r + \omega)} = \sum_{r=1}^{N} \frac{c_r}{(s - a_r)} = F_N(s) , \tag{23}$$

where

$$c_r = 4\delta_r/(1 - \alpha_r) < 0 , \qquad a_r = -4\alpha_r/(1 - \alpha_r) > 4 , \tag{24}$$

and so is of the required form (11). The transformations (20) are not the only ones that give approximants to $F(s)$ of the form (11) from Padé approximants to related series of Stieltjes, and a detailed discussion on this point has been given by Sweig (1968). It should be noticed that the constants δ_r, α_r and hence c_r and a_r depend on N the order of the approximation.

We have obtained a sequence of approximations to $F(s)$ which converge for $s < 0$, and want now to prove that the approximate solutions for D converge to the exact solution. Previously we did this by considering the Fredholm expansions of the solution of the approximate and exact integral equation for D (Common, 1967). It is probably more straightforward to use the concept of relatively uniform convergence as suggested by Sweig (1968). We give the definition of relatively uniform convergence and a related theorem that we need for the proof of convergence of our solutions (see, for example, Smithies, 1958):

DEFINITION. A sequence $\{k_N(s, x)\}$ of L^2 kernels is said to be relatively uniformly convergent to $k(s, x)$ if there exists an L^2 kernel $h(s, x)$ such that, given $\varepsilon > 0$, there is an integer $M(\varepsilon)$ for which

$$|k_N(s, x) - k(s, x)| \leq \varepsilon h(s, x)$$

for all $N > M(\varepsilon)$. The limit kernel is again in L^2.

THEOREM. If $\{k_N(s, x)\}$ is a relatively uniformly convergent sequence of degenerate approximations to $k(s, x)$ and if

$$f(s) = g(s) + \lambda \int k(s, x) f(x) \, dx$$

and

$$f_N(s) = g(s) + \lambda \int k_N(s, x) f_N(x) \, dx$$

where $g(s)$ is in L^2, then $f(s) = \lim_{N \to \infty} f_N(s)$ up to a possible addition of a solution of the homogeneous equation.

The foregoing theorem can be applied only to integral equations whose kernels are in L^2, and the integral equations for $N(s)$ and $D(s)/s$ do not have this property for all amplitudes in class (a). However, if we symmetrize

(7) and (8) by writing them, respectively, as

$$\frac{D(s)[\operatorname{Im} f(s)]^{1/2}}{s^{1/2}} = \frac{[\operatorname{Im} f(s)]^{1/2}}{s^{1/2}} + \frac{1}{\pi}\int_{-\infty}^{0} \frac{[sx]^{1/2}[\operatorname{Im} f(x) \operatorname{Im} f(s)]^{1/2}[F(x) - F(s)]}{(x - s)} \times \left\{\frac{D(x)[\operatorname{Im} f(x)]^{1/2}}{x^{1/2}}\right\} dx\,, \tag{25}$$

$$\frac{N(s)}{s^{1/2}} = \frac{B(s)}{s^{1/2}} + \frac{1}{\pi}\int_{4}^{\infty} \frac{\rho(x)}{(sx)^{1/2}(x - s)} [xB(x) - sB(s)] \times \frac{N(x)}{x^{1/2}}\, dx\,, \tag{26}$$

then these are integral equations with L^2 kernels for all amplitudes in class (a), as required.

Let $k(s, x)$ be the kernel of (25), and let $k_N(s, x)$ be the approximate kernel when $F(s)$ is replaced by the approximation (23). Then

$$\begin{aligned} k(s, x) - k_N(s, x) &= [sx \operatorname{Im} f(x) \operatorname{Im} f(s)]^{1/2} \frac{[F(x) - F_N(x) - F(s) + F_N(s)]}{(x - s)} \\ &= \frac{4}{(4 - x)(4 - s)} [sx \operatorname{Im} f(x) \operatorname{Im} f(s)]^{1/2} \\ &\quad \times \left\{\int_0^1 \frac{\rho[4/(1 - y)](1 - y)\, dy}{(1 + \omega y)(1 + \omega' y)} - \sum_{r=1}^{N} \frac{(1 - \alpha_r)\delta_r}{(1 + \alpha_r\omega)(1 + \alpha_r\omega')}\right\}, \end{aligned} \tag{27}$$

where ω' is related to x in the same way as ω is to s.

The function

$$f(\omega, \omega') = \int_0^1 \frac{(1 - y)\rho[4/(1 - y)]\, dy}{(1 + \omega y)(1 + \omega' y)} \tag{28}$$

appearing in (27) can be written in the general from·

$$f(\omega, \omega') = \int_0^\infty b(\omega, \omega', u)\, d\phi(u)\,, \tag{29}$$

where $d\phi(u)$ is a positive measure. This class of functions is very similar to the class of functions

$$g(\omega) = \int_0^\infty b(\omega, u)\, d\phi(u)\,, \tag{30}$$

whose properties have been discussed by Baker (1967). Many of these properties carry over immediately to functions $f(\omega, \omega')$ defined by (29). We shall state some results, without proof, which are useful in showing that the $k_N(s, x)$ tend relatively uniformly to $k(s, x)$.

Firstly,

$$\lim_{N\to\infty} \sum_{r=1}^{N} \frac{(1-\alpha_r)\,\delta_r}{(1+\alpha_r\omega)(1+\alpha_r\omega')} = f(\omega, \omega') \,, \tag{31}$$

and the convergence is uniform in particular for $0 \leq \omega, \omega' \leq \omega_0 < 1$, where ω_0 is fixed, or, equivalently, for $-s_0 \leq s,\ x \leq 0$, where s_0 depends on ω_0 through (20). Also, the approximants are less than $f(\omega, \omega')$ for these values of ω and ω'. Therefore,

$$0 \leq |k(s, x) - k_N(s, x)| \leq |k(s, x)| \,. \tag{32}$$

Since we are considering amplitudes in class (a),

$$|\mathrm{Im}\, f(s)| < C\,|s|^{-\mu}$$

for some $\mu > 0$ when $-\infty < s \leq s_0 < 0$.

Therefore, there exists $C_1 > 0$ such that

$$|k(s, x)| < \frac{C_1 \log |s|}{(xs)^{1/2+\mu}}$$

when $-\infty < s,\ x \leq s_0 < 0$.

Given $\varepsilon > 0$, there then exists $s_1 < 0$ such that, for all $s < s_1$ or all $x < s_1$,

$$|k(s, x) - k_N(s, x)| \leq |k(s, x)| < \varepsilon[sx]^{-(1/2)-(\mu/2)} \,. \tag{33}$$

Also, from the uniform convergence property, there exists $N_0(\varepsilon)$ such that, for all $s_1 \leq s,\ x \leq 0$,

$$|k(s, x) - k_N(s, x)| \leq \varepsilon\,|(s-1)(x-1)|^{-(1/2)-(\mu/2)} \,. \tag{34}$$

Since (33) is true for all N, $k_N(s, x)$ converges relatively uniformly to the exact kernel $k(s, x)$. The approximate solutions of (25) then converge relatively uniformly to the exact solution by the stated theorem. In fact, it can be proved from (33) and (34) that the convergence is uniform in any finite region of the negative axis. It then follows that

$$N_A(s) = \frac{1}{\pi} \int_{-\infty}^{0} \frac{\mathrm{Im}\, f(x) D_A(x)\, dx}{(x-s)} \,, \tag{35}$$

the approximate solutions of $N(s)$, converge in the whole complex s plane cut along the negative axis.

In exactly the same way, it may be shown that the approximate solution (18) of the integral eq. (26) converges uniformly to the exact solution $N(s)$

and that $D_A(s)$, given by (17), converges to $D(s)$, when Padé approximants are used for $B(s)$. The only difference with the previous case is that $B(s)$ is equivalent to the difference of two series of Stieltjes, since Im $f(s)$ may change sign on the left-hand cut. The decomposition of $B(s)$ in this case is

$$B(s) = \frac{(1+\omega)}{2\pi}\int_0^1 \frac{[|\mathrm{Im}\, f\{-1/(1-y)\}| - \mathrm{Im}\, f\{-1/(1-y)\}]\, dy}{(1+\omega y)} - \frac{(1+\omega)}{2\pi}\int_0^1 \frac{[|\mathrm{Im}\, f\{-1/(1-y)\}| + \mathrm{Im}\, f\{-1/(1-y)\}]\, dy}{(1+\omega y)}, \tag{36}$$

where

$$s = -\omega/(\omega+1)\,.$$

The approximants (16) are then obtained by taking the $[N, N-1]$ Padé approximants to each of the integrals in (36).

We have shown in this section that the Padé method gives approximate solutions of the N/D equations which converge to the exact solution for all amplitudes of class (a). In doing this, we have found not only the convergence properties of Padé approximants to series of Stieltjies and generalizations useful, but also the bounding properties as well. This will also be the case in the next section, when we consider amplitudes in class (b).

A particular example of a partial wave amplitude in class (a) is that for the neutral-scalar-meson bootstrap. Numerical results for this case have been obtained by Sweig (1968), and they show that the approximate solutions converge rapidly. For instance, if the pole approximation (16) for $B(s)$ is obtained from the [2, 1] Padé approximant to the related series of Stieltjes, the mass and coupling constant of the meson obtained from the corresponding approximate amplitude are correct to within an error of several percent.

IV. THE ρ BOOTSTRAP AND RELATED AMPLITUDES

In this section, we consider the N/D equations for amplitudes of class (b). As mentioned previously, this class includes the ρ bootstrap amplitude, when the force is assumed to be given by the exchange of a single ρ meson. We discuss the solution of the integral equation for N but not for D, since the former's solution has proved properties that we shall need in the proof of convergence of our approximate solutions.

The integral equation for $N(s)$ is (Atkinson, 1966)

$$\frac{N(s)}{(s-4)} = B(s) + \frac{1}{\pi}\int_4^\infty \frac{\rho(x)[xB(x) - sB(s)]N(x)\, dx}{(x-s)x}, \tag{37}$$

where in this case

$$B(s) = \frac{1}{\pi}\int_{-\infty}^{0} \frac{\operatorname{Im} f(x)\,dx}{(x-4)(x-s)}\ . \tag{38}$$

For $l = 0$, there is an arbitrary subtraction constant in (37), whereas for $l \geq 2$ there are extra conditions on $N(s)$ to ensure that the threshold behaviour is correct. The denominator function is given by

$$D(s) = 1 - \frac{s}{\pi}\int_{4}^{\infty} \frac{dx\,\rho(x)N(x)}{x(x-s)}\ , \tag{39}$$

where it is again normalized to unity at $s = 0$.

The integral eq. (37) does not have an L^2 kernel when the amplitude is in class (b). However, Atkinson and Contogouris (1965) have shown how it may be transformed to one with an L^2 kernel. We discuss briefly their method and give the properties of the exact solution of (37) which we shall need later.

The assumption is first made that

$$\operatorname{Im} f(s) = \lambda + r(s)\ , \tag{40}$$

where $r(s) = O(s^{-\delta})$, with $\delta > 0$. The integral eq. (37) can then be written as

$$\phi(s) = \phi_0(s) + \frac{\lambda}{\pi}\int_{4}^{\infty} dx\left[\frac{\log(x/s)}{\pi(x-s)} + K(s,x)\right]\phi(x)\ , \tag{41}$$

with $\phi(s) = N(s)/(s-4)$, $\phi_0(s) = B(s)$, and where $K(s, x)$ is an L^2 residual kernel. The preceding authors rewrote (41) as

$$\phi(s) = \psi(s) + \frac{\lambda}{\pi^2}\int_{4}^{\infty} \frac{\log(x/s)}{(x-s)}\,\phi(x)\,dx\ , \tag{42}$$

where

$$\psi(s) = \phi_0(s) + \frac{\lambda}{\pi}\int_{4}^{\infty} K(s,x)\phi(x)\,dx\ . \tag{43}$$

They were able to obtain an explicit form for the resolvent $R(s, x; \lambda)$ of (42) and showed that, when λ is in the complex plane cut from 1 to ∞, there is a unique L^2 solution of (42) which is an analytic function of λ in this cut plane. Arguments were given by these authors for choosing this solution to give the physical amplitude and rejecting all other solutions that are not in L^2 and not analytic at $\lambda = 0$. Lyth (1968) also reasons that the L^2 solution should be taken for the physical amplitude by showing that, if the exact discontinuity of the amplitude were used instead of being approximated by the ρ exchange term, the solution for $N(s)$ would be in L^2.

The final step in Atkinson's and Contogouris' method is to substitute from (43) in the solution of (42), getting

$$\begin{aligned}\phi(s) &= \psi(s) + \lambda \int_4^\infty R(s, x; \lambda)\psi(x)\, dx \\ &= \phi_0(s) + \lambda \int_4^\infty R(s, x; \lambda)\phi_0(x)\, dx \\ &\quad + \frac{\lambda}{\pi}\int_4^\infty \left\{K(s, x) + \lambda \int_4^\infty R(s, y; \lambda) K(y, x)\, dy\right\}\phi(x)\, dx\,. \end{aligned} \tag{44}$$

They showed that the kernel of this integral equation for $\phi(s)$ is in L^2, and it can therefore be solved using standard Fredholm methods.

As in the case of amplitudes of class (a), approximate solutions of (37) or, equivalently, (41) can be obtained by making pole approximations to $B(s)$. We shall show that Padé approximants can once again be used to give convergent sequences of approximations to $B(s)$ and that the corresponding solutions for $N(s)$ converge to the exact solution. In fact, they converge to the suggested "physical" solution i. e., the unique L^2 solution.

We start by writing

$$\begin{aligned}B(s) &= \frac{\lambda}{\pi}\int_{-\infty}^0 \frac{dx}{(x-4)(x-s)} + \frac{1}{\pi}\int_{-\infty}^0 \frac{r(x)\, dx}{(x-4)(x-s)} \\ &= \frac{\lambda}{\pi(s-4)}\log\left(\frac{s}{4}\right) - \frac{1}{2\pi}\int_{-\infty}^0 \frac{[|r(x)| - r(x)]\, dx}{(x-4)(x-s)} \\ &\quad + \frac{1}{2\pi}\int_{-\infty}^0 \frac{[|r(x)| + r(x)]\, dx}{(x-4)(x-s)} \end{aligned} \tag{45}$$

and, as in the previous section, use the Padé approximants to each of the foregoing three terms to give a pole approximation for $B(s)$. Consider the first term on the right-hand side of (45), which gives the kernel of (42), i. e.,

$$\frac{\log(s/4)}{(s-4)} = \int_{-\infty}^0 \frac{dx}{(x-4)(x-s)} = \frac{1}{4}\int_0^1 \frac{dy}{(1+\omega y)}\,, \tag{46}$$

where $\omega = (s-4)/4$. The $[N, N-1]$ Padé approximants to the right-hand side give the following pole approximation:

$$\frac{\log(s/4)}{(s-4)} = \sum_{r=1}^N \frac{\alpha_r}{(1+\beta_r s)}\,, \tag{47}$$

with $\alpha_r > 0$ and $\beta_r > 0$. The kernel $k_s(s, x)$ of (42) is then approximated by $k_N(s, x)$, where

$$k_N(s, x) = \frac{1}{\pi^2}\sum_{r=1}^N \frac{\alpha_r(1+4\beta_r)}{(1+\beta_r x)(1+\beta_r s)}\,. \tag{48}$$

We now show that, if this kernel is used to give approximate solutions of (42) for given $\phi(s)$, the approximate solutions converge to the unique L^2 solution for all λ in the complex plane cut from 1 to ∞. Let

$$\Phi(s) = \phi(s) - \phi_N(s) ,$$

where $\phi_N(s)$ is the approximate solution of (42) corresponding to kernel $k_N(s, x)$ and $\phi(s)$ is the L^2 solution of (42). Then

$$\Phi(s) = \lambda \int_4^\infty [k_s(s, x) - k_N(s, x)]\phi(x)\, dx + \lambda \int_4^\infty k_N(s, x)\Phi(x)\, dx . \tag{49}$$

Multiplying both sides of (49) by $\Phi^*(s)$ and integrating over s,

$$\int_4^\infty |\Phi(s)|^2\, ds = \lambda \int_4^\infty \int_4^\infty \Phi^*(s)[k_s(s, x) - k_N(s, x)]\phi(x)\, dx\, ds + \lambda \int_4^\infty \int_4^\infty \Phi^*(s) \sum_{r=1}^{N} \frac{\alpha_r(1 + 4\beta_r)}{(1 + \beta_r s)(1 + \beta_r x)} \Phi(x)\, dxds . \tag{50}$$

Using Schwartz's inequality,

$$\left| \frac{1}{\lambda} \int_4^\infty |\Phi(s)|^2\, ds - \sum_{r=1}^{N} \alpha_r(1 + 4\beta_r) \left| \int_4^\infty \frac{\Phi(x)\, dx}{(1 + \beta_r x)} \right|^2 \right|^2 \leq \int_4^\infty |\Phi(s)|^2\, ds \int_4^\infty \left| \int_4^\infty [k_s(s, x) - k_N(s, x)]\phi(x)\, dx \right|^2 ds . \tag{51}$$

The foregoing integrals exist since $\phi(s)$ and $\phi_N(s)$ are in L^2.

We shall now prove that the right-hand side of (51) tends to zero as $N \to \infty$. There again exists an inequality similar to (32), i. e.,

$$0 \leq |k_s(s, x) - k_N(s, x)| \leq |k_s(s, x)| . \tag{52}$$

Consider the functions

$$F_N(s) = \int_4^\infty [k_s(s, x) - k_N(s, x)]\phi(x)\, dx . \tag{53}$$

From (52),

$$|[k_s(s, x) - k_N(s, x)]\phi(x)| \leq |k_s(s, x)\phi(x)| . \tag{54}$$

It can be proved from (40) that $|\phi(x)| < C_1|x|^{-(\delta_1+1/2)}$ as $x \to +\infty$, where $\delta_1 > 0$ (Atkinson, 1966), so that

$$\int_4^\infty |k_s(s, x)\phi(x)|\, dx$$

exists. It follows from the "dominated convergence theorem" of Lebesgue [see, for example, Munroe (1953)] that

$$\lim_{N\to\infty} F_N(s) = 0\,. \tag{55}$$

We have that

$$|F_N(s)| \leq \int_4^\infty |k_s(s, x)\phi(x)|\, dx\,, \tag{56}$$

and the right-hand side is in L^2. Hence we can use the dominated convergence theorem again to give

$$\lim_{N\to\infty} \int_4^\infty \left| \int_4^\infty [k_s(s, x) - k_N(s, x)]\phi(x)\, dx \right|^2 ds$$

$$= \lim_{N\to\infty} \int_4^\infty |F_N(s)|^2\, ds = 0\,. \tag{57}$$

Going back to (51), we have proved that, given $\varepsilon > 0$, there exists N_0 such that, for all $N \geq N_0$,

$$\left| \frac{1}{\lambda} \int_4^\infty |\Phi(s)|^2\, ds - \sum_{r=1}^{N} \alpha_r(1 + 4\beta_r) \left| \int_4^\infty \frac{\Phi(x)\, dx}{(1 + \beta_r x)} \right|^2 \right|^2$$

$$\leq \varepsilon \int_4^\infty |\Phi(x)|^2\, dx\,. \tag{58}$$

If $-\infty < \lambda < 0$, it follows immediately that

$$\frac{1}{\lambda^2} \left[\int_4^\infty |\Phi(s)|^2\, ds \right]^2 \leq \varepsilon \int_4^\infty |\Phi(x)|^2\, dx\,,$$

i. e.,

$$\frac{1}{\lambda^2} \int_4^\infty |\Phi(s)|^2\, ds \leq \varepsilon\,. \tag{59}$$

Hence,

$$\lim_{N\to\infty} \int_4^\infty |\Phi(s)|^2\, ds = 0\,. \tag{60}$$

In the same way, if λ has a nonzero imaginary part, (60) again holds. Therefore, for λ in the complex plane cut from 0 to ∞, $\phi_N(s)$ converges to $\phi(s)$ in the mean.

The remaining case to consider is when $1 > \lambda \geq 0$. It has been shown by Lyth (1968) that $k_s(s, x)$ is a bounded operator in L^2, i. e.,

$$\sup_{f\in L^2} \frac{||kf||}{||f||} < \infty\,.$$

where

$$||f|| = \left\{ \int_4^\infty |f(x)|^2\, dx \right\}^{1/2}. \tag{61}$$

Since

$$0 \leq k_N(s, x) \leq k_s(s, x) , \qquad s, x \geq 4 , \tag{62}$$

one can prove that the kernel $k_s(s, x)$ and its approximant $k_N(s, x)$ satisfy certain conditions that allow us to use a theorem proved by Jones and Tiktopoulos (1966). We refer the reader to Appendix 3 of the paper by Lyth to see how this may be done. An immediate consequence of the aforementioned theorem is that, for $|\lambda| < 1$, the approximate solutions $\phi_N(s)$ converge to $\phi(s)$ in the mean. Therefore, we have convergence in the mean of our approximate solutions of (42) to the exact L^2 solution for all λ in the complex plane cut from 1 to ∞.

We consider now the approximate solutions $\phi_N(s)$ of (41) when we make the pole approximation to $B(s)$ by taking the $[N, N-1]$ Padé approximants to the three terms in (45). Let $K_N(s, x)$ be the approximant of $K(s, x)$, the L^2 part of the kernel of (41). It can be proved using the methods of Sect. III that $K_N(s, x)$ converges relatively uniformly to $K(s, x)$ as $N \to \infty$. Let $R(s, x)$ be the resolvent of (42) corresponding to the kernel $k_s(s, x)$ and $R_N(s, x)$ the resolvent corresponding to $k_N(s, x)$. Let ϕ now be the exact solution of (41) and ϕ_N the approximate solution when we make the pole approximation to $B(s)$. Then, in operator notation,

$$\phi = (1 + R)\phi_0 + (1 + R)K\phi \tag{63}$$

and

$$\phi_N = (1 + R_N)\phi_{0,N} + (1 + R_N)K_N\phi_N , \tag{64}$$

where $\phi_{0,N}$ is the approximant of ϕ_0.

We now prove that ϕ_N, given in (64), converges in the mean to ϕ given in (63). Firstly, $\chi_N = (1 + R_N)\phi_{0,N}$ converges in the mean to $\chi = (1 + R)\phi_0$, since they are solutions of

$$\chi_N = \phi_{0,N} + k_N\chi_N \tag{65}$$

and

$$\chi = \phi_0 + k_s\chi . \tag{66}$$

To prove this convergence, the same methods are used as those to prove the convergence of the approximate solution of (42) to the exact solution. The only difference is that here the inhomogeneous term $\phi_{0,N}$ in (65) also depends on N and converges to ϕ_0. The modification of Eq. (49)–(60) to take account of this difference is straightforward when use is made of the property that $|\phi_{0,N}(s)| < |a(s)|$ for all N, where $a(s)$ is an L^2 function independent of N. For the case when $0 \leq \lambda < 1$, the theorem of Jones and Tiktopoulos (1966) can again be used.

To show convergence of ϕ_N to ϕ, we shall use the relation

$$\lim_{N\to\infty} \|(1 + R)K - (1 + R_N)K_N\| = 0 , \tag{67}$$

which we now prove. We write

$$\|(1 + R)K - (1 + R_N)K_N\| \leq \|R(K - K_N)\| + \|(R - R_N)K_N\| . \tag{68}$$

The first term on the right-hand side tends to zero as $N \to \infty$, since K_N tends relatively uniformly to K. RK_N and $R_N K_N$ are, respectively, solutions of the integral equations

$$\Psi_N(x, y) = K_N(x, y) + \int_4^\infty k_s(x, z)\Psi_N(z, y)\, dz \tag{69}$$

and

$$\Phi_N(x, y) = K_N(x, y) + \int_4^\infty k_N(x, z)\Phi_N(z, y)\, dz . \tag{70}$$

We can prove convergence of $\Phi_N - \Psi_N$ to zero in the mean by carrying through the reasoning from (49)–(60) once again. The two differences are that here we have an extra variable to deal with, and again the inhomogeneous term depends on N. The latter difference is dealt with by observing that $|K_N(x, y)| \leq |A(x, y)|$, where $A(x, y)$ is in L^2 and is independent of N. The case where $0 \leq \lambda < 1$ is treated by using a simple modification of the theorem of Jones and Tiktopoulos, to take into account the extra variable. Therefore, $\|(R - R_N)K_N\|$ converges to zero as $N \to \infty$, and hence (67) is true.

Let T be the resolvent corresponding to the L^2 kernel $(1 + R)K$. Then, from (63) and (64),

$$\phi = (1 + T)\chi \tag{71}$$

and

$$\phi_N = (1 + T)\chi_N + (1 + T)\{(1 + R_N)K_N - (1 + R)K\}\phi_N . \tag{72}$$

Equation (72) is an integral equation for ϕ_N whose kernel H_N is such that

$$\lim_{N\to\infty} \|H_N\| = 0 ,$$

from (67). It can be seen from the Neumann series expansion of the corresponding resolvent that $\phi_N - (1 + T)\chi_N$ tends to zero in the mean as $N \to \infty$. But $(1 + T)\chi_N$ tends in the mean to $(1 + T)\chi = \phi$. Therefore, ϕ_N, the approximate solution of (41) when $B(s)$ is replaced by its pole approximation, converges in the mean to the exact solution as $N \to \infty$. Convergence in the mean does not imply pointwise convergence, unless ϕ_N and ϕ are sufficiently well behaved, e. g., if they are continuous. However, even if ϕ_N only converges to ϕ in the mean, the approximate solutions for $D(s)$ converge pointwise. This can be proved by using Schwartz's inequality in (39).

We have shown in this section that for amplitudes in class (b) the Padé method gives approximate solutions of the N/D equations which converge in some sense to an exact solution. In particular, for the ρ bootstrap they converge to the so-called "physical" solution. The approximants for N and D will be similar to N_A and D_A given in (18) and (17), respectively.

It would seem that the approximate solutions we have obtained for the ρ bootstrap equations are of a much simpler form to calculate than the Fredholm solution of (44) with its rather complicated kernel. It would be very interesting to compare the rate of convergence of our solutions with the rate of convergence of the Fredholm series.

Finally, we should like to stress again the usefulness of the bounding properties of the Padé approximants and generalizations in proving convergence of the approximate solutions of the N/D equations.

REFERENCES

Atkinson, D., (1966). *J. Math. Phys.* **7,** 1607.

Atkinson, D., and Contogouris, A., (1965). Nuovo Cimento **39,** 1082 and 1102.

Baker, G. A. Jr. (1965). *Advan. Theoret. Phys.* **1,** 1.

Baker, G. A. Jr., (1967). *Phys. Rev.* **161,** 434.

Bander, M., (1964). *J. Math. Phys.* **5,** 1427.

Chew, G. F., and Mandelstam, S., (1960). *Phys. Rev.* **119,** 467.

Common, A. K., (1967). *J. Math. Phys.* **8,** 1669.

Jones, C. E., and Tiktopoulos, G., (1966). *J. Math. Phys.* **7,** 311. (See Theorem 2 of the Appendix.)

Lyth, D. H., (1968). Nuovo Cimento **53A,** 969.

Martin, A. W., (1964). *Phys. Rev.* **135,** B967.

Munroe, M. E., (1953). "Measure and Integration," Addison-Wesley, Reading, Massochnsetls.

Pagels, H., (1965). *Phys. Rev.* **140,** B1599.

Smithies, F., (1958). "Integral Equations." Cambridge Univ. Press, London and New York.

Sweig, M. J., (1968). *Phys. Rev.* **165,** 1893.

CHAPTER 11 PADÉ APPROXIMANTS AS A COMPUTATIONAL TOOL FOR SOLVING THE SCHRÖDINGER AND BETHE–SALPETER EQUATIONS*

Richard W. Haymaker†

Department of Physics
University of California
Santa Barbara

Leonard Schlessinger‡

Department of Physics
University of Illinois
Urbana

I. INTRODUCTION

The notion of numerical continuation of analytic functions has led us to study novel methods of calculating scattering phase shifts for the Schrödinger and Bethe–Salpeter equations. We propose to solve the dynamical equations at unphysical energies (the bound state region) and subsequently continue the scattering amplitude back to scattering energies.

There are a number of circumstances that make this idea appealing. Most important is the fact that the inherent difficulties of solving the equations are greatly reduced by staying out of the scattering region. Accurate variational methods are readily adaptable to calculation at unphysical energies. The nature of the wave function in these two regions is very different, whereas the scattering amplitude changes smoothly from one region to the other. Padé approximants offer a method of continuing the amplitude back to the physical region in many cases. Any continuation scheme requires a knowledge of the analytic structure of the functions involved. The literature is rich indeed with such information on scatter-

* Work supported in part by the National Science Foundation.

† Present address: Laboratory of Nuclear Studies, Cornell University, Ithaca, New York 14850.

‡ Present address: Department of Physics and Astronomy, University of Iowa, Iowa City, Iowa 52240.

ing amplitudes. Finally, computers make it a simple matter to examine sequences of approximations to estimate errors in both the variational calculation and the continuation.

In this chapter, we should like to make a case for this point of view in the light of attempting difficult phase shift calculations. The two-body Schrödinger equation with Yukawa potentials provides us with the simplest problem to present our ideas and a "numerical laboratory" to study the ideas. Three additional examples are also presented with their individual difficulties: a two-channel Schrödinger equation (Sect. IV), a two-body Bethe–Salpeter equation (Sect. V), and finally the nonrelativistic scattering of a particle off a bound state, i.e., a three body problem (Sect. VI).

The continuation is performed by fitting the scattering amplitude to a Padé form. The input information consists of the values of the function at a set of points rather than the coefficients of a power series about one point. Since this differs from the other chapters of this book, we present (in Sect. III) various methods of determining the Padé coefficients. We illustrate the convergence with some examples of simple functions. Since there is no convergence theory for this method of continuation, we can judge the accuracy of our results for phase shifts only by examining stability of our answers for increasing orders of fitting functions.

This chapter is taken from five papers on this subject (Schlessinger and Schwartz, 1966; Haymaker, 1967, 1968; Schlessinger, 1968a, b). We have tried to unify those techniques and present the basic ideas with a simple example (Sect. II). In addition, we emphasize those techniques which were more successful and which have more promise for future applications.

II. BASIC PHILOSOPHY AND A SIMPLE EXAMPLE

Let us illustrate our method by outlining a sample calculation of the two-body Schrödinger scattering problem with a Yukawa potential. The equation for the transition operator is

$$T(q) = V + V(H_0 - q^2)^{-1}T(q) \,,$$

where q^2 is the center-of-mass energy. Taking matrix elements in plane wave states, we get the T matrix $T(\mathbf{k}', \mathbf{k}; q) = \langle \mathbf{k}' \mid T(q) \mid \mathbf{k} \rangle$. We do not want to impose the on-shell constraint $|\mathbf{k}| = |\mathbf{k}'| = q'$ but we shall define an off-shell continuation. Expanding T in partial waves,

$$T(\mathbf{k}', \mathbf{k}; q) = \sum_l (2l+1) P_l(\hat{k}' \cdot \hat{k}) T_l(k', k, q) \,,$$

and expressing T_l in terms of the wave function in coordinate space gives

$$T_l(k', k; q) = 4\pi \int_0^\infty j_l(k'r)\, V(r)\psi_{qk}^l(r) r^2\, dr\,,$$

where k' and k are the magnitudes of $\mathbf{k}'$ and $\mathbf{k}$. The wave function with scattering boundary conditions satisfies the Schrödinger integral equation

$$\begin{aligned}\psi_{qk}^l(r) &= j_l(kr) - iqh_l^{(1)}(qr) \int_0^r j_l(qr')\, V(r')\psi_{qk}^l(r') r'^2\, dr' \\ &\quad - iqj_l(qr) \int_r^\infty h_l^{(1)}(qr')\, V(r')\psi_{qk}^l(r') r'^2\, dr'\,, \qquad (1)\end{aligned}$$

where j_j and $h_l^{(1)}$ are spherical Bessel functions (Abramowitz and Stegun, 1965, Chapter 10). Our units are chosen such that $\hbar/2m = 1$. It is useful to separate the wave function into two parts:

$$\psi_{q,k}^l(r) = j_l(kr) + \chi_{q,k}^l(r)\,.$$

Finally we need a variational principle for numerical calculations:

$$\begin{aligned}[T_l(k', k; q)] = \int_0^\infty r^2\, dr \{ & j_l(k'r)\, V(r) j_l(kr) + j_l(k'r)\, V(r)\chi_{kq}^l(r) \\ & + \chi_{k'q}^{l*}(r)\, V(r) j_l(kr) - \chi_{k'q}^{l*}(r)[H_l - q^2]\chi_{kq}^l(r)\}\,, \qquad (2)\end{aligned}$$

where the hamiltonian H_l is

$$H_l = -\frac{1}{r}\frac{d^2}{dr^2} r + \frac{l(l+1)}{r^2} + V(r)\,.$$

This expression is stationary with respect to first-order variations of χ about its solution to Eq. (1).

These formulas define the problem to be solved, and we are now in a position to discuss our proposed solution. The values of the variables k and q are judiciously chosen to ensure a nice asymptotic behavior of the wave function for large r. Fix $k = k' = a$ physical scattering momentum, and choose q to be positive pure imaginary, i.e., the bound state region. Figure 1 displays the relevant q plane. The positive real axis is the upper lip of the scattering cut in the q^2 plane. The dominant asymptotic behavior can be read off from the integral equation:

$$\chi_{kq}^l(r) \to O(e^{-(\mathrm{Im}\,q)r}) \quad \text{or} \quad O(V(r))\,. \qquad (3)$$

For a Yukawa potential of range $1/\mu$, $V \sim e^{-\mu r}$. Such a wave function can be well represented by the following expansion in basis functions:

$$\chi_{k,q}^l(r) = \sum_{n=0}^{N} c_n r^{n+l} e^{-\alpha r}\,. \qquad (4)$$

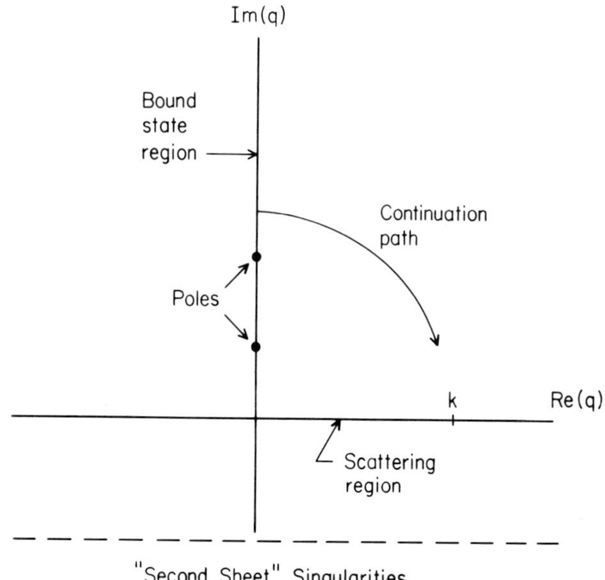

Fig. 1. The complex q plane, showing the region of calculation (the bound-state region) and the continuation path to a scattering energy $q = k$ for the example of Sect. II. $\mathrm{Im}(q) < 0$ is the "second sheet" of the q^2 plane.

TABLE I

Calculation of S-Wave Phase Shifts in a Yukawa Potential $V = -2e^{-r}/r$ for Momentum $k = 0.5$

Input accuracy		Output accuracy		
No. of basis functions	$t_l\ (q^2 < 0)$	Order of Padé	$(\tan \delta/k)$	Coefficient of $-i$
3	−0.642579435	[1, 1]	−10.135518	0.97273549
4	−0.642582974	[2, 2]	−11.132695	0.97840186
5	−0.642583099	[3, 3]	−16.861517	1.0003043
6	−0.642583709	[4, 4]	−16.894883	1.0000144
7	−0.642583752	[5, 5]	−16.894090	0.99999971
8	−0.642583755	[6, 6]	−16.893824	0.99999988
9	−0.642583755	[7, 7]	−16.893827	0.99999999

[a] The first column shows the convergence of the input, unphysical values of the T matrix for $W = -1.4571068$ as the number of trial functions in the variational principle is increased from 3 to 9. The second column shows the convergence of the fitting technique $(\tan \delta/k)$ as the order of the Padé approximant is increased. The third column indicates the convergence of the "unitarity coefficient" described in the text. This number should be equal to 1.

Putting this in the variational principle and treating the $\{c_n\}$ as variational parameters, we get linear equations for the c's and can very easily find T_l to a high accuracy. A sample of convergence is given in the first column of Table I for increasing N. The answer converges independent of α, but a "good" choice of α greatly enhances the convergence rate.

Great! However, we have T_l at peculiar unphysical values of q. To get the scattering amplitude at physical values, we fit $T_l(k, k, q)$ to a Padé form at a set of values of q.

$$T_l(k, k; q) = \frac{\sum_{j=0}^{n} a_j (iq)^j}{1 + \sum_{j=1}^{n} b_j (iq)^j} . \tag{5}$$

The $(2n + 1)$ coefficients a_j, b_j are found by multiplying through the denominator and evaluating the resulting expression at $(2n + 1)$ values of q. This results in linear equations for the coefficients. We continue in the variable q because the amplitude is regular at $q = 0$ in this variable. T_l is real for imaginary q, and thus the expansion coefficients are real. We choose a diagonal Padé form, since the amplitude goes to the first Born term for large q, and this term is independent of q. Finally, we evaluate the right-hand side of (5) at $q = k$. This is related to the phase shift through the formula

$$T_l^{-1} = -\frac{k}{4\pi} (\cot \delta - i) .$$

A sample of good convergence is given in Table I for $(\tan \delta)/k$, and the coefficient of $(-i)$. The latter gives a check on unitarity and thereby an independent measure of accuracy.

Compare this with the traditional calculation holding $k = k' = q =$ a scattering energy. The wave function is oscillatory for large r. The trial functions (4) are no longer adequate to represent the wave function. Functions must be chosen which have a variable phase of the large r oscillation. The calculation of T_l through the variational principle will be considerably more tedious. However, degree of tedium is not the only guiding principle in our work. For the Bethe–Salpeter problem that we present, the "traditional" method breaks down. The wave function is exponentially *growing* in the scattering region, and we see no practical alternative to staying out of the scattering region if we want to keep the "differential equation type" of variational principle (Schwartz and Zemach, 1966).

Our choice of values of k and q as presented previously is not the only choice. A more obvious choice is to stay on shell, i.e., $k = k' = q$, and make this variable pure imaginary. Because of the asymptotic behavior of the wave function [see Eq. (3)], our variational method will be applicable only for a finite range of q. More will be said about this in the section on the Bethe–Salpeter equation.

We would like to point out the close relationship between a bound state eigenvalue calculation and our calculation of T_l in the bound state region. The Rayleigh–Ritz variational principle for bound state energies is

$$\int_0^\infty r^2\,dr\,\chi^l(r)\,(H_l - [q^2])\chi^l(r) = 0\,.$$

Remembering the boundary conditions on bound state wave functions, we see that the expansion of $\chi^l(r)$ given in Eq. (8) gives a good representation of the wave function. Thus, the same matrix elements of $(H_l - q^2)$ are needed for both problems. The calculation of T_l in the bound state region requires some additional matrix elements, i.e., the second and third terms in Eq. (2), but they pose no special problems. Thinking of future more difficult calculation (e.g., the relativistic three-body problem), if a bound state problem is tractable using the Rayleigh–Ritz variational principle, the off-shell scattering amplitude can be found with no particular difficulty. No such statement can be made comparing the bound state problem and the traditional scattering problem, since totally different trial functions are needed for the two problems. For example, the Bethe–Salpeter bound state problem was first solved by Schwartz (1965) using Rayleigh–Ritz methods, but phase shifts were found from the integral equation, (Schwartz and Zemach, 1966) only after many attempts had been made to solve the differential equation.

A valid question at this point is why use variational methods? The Schrödinger equation can be easily integrated numerically, or solved by mesh point methods. Our answer is that variational methods are the only way to solve problems involving larger numbers of dimensions. Mesh point methods will require prohibitive matrix sizes, and the boundary conditions will become disastrous in trying to integrate the equation directly.

The weak link in our method is the continuation of the amplitude back to the scattering region. Although no one would complain about the results in Table I, it is an idealized problem. We are after all, extrapolating a function, and Padé fits seem to be the best way to do this. But if there are singularities near the extrapolated point, the continuation may converge very slowly or not at all. This can be counteracted by increasing the accuracy of the input numbers, but there is a limit to the accuracy that is practical to achieve. Our only test of this method to an individual problem is to do a sequence of fits and examine convergence.

III. RATIONAL–FRACTION FITTING

The scheme we have proposed depends on the possibility of using values of the scattering amplitude in an unphysical region to obtain a fit to that function which may then be analytically continued into the physical region. A rational fraction can provide such a representation, whereas a straightforward method of polynomial continuation generally cannot. The rational-fraction form allows the most important singularities of the amplitude to be accounted for by choice of expansion variable and also permits the asymptotic behavior of the amplitude to be directly incorporated in the representation. The crucial point, however, is that a ratio of polynomials is flexible enough to approximate well the poles and distant, less important, singularities of the amplitude. In this chapter, we discuss several methods that have been devised for obtaining a rational-fraction fit, given the values of a function at a set of points (Schlessinger, 1968a). Lacking a rigorous mathematical understanding of these techniques, the methods are discussed and compared in terms of some simple numerical examples.

A. Obtaining Rational Approximations

Given the value of a function $f(x)$ at a set of K input points $x_1 \cdots x_k$, we can obtain a rational approximation to $f(x)$ in three ways. We have called these techniques pointwise, moment, and norm fitting. Each method leads to a sequence of rational approximations to $f(x)$ of the form

$$f(x) \simeq R_{N,M}(x) = P_N(x)/Q_M(x)\,, \tag{6}$$

where P_N, Q_M are, respectively, Nth- and Mth-order polynomials in x. The techniques differ in that the norm and moment fitting use the values of $f(x)$ only to determine its moments over some range of x and in this way include all the values of f at each stage in the approximation, whereas the pointwise fitting uses the values of f directly in the representation and uses $N+M+1$ values of f to determine $R_{N,M}(x)$. The norm method, however, has not proved useful in numerical calculations, and we do not discuss it here (Schlessinger, 1968a).

One technique for obtaining a pointwise fit has already been described in Sect. II. Another method of generating the same type of representation makes use of continued fractions (Wall, 1948). Consider the continued fraction

$$C_N(x) = \frac{f(x_1)}{1+}\,\frac{(x-x_1)a_1}{1+}\,\cdots\,\frac{(x-x_N)a_N}{1}\,.$$

It is easy to see that $C_N(x_1) = f(x_1)$ and that the coefficients $a_1 \cdots a_N$ may be determined so that $C_N(x)$ is equal to $f(x)$ at the N points $x_1 \cdots x_N$. Requiring

$$C_N(x_{l+1}) = f(x_{l+1}) = \frac{f(x_1)}{1+} \frac{(x_{l+1} - x_1)a_1}{1+} \cdots \frac{(x_{l+1} - x_l)a_l}{1},$$

the a_l's may be shown to be

$$a_l = \frac{1}{(x_l - x_{l+1})} \left\{ 1 + \frac{(x_{l+1} - x_{l-1})a_{l-1}}{1+} \right.$$
$$\left. \times \frac{(x_{l+1} - x_{l-2})a_{l-2}}{1+} \cdots \frac{(x_{l+1} - x_1)a_1}{1 - [f(x_1)/f(x_{l+1})]} \right\}, \tag{7}$$

and

$$a_1 = \{f(x_1)/f(x_2) - 1\}/(x_2 - x_1) . \tag{8}$$

Equations (7) and (8) form a simple and efficient algorithm that may be used recursively to determine a pointwise rational-fraction fit to $f(x)$. This procedure yields a ratio of polynomials of order $((N - 1)/2, (N + 1)/2)$ for N odd and $(N/2, N/2)$ for N even. The procedure must be modified if different asymptotic behavior is desired.

To discuss moment fitting, we return to the notation introduced in Eq. (6), and we let

$$P_N(x) = \sum_{k=0}^{N} p_k u_k(x) ,$$

$$Q_M(x) = 1 + \sum_{k=1}^{M} q_k u_k(x) ,$$

where the $u_i(x)$ are ith-order orthogonal polynomials in x with the weight function $w(x)$ over the interval (a, b) which includes the set of input points x_i, $i = 1 \cdots k$. That is,

$$\int_n^b u_n(x) u_m(x) w(x)\, dx = \delta_{nm} A_n .$$

This method requires knowledge of the moments of $f(x)$ in the interval (a, b) in the basis $u_k(x)$. Accordingly, we assume the x_i are chosen so that integrals of $f(x)$ may be obtained by numerical quadrature, and we define

$$f_{nm} = \int_a^b u_n(x) u_m(x) f(x) w(x)\, dx \approx \sum_{i=1}^{K} u_n(x_i) u_m(x_i) f(x_i)\, W(x_i) ,$$

where $W(x_i)$ is an appropriate weight for the quadrature.

The moment technique determines a fit to $f(x)$ by requiring that the first $N + M + 1$ moments of the expression $f(x) Q_M(x) - P_N(x)$ are zero.

This requirement may be implemented by setting the moments of the following expression to zero.

$$\sum_{k=0}^{N} p_k u_k(x) - f(x) \sum_{k=0}^{M} q_k u_k(x) . \tag{9}$$

Here $q_0, u_0 = 1$. Multiplying (9) by $u(x)w(x)$ and integrating, we obtain

$$p_j = (A_j)^{-1} \sum_{k=0}^{M} f_{jk} q_k , \qquad j = 0 \cdots N, \tag{10}$$

and

$$\sum_{k=1}^{M} f_{jk} q_k = -f_{j0} N_0 , \qquad j = N + 1 \cdots N + M. \tag{11}$$

Equation (11) is a set of M linear equations for the M unknown q's. Using the solutions of (11), (10) then determines the p's. If the u_n are chosen to be $e^{in\theta}$ and the integration region is taken to be the perimeter of the unit circle, then this method (with the appropriate modifications for complex functions) gives the same result as the Padé (Wall, 1948) approximant method for a function that has a convergent power series in the unit circle.

The matrix elements of a function between orthogonal polynomials, e.g., f_{ij}, need not all be calculated from the integral. The standard recurrence relation on the particular set of polynomials chosen implies relations between the matrix elements. For example, if we choose Legendre polynomials, the recurrence relation for f_{ij} is

$$\begin{aligned} i(2j + 1)f_{ij} = {} & (2i - 1)(j + 1)f_{i-1,j+1} \\ & - (2j + 1)(i - 1)f_{i-2,j} + (2i - 1)jf_{i-1,j-1} . \end{aligned}$$

We should also like to point out some rather simple generalizations of the pointwise fitting techniques. Clearly, the first-mentioned method of pointwise fitting is not restricted to using only powers of x. That is, one could represent $f(x)$ as

$$f(x) \sim \frac{n_1 T_1(x) + n_2 T_2(x) + \cdots n_N T_N(x)}{1 + d_1 B_1(x) + d_2 B_2(x) + \cdots d_M B_M(x)} ,$$

where T_i, B_i are arbitrary functions of x. This procedure may be useful in approximately accounting for several different singularities of $f(x)$. We still get linear equations for the coefficients n_i, d_i.

Our fitting techniques are generally used to form a sequence of approximations to the desired function as the order of P_N and Q_M is increased keeping $N - M$ fixed. The moment method has the advantage of using all of the input points at each stage of approximation, but it is more susceptible to numerical errors because of the approximations made in representing the integrals by finite sums and because of the many manipulations of the moments f_{nm}. In our study of some simple examples, we have found that the moment and point fit usually give comparable results. We have not tried any examples of the more general pointwise fits.

We mentioned that, under certain conditions, moment fitting is equivalent to the Padé approximant method. However, the connection between the moment, norm, and various pointwise fitting techniques and their general relation to the Padé scheme is not clear. Perhaps determining these relationships would be a first step toward providing some badly needed mathematical rigor to the theory of pointwise fitting.

B. Comparison of the Methods with Examples

We attempt to establish here some empirical rules for the use of these methods by the consideration of some special examples. These examples were computed by taking the values of some simple functions at points along the positive z axis from $z = 0$ to $z = \infty$, applying the fits discussed previously, and evaluating the resulting rational fraction along the negative z axis as shown in Fig. 2.

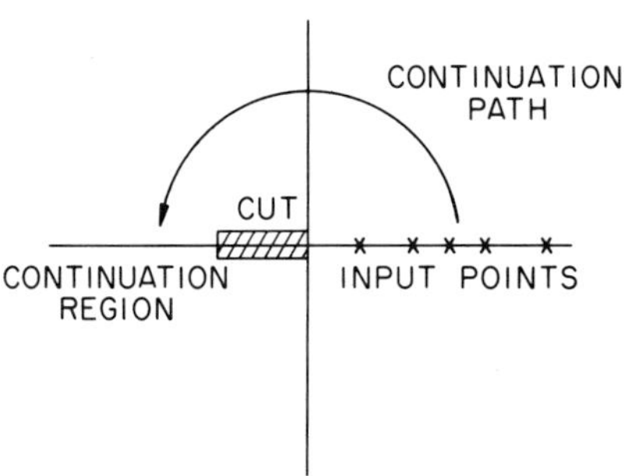

Fig. 2. Location of input points, branch cut, continuation path, and output region for the examples described in Sect. III.

We have found that this type of fit is able to represent a rational fraction exactly and gives a good approximation to a many-sheeted function on one sheet in the complex plane. For the functions studied, the "region of convergence" of the fit appears to be everywhere on one sheet of the complex plane except near the branch line, which is usually represented as a line of poles and zeroes.

Figure 3 is a comparison of the moment and pointwise fits to the function

$$f(z) = \left(\frac{z}{(1 + z)}\right)^{1/2},$$

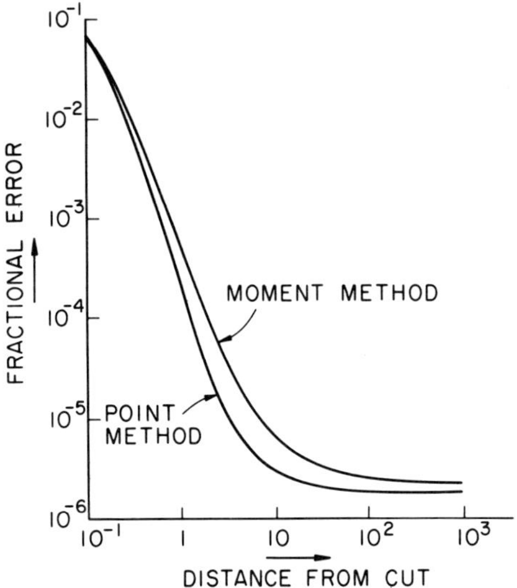

Fig. 3. Fractional error versus distance from $z = -1$ for moment and pointwise fit to $[z/(1+z]^{1/2}$. The order of approximation $N = 5$.

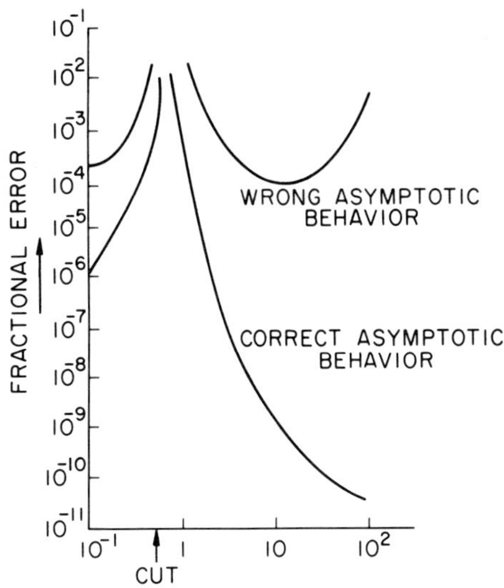

Fig. 4. Fractional error versus distance from $z = 0$ for pointwise fitting to $\{[(1 + z + z^2)(1 + 2z)]^{1/3} - 1\}/z$.

where we have plotted fractional error $[f(z) - R_{NN}(z)]/f(z)$ versus continuation distance from $z = -1$. In all the examples we tried, the pointwise and moment fits give comparable errors and rates of convergence.

We stressed the point that incorporating the correct asymptotic form directly into the rational-fraction fit can lead to more accurate continuation because the global bound imposed on the fitting function inhibits the growth of continuation error characteristic of polynomial expansions. This effect is demonstrated in Fig. 4, which is a graph of the fractional error $[f(z) - R_{NM}(z)]/f(z)$ versus continuation distance from $z = 0$ for a pointwise fit to the function $\{[(1 + z + z^2)(1 + 2z)]^{1/3} - 1\}/z$ for the [5, 5] and the [5, 4] approximations. The fit with the correct asymptotic behavior [5, 5] gives a good representation of the function, whereas the fit with the wrong asymptotic behavior [5, 4] exhibits the polynomial-like behavior at very large values of z.

Generally, the continuation $R_{N,N}$ converges rapidly as a function of N for both the point and moment fits. Figure 5 is an example of the rate of converegence for pointwise fit to the function

$$\left(\frac{1 + z}{2 - z}\right)^{-1/2}.$$

In Fig. 5, we have plotted $[R_{NN}(z) - f(z)]/f(z)$ versus the continuation distrance from $z = -1$ for $N = 5, 8, 10$. In this, as in all examples studied, the accuracy of the continuation is greatest at large distances from the singularity of the function. The preceding examples were calculated using

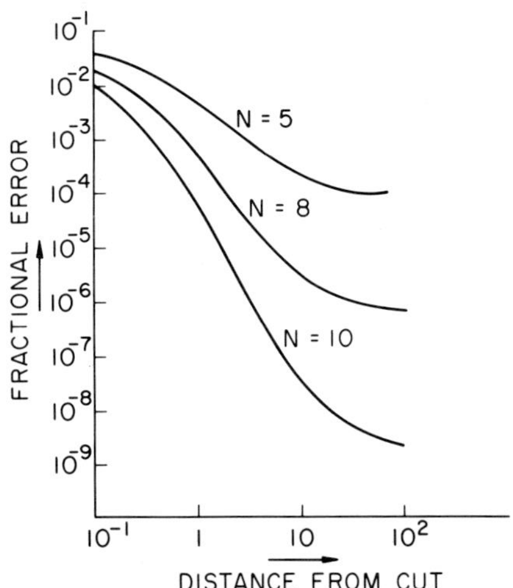

Fig. 5. Fractional error versus distance from $z = -1$ for pointwise fit to $[(1 + z)/(2 - z)]^{-1/2}$. The order of approximation $N = 5, 8, 10$.

input points distributed from $z = 0$ to $z = \infty$ as needed for the evaluation of the moments of the functions according to Laguerre quadrature. This last result suggests, however, that, to obtain an accurate fit to a function at a particular point, $z = a$, one should distribute the points "evenly" over the transformed input region in the plane where the point $z = a$ is infinitely far from the singularities of the function. To get a good fit to a function at $z = a$, we take points evenly distributed in the y plane, where

$$y = z/(z - a) .$$

In the y plane, the singularities are infinitely far from $z = a$ and the input points $z = 0$ to $z = \infty$ are mapped into $y = 0$ to $y = 1$ (for $a < 0$). We have used this method for picking the input points in the nonrelativistic scattering calculations, and it does, in most cases, lead to more accurate continuations. In this respect, our methods differ from the Padé approximant techniques, since the Padé approximants are invariant under such transformations.

The physical example discussed in Sect. II was the calculation of the scattering amplitude for the S-wave Schrödinger equation. For this example, we used a Yukawa potential of unit range strong enough to produce a bound state near $q^2 = 0$. The input points were chosen to be evenly distributed in the y plane, where

$$y = iq/(iq - k) ,$$

and k was taken to be 0.5. The T matrix has a very complicated cut structure in the lower half-q plane. Table I is evidence that the effect of this cut structure as well as the bound state pole can be approximated quite well in the scattering region by the rational-fraction form.

An important and interesting question is the effect of the accuracy of input values on the accuracy of the output values of the function to be continued. We have not been able to arrive at any useful conclusions on this point either theoretically or by the consideration of simple examples. At present, this effect must be examined in the context of the particular problem to which these fitting methods are applied. Accordingly, we shall discuss the effects of the input accuracy on the output accuracy, along with the results of each physical calculation.

IV. TWO-CHANNEL NONRELATIVISTIC SCATTERING

To demonstrate one area of usefulness of the foregoing techniques, we compute scattering amplitudes for a two-channel system (Schlessinger, 1968a).

The system we wish to describe is the scattering of a particle of mass

$m = \frac{1}{2}$, momentum p, from a system that can exist in two states of energy: E_1 and E_2. The wave function for such a process is the solution of the two-channel Schrödinger equation, which we write as

$$(E - E_1 - H_0)\psi_1 = V_{11}\psi_1 + V_{12}\psi_2\,,$$
$$(E - E_2 - H_0)\psi_2 = V_{21}\psi_1 + V_{22}\psi_2\,,$$

where we have taken $H_0 = -\nabla^2$ and $E - E_i = p_i^2$, with p_i the momentum in the ith channel.

The discussion of the scattering operator for the single-channel case was quite general, and we may apply the same arguments here to obtain

$$T_{ij}(W) = V_{ij} + V_{ik}[W - H_0]_{kl}^{-1}T_{lj}\,.$$

The physical value of T_{ij} is obtained as the limit as $W \to E + i\varepsilon$ of the matrix element of $T_{ij}(W)$ between the asymptotic states φ_i satisfying

$$(p_i^2 - H_0)\varphi_i = 0\,.$$

The T matrix elements are then

$$T_{ij}(p_i, p_j, E) = \lim_{W \to E + i\varepsilon} [\varphi_i, T_{ij}(W)\varphi_j]\,.$$

The momenta in the two channels are related by

$$p_2^2 = p_1^2 + E_1 - E_2\,,$$

and, for a channel that is closed at a given energy ($p_2^2 < 0$), the asymptotic wave function is zero. Just as in the single-channel case, unitarity of the T operator gives us a useful check on the numerical work. In this case, the statements of two-channel unitarity are

$$\operatorname{Im}(T^{-1})_{11}/p_1 = 1\,,$$
$$\operatorname{Im}(T^{-1})_{22}/p_2 = 1\,,$$
$$\operatorname{Im}(T^{-1})_{12} = \operatorname{Im}(T^{-1})_{21} = 0\,.$$

We use the form

$$T(W) = V + V[W - H]^{-1}V$$

to obtain the analytic structure of T as a function of W. From knowledge of the spectrum of H, we conclude $T(W)$ may have simple poles for $W < E_1$, corresponding to bound states of H, a branch point at $W = E_1$ at the threshold for scattering states in the first channel, and similarly a threshold branch point at $W = E_2$ corresponding to scattering states of H in the second channel. Everywhere else on the first sheet of the W plane, $T(W)$ is analytic, as shown in Fig. 6.

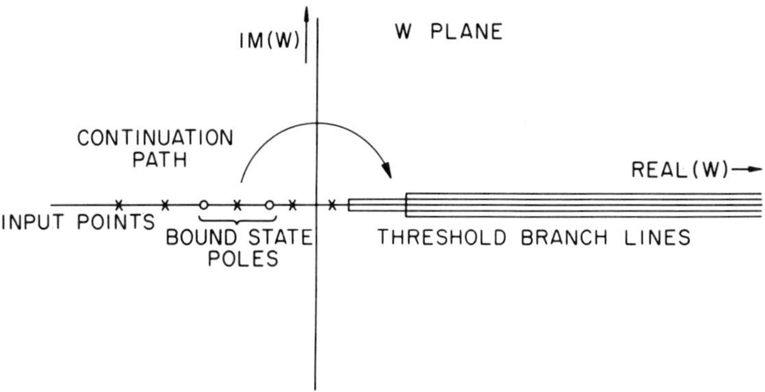

Fig. 6. Analytic structure of the two-channel scattering amplitude in the complex energy plane.

To calculate physical scattering amplitudes, we first find $T_{ij}(W)$ for several values of $W < E_1$ (shown as x's in Fig. 6) and then fit these values to a rational fraction that is evaluated at the physical energy. The input values of $T_{ij}(W)$ may again be calculated using variational principles.

The two-channel problem differs from the single-channel case in that it has two scattering thresholds. As these are just two-body thresholds, $T_{ij}(W)$ is analytic in the variable $(W - E_i)^{1/2}$ near the ith threshold. Fitting $T_{ij}(W)$ to a ratio of polynomials in the variable $y = (-W + E_1)^{1/2}$, we take into account the first threshold, but the second threshold then occurs at $y = \pm i(E_2 - E_1)^{1/2}$. The transformation

$$z = \frac{y + i(E_2 - E_1)^{1/2}}{y - i(E_2 - E_1)^{1/2}}$$

maps these branch points to $z = 0$ and $z = \infty$. In the variable $z' = z^{1/2}$, the second threshold is included correctly, and an everywhere analytic function of $(W - E_1)^{1/2}$ and $(W - E_2)^{1/2}$ is an everywhere analytic function of z'. This process, known as uniformization (Newton, 1966), cannot be applied to problems involving more than two channels. For many channels, one could try to include the effect of each threshold approximately by representing T as a rational fraction in the many variables $(E - E_i)^{1/2}$. We have not tried this type of representation.

The numerical examples we studied were of S wave scattering in the Yukawa potentials

$$V_{ij} = \lambda_{ij} \frac{e^{-r}}{r}.$$

The pointwise fitting technique was used to calculate successive approximations to the scattering amplitude for the sequence $N = 1, 2 \ldots,$ where

TABLE II

Successive Approximations to the Three Unitarity Conditions for the Two-Channel T Matrix for the Potentials $V_{11} = -2e^{-r}/r$, $V_{22} = e^{-r}/r$, $V_{12} = V_{21} = -0.5e^{-r}/r$ for the Momenta $p_1 = 0.3$, $p_2^2 = 0.5$, and $E_2 = 0.04$ (Both Channels Open)

	Second threshold not included			Second threshold included		
N	$\mathrm{Im}(T_{11}^{-1}/p_1)$	$\mathrm{Im}(T_{22}^{-1}/p_2)$	$\mathrm{Im}(T_{12}^{-1})$	$\mathrm{Im}(T_{11}^{-1}/p_1)$	$\mathrm{Im}(T_{22}^{-1}/p_2)$	$\mathrm{Im}(T_{12}^{-1})$
1	0.998	0.0407	2×10^{-3}	0.166	0.036	4×10^{-3}
2	0.993	0.204	8×10^{-3}	1.015	0.276	3×10^{-2}
3	0.9994	0.697	5×10^{-3}	1.0091	1.039	8×10^{-4}
4	0.995	0.961	3×10^{-3}	0.9964	1.0084	4×10^{-4}
5	1.001	1.128	3×10^{-3}	1.00007	1.0041	2×10^{-4}
6	0.99998	1.01	2×10^{-4}	0.9999916	0.99994	5×10^{-6}
7	0.999994	1.01	2×10^{-4}	0.9999913	1.0000004	5×10^{-6}

$2N + 1$ is the number of input points at each stage, and N also equals the order of numerator and denominator of the fraction. In these examples, $E_1 = 0$.

We have found that the calculations done for processes with the second channel closed are insensitive to the inelastic threshold. For this energy region, the results of the fits including the inelastic threshold and those including only the elastic threshold agree to within the errors of the calculation. We conclude then that below the second threshold keeping only the elastic singularity in the fit yields an excellent approximation. Above the inelastic threshold, however, the improvement obtained by including the second threshold is marked. In Table II, we compare the three unitarity conditions for the T-matrix elements evaluated by these two methods for the parameters $\lambda_{11} = -2$, $\lambda_{22} = 1$, $\lambda_{12} = -0.5$, $p_1 = 0.3$, $p_2^2 = 0.05$. The first three columns are the results of the fit ignoring the second threshold, and the last three are the results obtained when the second threshold is included. The errors incurred by the neglect of the second threshold are not large, but inclusion of the threshold does increase the accuracy. Our input values for these calculations were accurate to 7–10 decimal places; thus we appear to lose about two decimal places of accuracy in the continuation.

V. BETHE–SALPETER EQUATION

These techniques are now applied to the Bethe–Salpeter equation (Haymaker, 1967, 1968). As mentioned earlier, traditional differential equation methods for calculating phase shifts break down. Again the asymptotic behavior of the wave function is the overriding issue. We shall examine the asymptotic behavior and then present a sample calculation. We depart from the earlier examples in this chapter in that the on-shell amplitude is continued to the scattering region.

Consider the scattering of two mass m particles with initial four-momentum k_1, k_2 to final four-momentum k_1', k_2'. The off-shell T matrix is a function of the three four-vectors.

$$k = \frac{k_1 - k_2}{2}, \qquad k' = \frac{k_1' - k_2'}{2}, \qquad K = k_1 + k_2 .$$

The on-shell constraints are $k = (0, \mathbf{k})$, $\mathbf{k}' = (0, \mathbf{k})$, $|\mathbf{k}| = |\mathbf{k}'| = q'$ where $-K^2 = s = 4(q^2 + m^2)$. We again calculate below threshold, i.e., $s < 4m^2$, but now maintain the on-shell constraints. The unitarity constraint is simple and can be incorporated directly in the continuation. This is then a sequel to the preceding discussion titled "analyticity and unitarity as a useful computational tool."

A. Wick Rotation and Asymptotic Behavior

The asymptotic behavior of the wave function can be found fróm the integral equation (Haymaker, 1968; Schwartz and Zemach, 1966)

$$\psi_{k,s}(x) = \psi_k{}^0(x) + \int d^4x'\, G_s(x - x')\, V(x')\psi_{k,s}(x') ,$$

where

$$\psi_k{}^0(x) = e^{ik\cdot x} ,$$

$$G_s(x - x') = -i \int \frac{d^4p}{(2\pi)^4}\, e^{ip\cdot(x-x')}\, [(p^2 - s/4 + m^2 - i\varepsilon)^2 - sp_0{}^2]^{-1} ,$$

$$V(x) = \frac{4\mu\lambda}{|x|} K_1(\mu\,|x|) , \tag{13}$$

and K_1 is the modified Bessel function (Abramowitz and Stegun, 1965, Chapter 9). This potential corresponds to a spin zero exchange of mass μ. We have gone to the center-or-mass system.

As long as the integral converges, we see that the large x behavior is governed by the free-wave term and the Green's function. The free-wave term can be easily subtracted out, leaving the Green's function as our main concern.

It is well known that the Wick rotation can be performed on the Bethe–Salpeter equation for s below threshold. This has the practical advantage of allowing the use of expansions in four-dimensional spherical harmonics in the subsequent work. Henceforth, we shall work with the rotated forms of the equations, where now all relative momentum vectors and coordinate vectors are Euclidean:

$$k = (\mathbf{k}, k_4), \qquad x = (\mathbf{x}, x_4),$$
$$k_4 = k_0 e^{-i\pi/2}, \qquad x_4 = x_0 e^{i\pi/2}.$$

Turning now to the integral representation of the Green's function in x space, Eq. (13), and holding $s < 4m^2$, the Wick rotation can be carried out without passing any singularities. In the rotated form, the exponential components of the Green's function for large x are (Schwartz and Zemach, 1966; Schwartz, 1965)

$$G_E(x) \sim \exp[iq\,|\mathbf{r}|] + \exp \tfrac{1}{2}[s^{1/2}\,|x_4| - 2mR],$$

where $R = (|\mathbf{r}|^2 + x_4^2)^{1/2}$. This clearly shows that, for $s < 4m^2$ on the physical sheet, the Green's function is exponentially damped in all directions of the Euclidean four-space $(\mathbf{r}, x_4)$ whereas for $s > 4m^2$ the second term is a growing exponential in x_4.

The scattered part of the wave functions is also damped for $s < 4m^2$ as long as the integral in Eq. (12) converges. For the off-mass-shell case in which s and k are independent, the integral does indeed converge for the full region of s stated. However, by going to the mass shell, $s = 4(|\mathbf{k}|^2 + m^2)$, below threshold the free-wave term $\psi_k{}^0(x)$ in Eq. (12) becomes a real exponential. This exponential competes with the damped exponential under the integral, giving only a finite region of s where the scattered part of the wave function is exponentially damped. This effect signals the onset of the second Born contribution to the left-hand cut and will be discussed again below.

B. The Differential Equation

We can obtain the differential equation from the integral equation, Eq. (12) (where now all relative vectors are Euclidean), by operating on the equation with differential operator with the property

$$D(\partial_\nu)G_s(x - x') = \delta^4(x - x'), \tag{14}$$

giving the wave-function equation in the center-of-mass system:

$$D(\partial_\nu)\psi_{k,s}(x) = V(x)\psi_{k,s}(x) + D(ik_\nu)\psi_k{}^0(x),$$

where

$$D(\partial_\nu) = [-\Box - (s/4) + m^2]^2 - s(\partial/\partial x_4)^2$$

and

$$\Box = \sum_{\nu=1}^{4} \frac{\partial}{\partial x_\nu}\frac{\partial}{\partial x_\nu}.$$

The last term comes from the free-wave term in Eq. (12) and is zero for k_ν on the mass shell.

The scattered part of the wave function is

$$\chi_{k,s}(x) = \psi_{k,s}(x) - \psi_k{}^0(x) .$$

The differential equation for $\chi_{k,s}(x)$ is

$$D(\partial_\nu)\chi_{k,s}(x) = V(x)[\chi_{k,s}(x) + \psi_k{}^0(x)] . \tag{15}$$

C. Variational Principle

A Kohn-type variational principle based on the differential Eq. (15) which gives a stationary expression for T is

$$[T(k', k, s)] = \int d^4x[\chi^*_{k's}(x)\{D(\partial_\nu) - V(x)\}\chi_{k,s}(x) + \chi^*_{k's}(x)\,V(x)\psi_k{}^0(x)$$
$$+ \psi^{0*}_{k'}(x)\,V(x)\chi_{k,s}(x) + \psi^{0*}_{k'}(x)\,V(x)\psi_k{}^0(x)] . \tag{16}$$

The last term of the variational principle is just the first Born term. This is considerably simpler to apply in practice than the Schwinger variational principle used by Schwartz and Zemach (1966).

This variational principle is an exact statement and holds for arbitrary k, k', and s. However, it may be that the integrals ostensibly diverge and have a meaning only through analytic continuation. Also, when introducing a set of expansion functions for χ which are capable of representing the true wave function, it may well be that the matrix elements in such a basis will diverge. This happens for certain values of s, k, and k' and severely limits the applicability of this variational principle in practice. The off-mass-shell and on-mass-shell cases will be consisdered separately.

1. *Off-Mass Shell*

For this case, s is independent of k and k'. Fix k and k' to represent true scattering states: $k = (\mathbf{k}, 0)$, $k' = (\mathbf{k}', 0)$, $|\mathbf{k}'| = |\mathbf{k}|$. For $s < (2m)^2$, $\chi(x)$ is a decaying exponential for large x which may be represented well by a set of trial functions with a decaying exponential behavior, and all the integrals in Eq. (16) converge. For $s > (2m)^2$, however, $\chi(x)$ contains a growing exponential component in x_4, and it defies one's imagination to find a *simple* set of trial functions with the growing exponential which will yield convergent integrals. The derivative term in Eq. (16) is the troublesome one; all other terms have a potential present which is a decaying exponential in all directions. A possible method of circumventing this difficulty is to introduce an integral representation of the wave function:

$$\chi_{k,s}(x) = \int d^4x' G_s(x - x')\varphi(x') .$$

An examination of the integral equation for $\chi(x)$ shows that $\varphi(x)$ contains a decaying exponential in the elastic-scattering region. Using the property (14), we see that the derivative term is a convergent integral. But now we have integrals to evaluate of the same difficulty as those of the Schwinger method (Schwartz and Zemach, 1966).

2. *On-Mass Shell*

The additional constraints $s = 4(|\mathbf{k}|^2 + m^2) = 4(|\mathbf{k}'|^2 + m^2)$ limit further the applicability of Eq. (16). The free-wave $\phi_k{}^0(x)$, which is oscillatory for the foregoing case, becomes a real exponential below threshold, and for $\mathrm{Im}(k) > \mu$ the integrals again diverge. Also at this point our estimate of the asymptotic behavior of $\chi(x)$ is no longer valid, as discussed previously. So finally the domain of applicability in s is

$$4(m^2 - \mu^2) < s < (2m^2) .$$

This is the region between the elastic threshold and the second Born contribution to the left-hand cut. (The first Born tern was included explicitly.) These considerations might suggest that the full amplitude has a singularity at $s = 4(m^2 - \mu^2)$. Although the variational principle for the full amplitude breaks down at this point, only the partial-wave amplitude in fact has this singularity.

D. Numerical Method

To solve the Bethe–Salpeter equation, we first introduce a linear parameterization of the wave function $\chi_{k,s}(x)$:

$$\chi_{k,s}(x) = \sum_{\substack{l=0\\ n=l\\ j=n}}^{\infty} a_{l,n,j} Y_{nl0}(\hat{x}) \varphi_j(R) ,$$

where Y_{nlm} are four-dimensional spherical harmonics on the four-sphere, and $\hat{x}$ is the unit four-vector in the Euclidean space. The four-dimensional spherical harmonics can be expressed in terms of the usual three-dimensional spherical harmonics Y_{lm} and Gegenbauer polynomials $C_\beta{}^\alpha$ (Schwartz and Zemach, 1966):

$$Y_{nlm}(\hat{x}) = Y_{lm}(\hat{r}) \left[\frac{2^{2l+1}(n+1)(n-1)!\, l!^2}{\pi(n+l+1)!}\right]^{1/2} C_{n-l}^{l+1}(\cos\theta) \sin^l(\theta) ,$$

where $\cos\theta = x_4/R$, and $\hat{r}$ is the three-space unit vector. The functions $\varphi_j(R)$ are radial functions depending only on R and were chosen to be

$$\varphi_j(R) = R^j e^{-\alpha R} .$$

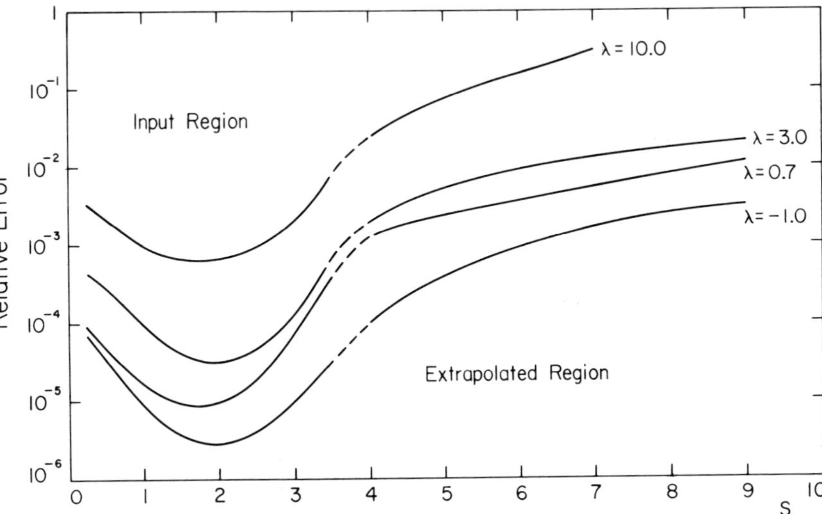

Fig. 7. Relative errors incurred in the calculation of S-wave phase shifts using the on-mass-shell continuation. The variational calculation was done in the region $0 < s < 3.5$, and the errors indicate the quality of input numbers for the extrapolation. The extrapolated region is $s > 4$.

We can expand $\chi^*_{k'}(x)$ in the same set of functions, but now with expansion parameters $a'_{l',n',j'}$. These functions have been used before by Schwartz and have been applied directly to our problem. The free-wave term has the following expansion in these angular functions:

$$\psi_k{}^0(x) = (2\pi)^2 \sum_{n=0}^{\infty} i^n \frac{J_{n+1}(|k|R)}{|k|R} \sum_{l=0}^{n} \sum_{m=-l}^{l} Y^*_{nlm}(\hat{x})\, Y_{nlm}(\hat{k}) \,,$$

where $\hat{k}$ is the unit four-vector of length $|k|$ in the direction k, and J is the ordinary Bessel function (Abramowitz and Stegun 1965, Chapter 9).

Substituting these expansions in the variational principle, treating the a's as variational parameters, we can proceed to calculate $T(k, k, s)$. Defining the partial wave amplitude through the expansion

$$T(k', k, s) = \sum_{l=0}^{\infty} (2l + 1) t_l(|\mathbf{k}'|, k_0', |\mathbf{k}|, k_0; q) P_l(\cos\theta) \,,$$

where $\cos\theta = (\mathbf{k} \cdot \mathbf{k}')/|\mathbf{k}|\,|\mathbf{k}'|$, we can calculate t_l and relate it to the phase shift

$$t_l(q, 0, q, 0; q) = \rho^{-1} e^{i\delta_l} \sin\delta_l \,, \qquad \rho = q/(8\pi s^{1/2}) \,.$$

We calculate t_l for a set of values of s, between $4m^2 - 4\mu^2$ and $4m^2$, in preparation for the continuation. A sample of accuracy for various potentials is shown in Fig. 7.

E. Continuation to the Scattering Region

We now continue the on-shell amplitude to the scattering region. This is a variation on the foregoing theme. We had dubious success applying the off-shell methods to this problem. A search was made for a smoother function with fewer nearby singularities. The study of the singularity structure of amplitudes leads us to believe that the on-shell K matrix is the optimum function

$$K_l(s) = \frac{t_l(q, 0, q, 0; q)}{2 - 2i\rho t_l(q, 0, q, 0; q)} .$$

It is well known that $K_l(s)$ is analytic in s at threshold and has branch points only on the real s axis. This follows from unitarity.

The calculation proceeds as follows. We calculate $t_l(q)$ (suppressing the first four arguments) as described in Sect. D, where

$$t_l(q) = B_l(q) + \tilde{t}_l(q) .$$

The first Born term $B_l(q)$ is known exactly, and $\tilde{t}_l(q)$ is calculated numerically. The amplitude t_l and thus $K_l(s)$ can be found for $4(m^2 - \mu^2) < s < 4m^2$. It is a real analytic function of s with a branch point in this

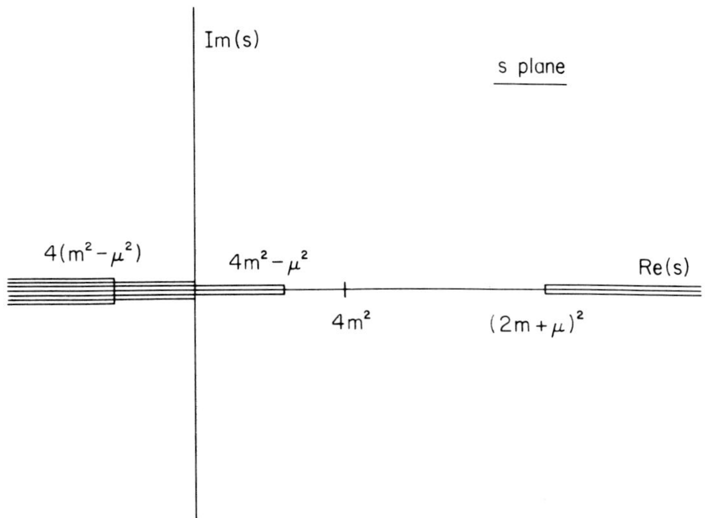

Fig. 8. Cut structure of $K_l(s)$ showing the first inelastic threshold, the first two contributions to left-hand cut, and the kinematical branch point at $s = 0$.

region at $s = 4m^2 - \mu^2$ coming from $B_l(q)$. There is also a kinematical branch point at $s = 0$ coming from ρ. These singularities are shown in Fig. 8. We can then expand the domain of analyticity further by removing the cut contribution to $K_l(s)$ for $4(m^2 - \mu^2) < s < 4m^2 - \mu^2$. Define

$$\tilde{K}_l(s) = K_l(s) - K_l^{\text{cut}}(s) ,$$

where

$$K_l^{\text{cut}}(s) = \frac{1}{\pi} \int_{4(m^2-\mu^2)}^{4m^2-\mu^2} ds' \frac{\Delta K_l(s')}{(s' - s)}$$

and

$$\Delta K_l(s) = (2i)^{-1} [K_l(s + i\varepsilon) - K_l(s - i\varepsilon)] .$$

The function $\tilde{K}_l(s)$ is analytic on the real axis in the region $4(m^2 - \mu^2) < s < (2m + \mu)^2$. This function was continued to the scattering region using the Legendre fit, Eqs. (9)–(11), in the variable s, and then the cut contribution $K_l^{\text{cut}}(s)$ was added back in. The phase shift was found from the formula

$$K_l(s) = \frac{4\pi s^{1/2}}{q} \tan \delta_l .$$

The results compared quite favorably with those of Schwartz and Zemach for moderately strong potentials. The relative errors in the phase shifts for various potentials are plotted in Fig. 7. In the region of s between zero and four, the relative error of t_l is plotted. In both cases, the error was estimated from the convergence of sequences of approximations. We see that, on the average, about two significant digits are lost in the extrapolation. The errors in both the input numbers and phase shifts increase for stronger potentials. Historically, Schwartz and Zemach (1966) were the first to calculate phase shifts. Our calculation agreed with their calculation within our error except for our very accurate results (e.g., $\lambda = -1$, Fig. 7). There was as much as a 10 standard deviation discrepancy, and it remained unexplained for some time. Recently, a small error was discovered in their calculation, and now we both agree within errors (McInnis and Schwartz, 1969).

This calculation of phase shifts was attempted using the off-shell ideas presented for the Schrödinger equation without much success. We were able to do short extrapolations to get low-energy behavior. This is included in Haymaker (1968). Why should this be after obtaining such nice results from the Schrödinger equation? There are two factor working against us:

(1) The Bethe–Salpeter equation is two-dimensional in $|\mathbf{r}|$, x_4, compared to the one-dimensional Schrödinger equation in $|\mathbf{r}|$. This limits the accuracy that is practical to achieve for the input numbers to the extrapolation.

(2) The analytic structure of the off-shell Bethe–Salpeter T matrix is much more complicated than the off-shell Schrödinger T matrix. There are a maze of thresholds for real positive s in addition to the inelastic thresholds that are present on-shell. By going to the on-shell K matrix, we have coveniently put the cuts, the fitting region, and the scattering region all on the real line. We believe this has an advantage in that fewer poles of the Padé fit are necessary to get a good representation of the cuts. This is important, since we must see convergence of the Padé fits in low orders before roundoff error obscures the results.

VI. THREE-BODY SCATTERING

Recently there has been considerable interest in the nonrelativistic three-body problem. Many of the more recent numerical calculations have been concerned with the solution of the Faddeev equation (Faddeev, 1960), usually with separable potentials (Watson and Nuttall, 1967). There have, however, been some calculations done with local potentials, and these works generally make use of variational principles applied to the Schrödinger equation (Watson and Nuttall, 1967). Although the application of some of the ideas we have discussed provides a considerable simplification in numerical calculations, we can claim only partial success in dealing with the three-body problem. Using the techniques of rational-function fitting on the off-energy-shell scattering amplitude, we are able to calculate phase shifts and cross sections for the process where one particle scatters elastically from a bound state of the other two (Schlessinger, 1968b); we have not been able to calculate the amplitude for the breakup process.

In this section, after a brief review of the notation and conventions associated with the three-body problem, we describe the application of our methods to some simple systems. We discuss in detail how to obtain the relative S-wave part of the $J = 0$ scattering amplitude for bound-state elastic three-body scattering with local potentials and compare our results to those obtained in other ways. We also describe the problems associated with the breakup reaction.

A. Formalism

In the scattering of three particles of mass m_1, m_2, m_3, and coordinates $\mathbf{r}_1$, $\mathbf{r}_2$, $\mathbf{r}_3$, many different processes may occur. Since there is no generally accepted notation, we label the transition operator by the bound-state pair in the initial and final states. The scattering of particle 1 from a bound state of the (2, 3) pair resulting in a bound state of the (1, 3) pair with particle 2 free is then described as

$$1 + (2, 3) \to (1, 3) + 2$$

and has the transition operator $T_{13,23}$. The general process is described by the operator $T_{\alpha,\beta}$, where α, β may have the values 12, 13, 23, or 0; 0 corresponds to three particles free. For this discussion, we consider only pairwise local interactions, and the potentials between the ith and jth particle have the form

$$V_{ij} = V_{ij}(|\mathbf{r}_i - \mathbf{r}_j|) .$$

The total Hamiltonian is given by

$$H = H_0 + V_{12} + V_{13} + V_{23} = H_0 + V ,$$

where H_0 is the total kinetic energy. We define the Hamiltonian that describes the interaction of two particles in the three-particle system as

$$H_\alpha = H_0 + V_\alpha ,$$

and for future use we also define

$$\hat{V}_\alpha = V - V_\alpha , \qquad V_0 = 0 , \qquad \hat{V}_0 = V .$$

Armed with these convention, we can now write down the scattering operators. The transition from the asymptotic state α to the asymptotic state β is described, as usual, by the matrix element

$$T(\beta, \alpha, E) = \lim_{W \to E + i\varepsilon} [\Phi_\beta(E), T_{\beta\alpha}(W)\Phi_\alpha(E)] . \tag{17}$$

Φ_β is the asymptotic state in which the pair β is bound and the third particle is free. Accordingly, $\Phi_\beta(E)$ satisfies the Schrödinger equation

$$(E - H_\beta)\Phi_\beta(E) = 0 , \tag{18}$$

with the boundary conditions pair β bound and the third particle free at large distances. Equation (17) may also be written as

$$\begin{aligned} T(\beta, \alpha, E) &= \lim_{W \to E + i\varepsilon} [\Psi_\beta(W), \hat{V}_\alpha \Phi_\alpha(E)] \\ &= \lim_{W \to E + i\varepsilon} [\Phi_\beta(E), \hat{V}_\beta \Psi_\alpha(W)] , \end{aligned} \tag{19}$$

where $\Psi_\alpha(W)$, for instance, satisfies the inhomogeous (for $W \neq E$) Schrödinger equation

$$(W - H)\Psi_\alpha(W) = (W - H_\alpha)\Phi_\alpha(E) . \tag{20}$$

Combining (19), (20), and (17), we arrive at the usual formula for the transition operator:

$$T_{\beta\alpha}(W) = \hat{V}_\beta + \hat{V}_\beta[W - H]^{-1}\hat{V}_\alpha\,. \tag{21}$$

The physical scattering amplitude $T(\beta, \alpha, E)$ is then obtained as the matrix elements of $T_{\beta\alpha}(W)$ between the states Φ_β, Φ_α as the energy approaches its physical value $W \to E + i\varepsilon$ as in Eq. (17).

B. Rational-Fraction Fitting

To apply the methods of rational-fraction fitting to this physical problem, one must first determine the analytic structure of the off-energy-shell T matrix defined by Eq. (21). This discussion is similar to that for the two-body case. Knowing the spectrum of the operator H, we can, from Eq. (21), conclude that $T(\beta, \alpha, W)$ has poles for negative values of W corresponding to the three-body bound states of H. Furthermore, $T(W)$ has several branch points, each beginning at $W = E_i$, where E_i is the bound-state energy of the two-particle bound states of H. Finally, $T(W)$ has a branch point at $W = 0$ corresponding to the continuum of scattering states of H. Everywhere else on the first sheet of the W plane, $T(W)$ is analytic, as shown in Fig. 9.

Since the physical process occurring at the thresholds $W = E_i$ is really two-body scattering, with one of the particles a two-body bound state of H, $T(W)$ is analytic in the variable $(W - E_i)^{1/2}$ near each of these thresholds. This behavior may be verified by considering the Born series for $T(W)$. In the same way, one can show that T behaves as $W^2 \ln(W)$ near the breakup threshold at $W = 0$. This, admittedly, nonrigorous discussion of the singularities of $T(W)$ is, we believe, quite adequate for our purposes for bound-state elastic scattering. For breakup reactions, however, one must be more careful. Since there are also other difficulties involved in

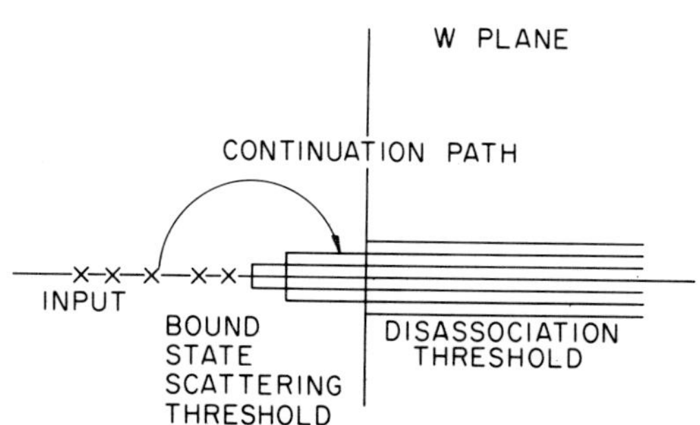

Fig. 9. Analytic structure of the elastic three-body scattering amplitude in the complex energy plane.

the breakup calculation, we shall limit this discussion to elastic bound-state scattering and consider the inelastic reactions in a later section.

Knowing the important features of the analytic structure of $T(W)$, the application of the fitting technique proceeds as in the two-body case. That is, the calculation is done in two steps. First several values of $T(\alpha, \alpha, W)$ are calculated for $W < E_0$, the lowest two-body bound-state energy of H. These numerical values are then fit to a rational fraction that is evaluated at the physical energy to give scattering phase shifts. We should emphasize here that $T(\alpha, \alpha, W)$ is originally calculated, as in the two-body case, for fixed and physical values of the external momenta, and for unphysical values of W. As we shall show, the first part of the calculation is considerably easier than the direct solution of the Schrödinger equation. The reason for this simplification is, essentially, that for $W < E_0$ the expression $[W - H]$ in (17) can vanish only at the three-body bound-state energies. In coordinate space, this property is reflected in the large distance behavior of the wave functions, which are exponentially damped in all directions. This damping restricts the calculation to a finite region and provides a great simplification in numerical work. To help clarify these notions, we discuss, in the next section, some model calculations.

C. Model Calculations

Perhaps the major difficulty in any calculation of a three-body problem is the large number of coordinates and amplitudes which must be handled. Since these complications are not essential to the dynamical aspects of the problem, we have chosen models that eliminate many of these geometrical details. We consider S-wave elastic scattering of particle 1 from a bound state of the (2, 3) pair. Furthermore, we chose the mass of particle 3 to be infinite and set the masses of particles 1 and 2 equal. These restrictions allow the reaction to be described in terms of three variables: r_1, the distance between particles 1 and 3; r_2, the distance between particles 2 and 3; and $\cos\theta_{12}$, the angle between r_1 and r_2. Ψ, the wave function for such a process, may be expressed as

$$\Psi(r_1, r_2, \cos\theta_{12}) = \sum_l \Psi^l(r_1, r_2) P_l(\cos\theta_{12}) . \tag{22}$$

Our final simplification is just to consider relative S-wave interaction between the equal mass particles, that is, we only use the $l = 0$ term in Eq. (22).

A good method for obtaining the value of the T matrix for $W < E_0$ is a variational principle based on the coordinate space differential equation for the unphysical wave function $\Psi_\alpha(W, r_1, r_2)$. It is important to notice that $\Psi_\alpha(W)$ depends on two independent energy parameters, W and E. The dependence of $\Psi_\alpha(W)$ on E, the energy of the asympotic state α, is implied in our notation by the subscript α. In these calculations, we determine $\Psi_\alpha(W)$ as a function of W for E fixed at its physical value determined by the momenta of the particles in the initial or final state. With the simplifications we have made, the equation for $\Psi_\alpha(W, r_1, r_2)$ in Yukawa potentials is

$$(W - H)\Psi_{23}(W, r_1, r_2) = (W - H_{23})\Phi_{23}(E, r_1, r_2) \,. \tag{23}$$

Here

$$\begin{aligned} H = & -\frac{1}{2m_1}\frac{1}{r_1}\frac{d^2}{dr_1^2}r_1 \\ & -\frac{1}{2m_2}\frac{1}{r_2}\frac{d^2}{dr_2^2}r_2 + V_{12}(r_1, r_2) + V_{13}(r_1) + V_{23}(r_2) \end{aligned}$$

and

$$V_{12}(r_1) = \lambda_{13}\frac{e^{-\mu_1 r_1}}{r_1}\,, \qquad V_{23}(r_2) = \lambda_{23}\frac{e^{-\mu_2 r_2}}{r_2}\,,$$

$$V_{12}(r_1, r_2) = \frac{\lambda_{12}}{\mu}\frac{\sinh(\mu r_<)e^{-\mu r_>}}{r_< r_>}\,,$$

where $r_<$ ($r_>$) is the lesser (greater) of r_1 and r_2. Equation (23) for $V_{12}(r_1, r_2)$ is the relative S-wave projection of

$$V_{12}(r_1, r_2) = \lambda_{12}\frac{e^{-\mu|r_1 - r_2|}}{|r_1 - r_2|}\,.$$

Since $\Phi(E, r_1, r_2)$ satisfies $(W - H_{23})\Phi_{23}(E, r_1, r_2) = 0$, we may write it as

$$\Phi_{23}(E, r_1, r_2) = (\sin kr_1/kr_1)\varphi_{23}(r_2)\,,$$

where $\varphi_{23}(r_2)$ is the two-body bound-state wave function of the (2, 3) pair, and k is the momentum of particle 1. If we represent $\Psi_{23}(W, r_1, r_2)$ as

$$\Psi_{23}(W, r_1, r_2) = \chi_{23}(W, r_1, r_2) + \Phi_{23}(E, r_1, r_2)\,,$$

then the differential Eq. (23) becomes

$$(W - H)\chi_{23}(W, r_1, r_2) = \hat{V}_{23}\Phi_{23}(E, r_1, r_2)\,, \tag{24}$$

which is, in the general case,

$$(W - H)\chi_\alpha = \hat{V}_\alpha\Phi_\alpha\,. \tag{24$'$}$$

The simplification obtained by calculating with $W < E_0$ can be seen from Eq. (24), for in this energy region $\chi_{23}(W, r_1, r_2)$ decays exponentially at large values of either r_1 or r_2. In the physical energy region, $\chi_{23}(W)$

decays in this way only at large values of r_2 but has oscillatory behavior at large r_1; this behavior complicates the on-energy-shell calculation considerably.

The variational principle we used can be written, for the general process, as

$$T(\beta, \alpha, W) = (\Phi_\beta, \hat{V}_\beta \Phi_\alpha) + (\chi_\beta, \hat{V}_\alpha \Phi_\alpha) + (\Phi_\beta, \hat{V}_\beta \chi_\alpha) - (\chi_\beta [W - H] \chi_\alpha) . \qquad (25)$$

This expression is equal to $T(\beta, \alpha, W)$ when χ_β or χ_α satisfies Eq. (24) and is stationary under small variations of χ_α, χ_β around those values. The usual procedure for making use of such an expression is to represent $\chi(W, r_1, r_2)$ as a finite sum of some elements of a complete set of functions:

$$\chi(W, r_1, r_2) = \sum_{n=1}^{N} \sum_{m=1}^{M} C_{nm} U_{nm}(r_1, r_2) .$$

The U_{nm} are chosen so that they can, in some sense, easily approximate χ. This expression is then substituted into (25), and the variation of the parameters C_{nm} leads to a set of linear equations for the C_{nm}. Solving those equations determines χ, which, when substituted into Eq. (25), determines the variational approximation to $T(\beta, \alpha, W)$. For our choice of potentials, the functions

$$U_{nm}(r_1, r_2) = r_<^{n-1} r_>^{m-1} e^{-\alpha(r_1+r_2)}$$

and

$$V_{nm}(r_1, r_2) = (r_1 - r_2) U_{nm}(r_1, r_2)$$

form an especially good set of trial functions, since they not only have the correct behavior (that is, like χ) at both small and large values of r_1, r_2, but they can also easily approximate the distinctive behavior of χ at $r_1 = r_2$ caused by the potential $V_{12}(r_1, r_2)$. Furthermore, with this representation for χ, the integrals appearing in Eq. (25) can easily be done. We have obtained results, in this way, both for scattering by short-range Yukawa potentials and for Coulomb scattering (Schlessinger, 1968b).

D. Coulomb Scattering: Electron Hydrogen Scattering

For Coulomb scattering with the assumption of relative S-wave interactions, the Hamiltonian, with the units of length of $a_0 = \hbar^2/me^2$, is

$$H = -\frac{1}{2r_1}\frac{d^2}{dr_1^2} r_1 - \frac{1}{2r_2}\frac{d^2}{dr_2^2} r_2 - \frac{1}{r_1} - \frac{1}{r_2} + \frac{1}{r_>} .$$

The fact that the electrons are identical particles is taken into account by making the wave functions symmetric (singlet) or antisymmetric (triplet) under the exchange $r_1 - r_2$. The trial functions are similarly chosen, and only the U_{nm} are used for the singlet and the V_{nm} for the triplet states. The asymptotic wave function Φ_{23} for S-wave scattering from the hydrogen atom in its ground state is

$$\Phi_{23} = \sqrt{2}\left[\frac{\sin pr_1}{pr_1}e^{-r_2} \pm \frac{\sin pr_2}{pr_2}e^{-r_1}\right].$$

In these units, $2e^{-r}$ is the S-wave ground-state wave function and p is the momentum of the incident electron.

Table III shows a comparion between some values of $\tan\delta/p$ obtained using our method and those obtained in other ways for singlet and triplet S-wave e-H scattering with the approximations discussed previously. In all cases, the agreement is quite good. Our triplet results are more accurate than our singlet results because the input to the continuation was more precise (five to eight decimal places) than for the singlet (four to seven decimal places). To obtain these results, we used 11 input points distributed along the negative W axis from $W = -E_0$ to $W = -\infty$ as discussed for two-body scattering.

We also calculated the triplet elastic amplitude above the first excitation threshold but below the breakup threshold. These results are also in good agreement with those obtained by other methods. The calculation using short-range potentials was also successful and yielded answers of accuracy comparable to the accuracy of the electron hydrogen phase shifts.

There are several complications involved in using these techniques for the breakup process. One problem deals with the so-called disconnected pieces of the scattering amplitude that describe the scattering of two of the particles of the system, with the third particle a spectator. These discon-

TABLE III

$\tan\delta/pa_0$ for Singlet and Triplet S-Wave Elastic ($1s \to 1s$) e-H Scattering

pa_0	$\tan\delta/pa_0$ (continuation)	$\tan\delta/pa_0$*
	Singlet	
0.0	−7.842	−7.815
0.6	2.15	2.147
	Triplet	
0.0	−2.348	−2.3482
0.6	−4.8486	−4.8486

* Schwartz (1961, 1962); Kyle and Temkin (1964).

nected graphs manifest themselves in this formalism in an oscillatory behavior of the wave functions at large distances. This behavior may be eliminated, however, by isolating the two-body contributions to the wave functions and treating them separately.

Having made this separation and thus limiting the input calculation to a finite spatial region, we obtained convergent values for the breakup T matrix for $W < E_0$. We were, however, not able to continue these amplitudes to the physical region. That is, our fitting procedure did not converge. This failure is probably caused by our inability to include the threshold at $W = 0$ in the correct way or by the presence of other important singularities of the amplitude which we had not included. The precise cause for the failure of the method in this case is uncertain, and the applicability of rational fraction fitting to the calculation of the breakup amplitude is still an open question.

ACKNOWLEDGMENTS

This work was stimulated by Professor Charles Schwartz, and we wish to express our sincere appreciation to him for his encouragement and numerous suggestions throughout the course of this work.

REFERENCES

Abramowitz, M., and Stengun, I. (1965). "Handbook of Mathematical Functions." Dover, New York.

Baker, G. A., Jr. (1965). "Advances in Theoretical Physics" (K. A. Brueckner, ed.), Vol. 1, pp. 1-58. Academic Press, New York.

Faddeev, L. D. (1960). *Zh. Eksperim. Teor. Fiz.* **39,** 1459. English transl. *Soviet Phys.—JEPT* **12,** 1014.

Haymaker, R. (1967). *Phys. Rev. Letters* **18,** 968.

Haymaker, R. (1968). *Rhys. Rev.* **165,** 1790.

Kyle, H. L., and Temkin, A. (1964). *Rhys. Rev.* **139,** A600.

McInnis, B. C., and Schwartz, C. (1969). *Phys. Rev. Comments and Annenda* **177,** 2621.

Newton, R. G. (1966). "Scattering Theory of Waves and Particles." McGraw-Hill, New

Schlessinger, L. (1968a). *Phys. Rev.* **167,** 1411.

Schlessinger, L. (1968b). *Phys. Rev.* **171,** 1523.

Schlessinger, L., and Schwartz, C. (1966). *Rhys. Rev. Letters* **17,** 1173.

Schwartz, C. (1961). *Rhys. Rev.* **124,** 1468.

Schwartz, C. (1962). *Phys. Rev.* **126,** 1015.

Schwartz, C. (1965). *Phys. Rev.* **137,** B717.

Schwartz, C., and Zemach, C. (1966). *Phys. Rev.* **141,** 1454.

Wall, H. S. (1948). "Continued Fractions." Van Nostrand, Princeton, New Jersey.

Watson, K. M., and Nuttall, J. (1967). "Topics in Several Particle Dynamics." Holden-Day, San Francisco, California.

CHAPTER 12 SOLUTION OF THE BETHE–SALPETER EQUATION BY MEANS OF PADÉ APPROXIMANTS

J. A. Tjon

University of Utrecht
The Netherlands

H. M. Nieland

University of Nijmegen
The Netherlands

I. INTRODUCTION

During the past few years, there has been some interest in the solution of the Bethe–Salpeter (BS) equation in the ladder approximation as a relativistic model for two-body scattering. Rather interesting features appear in this model, such as the violation of the unitarity relation above inelastic threshold and the occurence of Regge daughters due to the $O(4)$ symmetry in this equation.

Historically, the equation was considered mainly in connection with the bound-state problem; the interest in the scattering region depended essentially on the possibility of a numerical solution.

The first work in this direction was done by Schwartz and Zemach (1966), who considered scalar particles with equal masses and obtained phase shifts in the elastic scattering region using variational techniques.

A possibility of extending the calculation above the inelastic threshold was demonstrated by Levine *et al.* (1967), who considered also scalar particles with equal masses. They regularized the equation by a technique due to Noyes (1965) and Kowalski (1965) and obtained after the usual Wick rotation a solution, which, however, for certain values of the parameters, violated unitarity above inelastic threshold. This could be cured by inserting a "bubble" in one of the Green's function propagators in order to satisfy three-particle unitarity.

The study of the case of scalar particles with equal masses is, of course, useful for understanding the general properties of the equations and their solutions. However, it is of some interest to extend the calculation to a more realistic situation. The simplest case is then the πN system. Two new features appear: the nucleon spin and the unequal masses of the external particles. The unequal masses change the locations of the singularities, but do not affect the method of solution in an essential way. However, it turns out that the presence of spin makes it technically rather difficult to solve the integral equations by matrix inversion, because of the dimensions of the kernel. As another possible method, we have considered the use of Padé approximants. It turned out that this method works quite well, although it is found that the lowest-order diagonal approximant [1, 1] can be a bad approximation, which makes it necessary to go to higher-order approximants (Nieland and Tjon, 1968).

In Sect. II, the integral equations for πN scattering will be written down using the ladder approximation with nucleon exchange as potential. Any attempt to reproduce the main features of low-energy πN scattering would require a more sophisticated potential, for example, the inclusion of N^* and/or ρ exchange. Although this can be done without further difficulties, we shall restrict ourselves here to N exchange.

Section III presents some details of the method of solution. Finally, the results are given in Sect. IV. The Appendices take care of some formulas.

II. THE INTEGRAL EQUATIONS

Our model will be the BS equation for πN scattering in the ladder approximation with nucleon exchange as potential, and pseudoscalar coupling. This is shown in diagrammatic form in Fig. 1.

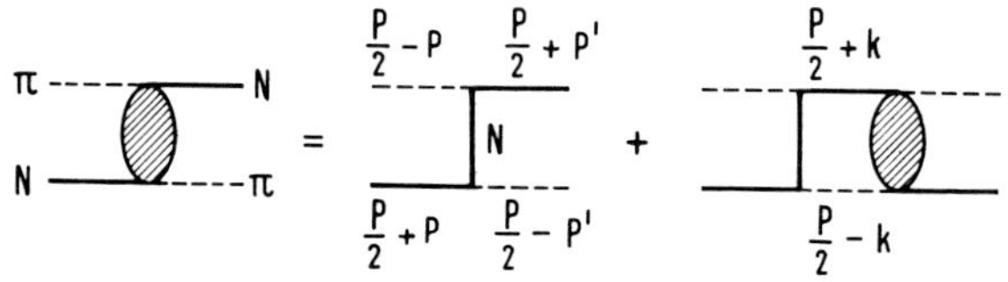

Fig. 1. The Bethe–Salpeter equation in diagrammatic form. P is the total four-momentum; in the center-of-mass system, $P = (0, 0, 0, s^{1/2})$.

The corresponding integral equation for the scattering amplitude T in momentum space is given by

$$T(p, p') = V(p, p') + \int \frac{d^4k}{(2\pi)^4} G(k) K(p, k) T(k, p') \,, \tag{1}$$

with

$$\begin{aligned}
V(p, p') &= -4\pi g^2[i\gamma(p + p') + M]\cdot\Delta(p + p')\,, \\
K(p, k) &= i\cdot 4\pi g^2[i\gamma(p + k) + M]\cdot\Delta(p + k)\cdot[i\gamma(\tfrac{1}{2}P + k) - M]\,, \\
\Delta(k) &= [k^2 + M^2 - i\varepsilon]^{-1}\,, \\
G(k) &= [(\tfrac{1}{2}P + k)^2 + M^2 - i\varepsilon]^{-1}\cdot[(\tfrac{1}{2}P - k)^2 + 1 - i\varepsilon]^{-1}\,.
\end{aligned}$$

M is the nucleon mass in units of the poin mass $m_\pi = 1$, and g^2 is the coupling constant whose experimental value is given by 14.6. Compared with the case of scalar particles, the kernel of the integral equation contains an extra factor k^2 for large k due to the spin $\frac{1}{2}$ propagators which destroys the compactness. In coordinate space, this corresponds to a marginal singular behavior of the potential near the origin and will make it convenient to use a form factor. Therefore, we introduce a cutoff in the integration for high momenta; this is done in a relativistic invariant way by making the substitution

$$\Delta(k) \to \Delta_\Lambda(k) = [k^2 + M^2 - i\varepsilon]^{-1} - [k^2 + \Lambda^2 - i\varepsilon]^{-1}\,.$$

The integrand will then contain an extra damping $(k^2)^{-1}$ for large k, which is sufficient in this case. It may be worthwhile to note that even without cutoff each term of the ladder series is finite, as can be shown with power-counting arguments.

In order to obtain equations for the partial wave amplitudes, a complete set of helicity spinors is introduced. The set belonging to a certain four-momentum $q = (\boldsymbol{q}, q_0)$, which is in general off-shell, is chosen as

$$u_\lambda(q) = N_q\begin{pmatrix}(E_q + M)\chi_\lambda(\boldsymbol{q}) \\ 2\lambda\,|\boldsymbol{q}|\,\chi_\lambda(\boldsymbol{q})\end{pmatrix}, \qquad v_\lambda(q) = N_q\begin{pmatrix}|\mathbf{q}|\,\chi_\lambda(\mathbf{q}) \\ 2\lambda(E_q + M)\chi_\lambda(\mathbf{q})\end{pmatrix},$$

where $E_q = (|\mathbf{q}|^2 + M^2)^{1/2}$, $N_q = [2M(E_q + M)]^{-1/2}$, and $\chi_\lambda(\boldsymbol{q})$ is an eigenfunction of the helicity operator $(\boldsymbol{\sigma}\cdot\boldsymbol{q})/|\boldsymbol{q}|$ with eigenvalues $2\lambda = \pm 1$.

Inserting in the integrand the completeness relation

$$\sum_{\sigma=\pm 1/2} [u_\sigma(\mathbf{q})\bar{u}_\sigma(\mathbf{q}) - v_\sigma(\mathbf{q})\bar{v}_\sigma(\mathbf{q})] = 1\,,$$

with $q = \frac{1}{2}P + k$, we get two coupled equations for T:

$$\begin{aligned}
T^{uu}_{\lambda\mu}(p, p') &= V^{uu}_{\lambda\mu}(p, p') + \int \frac{d^4k}{(2\pi)^4} G(k) \sum_\sigma [K^{uu}_{\lambda\sigma}(p, k)\, T^{uu}_{\sigma\mu}(k, p') \\
&\quad - K^{uv}_{\lambda\sigma}(p, k)\, T^{vu}_{\sigma\mu}(k, p')]\,, \qquad (2) \\
T^{vu}_{\lambda\mu}(p, p') &= V^{vu}_{\lambda\mu}(p, p') + \int \frac{d^4k}{(2\pi)^4} G(k) \sum_\sigma [K^{vu}_{\lambda\sigma}(p, k)\, T^{uu}_{\sigma\mu}(k, p') \\
&\quad - K^{vv}_{\lambda\sigma}(p, k)\, T^{vu}_{\sigma\mu}(k, p')]\,,
\end{aligned}$$

where, for example, $T^{vu}_{\lambda\mu}(p, p') = \bar{v}_\lambda(p)T(p, p')u_\mu(p')$, etc. This coupling is a consequence of the off-mass shell character of the intermediate state, because the fermion propagator $\{i\gamma[(P/2) + k] + M\}^{-1}$, operating on the inserted completeness relation, will then give a nonvanishing contribution of the v-spinor part.

Since the final state momentum p' is a dummy variable in the integral equations, it may as well be taken completely on shell (physical masses and energy) and therefore be suppressed in the notation. Thus,

$$T^m_{\lambda\mu}(p) = V^m_{\lambda\mu}(p) + \sum_{n=1}^{2} \int \frac{d^4k}{(2\pi)^4} G(k) \sum_\sigma K^{mn}_{\lambda\sigma}(p, k) T^n_{\sigma\mu}(k) ,$$

with the notation $T^1 = T^{uu}$, $T^2 = T^{vu}$, similarly for V, and the 11, 12, 21, and 22 components of K corresponding to uu, uv, vu, and vv, respectively. In the remainder, this two-dimensional character will be exhibited only if necessary.

The partial wave amplitudes will satisfy an equation of the form

$$T(p, p_0) = V(p, p_0) - (i/\pi) \int_0^\infty dk \int_{-\infty}^{+\infty} dk_0 \, G(k, k_0) K(p, p_0; k, k_0) T(k, k_0) ,$$

where p and k are now the absolute values of center-of-mass three-momenta. The isospin may be taken into account by multiplication of the coupling constant g^2 with the appropriate isospin factor; this factor is here $+2$ for $I = \frac{3}{2}$ and -1 for $I = \frac{1}{2}$.

The kernel $K(p, p_0; k, k_0)$ is a linear combination of two Legendre functions of the second kind $Q_l(z)$ and $Q_{l\pm1}(z)$ with

$$z = \frac{(p_0 + k_0)^2 - p^2 - k^2 - M^2 + i\varepsilon}{2pk} ,$$

and is given in Appendix A.

The reduction of this integral equation for $s^{1/2} < 2M + 2$ to a nonsingular form can be done in a way, that is slightly different from that of the scalar case with equal masses. Let us consider the singularities of the Green's function $G(k, k_0)$ in some detail (see Fgi. 2). The poles of $G(k, k_0)$ located at

$$k_0 = -\tfrac{1}{2}s^{1/2} + (k^2 + M^2)^{1/2} - i\varepsilon \equiv \omega_3 ,$$
$$k_0 = +\tfrac{1}{2}s^{1/2} - (k^2 + 1)^{1/2} + i\varepsilon \equiv \omega_1$$

do not pinch at the origin of the k_0 plane, as in the case of equal masses, but at a given $k_0 = \frac{1}{2}(\hat{p}^2 + M^2)^{1/2} - \frac{1}{2}(\hat{p}^2 + 1)^{1/2}$, where $\hat{p}$ is the absolute value of the center-of-mass three-momentum on the energy shell, satisfying

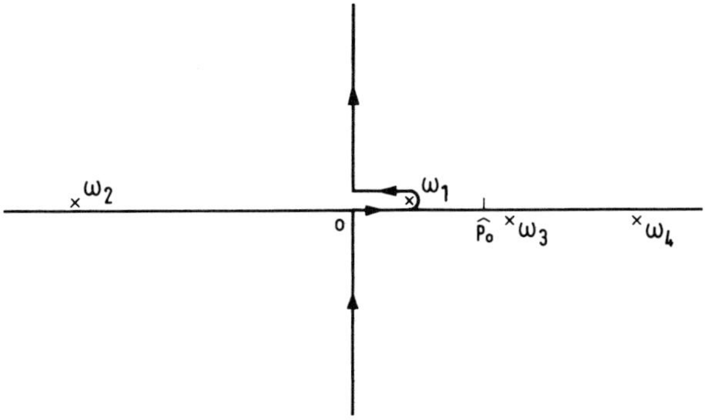

Fig. 2. The Wick-rotated path of integration in the k_0 plane. $\hat{p}_0 = \frac{1}{2}(\hat{p}^2 + M^2)^{1/2} - \frac{1}{2}(\hat{p}^2 + 1)^{1/2}$ is the pinching point of the poles ω_1 and ω_3.

$s^{1/2} = (\hat{p}^2 + M^2)^{1/2} + (\hat{p}^2 + 1)^{1/2}$. As a consequence, we do not have to use the Noyes–Kowalski trick for regularizing the kernel, but can directly apply the Wick rotation $k_0 \to e^{i\alpha}k_0$, $p_0 \to e^{i\alpha}p_0$, $0 \leq \alpha \leq \pi/2$, to deform the contour in the k_0 plane to the imaginary axis.

The two other poles of $G(k, k_0)$ are always in the second and fourth quadrant and will give no extra contributions.

The other singularities of the integrand are the branchpoints of the Q_l functions in $K(p, p_0; k, k_0)$. Since their location depends on p_0, they will be rotated away from the path of integration.

Thus, we get only contributions from the poles ω_1 and ω_3 for those values of k for which they are located in the first and third quadrants, respectively. Their residues must then be added to the rotated equation; this introduces an extra coupling with the values of the T matrix at the poles $T[p, \omega_1(p)]$ and $T[p, \omega_3(p)]$. The pole $\omega_3 = -\frac{1}{2}s^{1/2} + (k^2 + M^2)^{1/2} - i\varepsilon$ can be in the third quadrant only if $s^{1/2} > 2M$. Since we shall restrict our study to energies $s^{1/2} < 2M$, the contribution of ω_3 may be omitted.

The other pole $\omega_1 = \frac{1}{2}s^{1/2} - (k^2 + 1)^{1/2} + i\varepsilon$ is located in the first quadrant for $0 < k < k_{\max}$, with $k_{\max} = \frac{1}{2}(s - 4)^{1/2}$. This leads to the following Wick-rotated equations:

$$\begin{aligned} T(p, ip_0) &= V(p, ip_0) + \pi^{-1} \int_0^\infty dk \int_{-\infty}^{+\infty} dk_0\, G(k, ik_0) K(p, ip_0; k, ik_0) \\ &\quad \cdot T(k, ik_0) + \int_0^{k_{\max}} dk\, G_\omega(k) K[p, ip_0; k, \omega_1(k)] T_\omega(k)\,, \\ T_\omega(p) &= V_\omega(p) + \pi^{-1} \int_0^\infty dk \int_{-\infty}^{+\infty} dk_0\, G(k, ik_0) K[p, \omega_1(p); k, ik_0] \\ &\quad \cdot T(k, ik_0) + \int_0^{k_{\max}} dk\, G_\omega(k) K[p, \omega_1(p); k, \omega_1(k)] T_\omega(k)\,, \end{aligned} \tag{3}$$

with

$$V_\omega(p) = V[p, \omega_1(p)], \qquad T_\omega(p) = T[p, \omega_1(p)],$$
$$G_\omega(k) = (k^2 + 1)^{-1/2}\{k^2 + M^2 - i\varepsilon - [\tfrac{1}{2}s^{1/2} + \omega_1(k)]^2\}^{-1}.$$

What are now the remaining singularities of the integrands? The function in the two-dimensional integral

$$G(k, ik_0) = [k^2 + M^2 - (\tfrac{1}{2}s^{1/2} + ik_0)^2]^{-1}\cdot[k^2 + 1 - (\tfrac{1}{2}s^{1/2} - ik_0)^2]^{-1}$$

is singular only if simultaneously $k_0 = 0$ and $k = \frac{1}{2}(s - 4)^{1/2}$, but such a singularity is integrable, and can be taken into account numerically.

The other Green's function singularity occurs in the single integral:

$$k^2 + M^2 - [\tfrac{1}{2}s^{1/2} + \omega_1(k)]^2 = 0 \qquad \text{for} \quad k = \hat{p}.$$

This is the Lippmann–Schwinger type of singularity, related to the elastic unitarity for $M + 1 \leq s^{1/2} \leq M + 2$. Branch points of the Q_l functions in the potential V and the kernel K occur if the exchanged particle is on its mass shell.

In the Wick-rotated form, p_0 can have two types of values: the pole contribution value $p_0 = \omega_1(p) = \frac{1}{2}s^{1/2} - (p^2 + 1)^{1/2}$, which can be rewritten as the on-mass shell condition for the initial meson $(\frac{1}{2}P - p)^2 + 1 = 0$, or purely imaginary values $p_0 = iy$ $(-\infty < y < +\infty)$.

It is then easy to see that the inhomogeneous term V cannot have branch points: since the final state is already on shell, an exchanged particle with physical mass would be incompatible with conservation of four-momentum on the vertex with the incoming meson. For $p_0 = iy$, the condition $(p + \hat{p})^2 + M^2 = 0$ can be satisfied only if $p_0 = 0$, which implies $\hat{p}_0 > M$. However, we have $\hat{p}_0 = \frac{1}{2}(\hat{p}^2 + M^2)^{1/2} - \frac{1}{2}(\hat{p}^2 + 1)^{1/2} < (M - 1)/2$. Thererfore, V has no branch points. In the kernel K, the condition $(p + k)^2 + M^2 = 0$ leads to logarithmic cuts in the k_0 plane:

$$k_0 = -p_0 \pm [(p + \lambda k)^2 + M^2]^{1/2}, \qquad -1 \leq \lambda \leq +1.$$

The first inelastic threshold occurs whenever the branch points of these cuts pinch with the singularities of the Green's functions. A study of all possible combinations of the pole values and the values for k_0 and p_0 shows that the pinching occurs only in the one-dimensional integral in Eq. (3) at the values where

$$\omega_1(k) = -\omega_1(p) \pm [(p + \lambda k)^2 + M^2]^{1/2}.$$

Since $0 < \omega_1(p) + \omega_1(k) < 2(\frac{1}{2}s^{1/2} - 1)$, this happens only if

$$2(\tfrac{1}{2}s^{1/2} - 1) > M \qquad \text{or} \qquad s^{1/2} > M + 2.$$

The next inelastic threshold is located at $s^{1/2} = 3M$.

In the numerical calculations we have performed, we have restricted ourselves to energies $s^{1/2} < 2M$. As a consequence, we do not have to worry about these extra singularities, which give rise to the second inelastic threshold. The presence of cutoff terms introduces, of course, also a branch point at $s^{1/2} = \Lambda + 2$, but this will give no extra problems as long as we choose $\Lambda > 2M$.

To summarize, the only important singularities of the integrands are the pole in $G_\omega(k)$ at $k = \hat{p}$, and a branch point in $K[p, \omega_1(p); k, \omega_1(k)]$, if $s^{1/2} > M + 2$. These singularities, occurring in the one-dimensional integration of the Wick-rotated equation, can easily be removed by subtractions.

III. THE NUMERICAL METHOD

First of all, the equations must be written in a form suitable for a numerical solution. This is done as follows.

Consider for the moment the elastic region; the imaginary part of the right-hand side of Eq. (3) comes from the one-dimensional integral and is proportional to V. It may therefore be included in the inhomogeneous term. According to Eq. (3), this imaginary part is given by

$$R_j(p, p_0) = \frac{i\pi}{2\hat{p}s^{1/2}} \sum_{n=1}^{2} K_{jn}(p, p_0; \hat{p}, \hat{p}_0)\, T_n(\hat{p}, \hat{p}_0) ,$$

where the u, v indices are again written explicitly.

Now it should be remembered that, using for a moment p and k as four-momenta again, $K(p, k)$ is essentially the product of $V(p, k)$ and $(i\gamma k - M)$, which, divided by $-2M$, is a projection operator on the u spinors, if k is on the mass shell. Therefore, $K_{jn}(p, \hat{p})$ will be proportional to $\delta_{1n} V_j(p)$; the factor turns out to be $M/8\pi^2$. Thus,

$$\begin{aligned} R_j(p, p_0) &= i\frac{M}{16\pi\hat{p}s^{1/2}} T_1(\hat{p}, \hat{p}_0)\, V_j(p, p_0) \\ &= if\hat{T}V_j(p, p_0) . \end{aligned}$$

$\hat{T}$ is related to the phase shift by $\hat{T} = f^{-1}e^{i\delta} \sin\delta$. Including R in V, this leads to a new potential αV, with $\alpha = 1 + if\hat{T}$. Furthermore, the functions V, G, and K exhibit as functions of the imaginary variable the symmetry property $F^*(z) = F(z^*)$, which can be used to reduce the range of the variables p_0 and k_0 from $(-\infty, +\infty)$ to $(0, \infty)$.

Introducing the notation $\alpha^{-1}T(p, \pm ip_0) = T^\pm(p, p_0)$ and $\alpha^{-1}T_\omega(p) =$

$t(p)$, Eq. (3) is reduced to the following four equations:

$$T^+(p, p_0) = V(p, ip_0) + \pi^{-1}\int_0^\infty dk \int_0^\infty dk_0[G(k, ik_0)K(p, ip_0; k, ik_0)T^+(k, k_0) + G^*(k, ik_0)K(p, ip_0; k, -ik_0)T^-(k, k_0)] + \int_0^{k_{\max}} dk\, G_\omega(k)\{K[p, ip_0; k, \omega_1(k)]t(k) - (k/\hat{p})K(p, ip_0; \hat{p}, \hat{p}_0)t(\hat{p})\} + \hat{p}^{-1}K(p, ip_0; \hat{p}, \hat{p}_0)t(\hat{p}) \cdot I(s) , \tag{4}$$

$$T^-(p, p_0) = V^*(p, ip_0) + \pi^{-1}\int_0^\infty dk \times \int_0^\infty dk_0[G(k, ik_0)K^*(p, ip_0; k, -ik_0)T^+(k, k_0) + G^*(k, ik_0)K^*(p, ip_0; k, ik_0)T^-(k, k_0)] + \int_0^{k_{\max}} dk\, G_\omega(k)\{K^*[p, ip_0; k, \omega_1(k)]t(k) - (k/\hat{p})K^*(p, ip_0; \hat{p}, \hat{p}_0)t(\hat{p})\} + \hat{p}^{-1}K^*(p, ip_0; \hat{p}, \hat{p}_0)t(\hat{p}) \cdot I(s) , \tag{5}$$

$$t(p) = V_\omega(p) + \pi^{-1}\int_0^\infty dk \int_0^\infty dk_0\{G(k, ik_0)K[p, \omega_1(p); k, ik_0]T^+(k, k_0) + G^*(k, ik_0)K^*[p, \omega_1(p); k, ik_0]T^-(k, k_0)\} + \int_0^{k_{\max}} dk\, G_\omega(k)\{K[p, \omega_1(p); k, \omega_1(k)]t(k) - (k/\hat{p})K[p, \omega_1(p); \hat{p}, \hat{p}_0]t(\hat{p})\} + \hat{p}^{-1}K[p, \omega_1(p); \hat{p}, \hat{p}_0]t(\hat{p}) \cdot I(s) , \tag{6}$$

and

$$t(\hat{p}) = V_\omega(\hat{p}) + \pi^{-1}\int_0^\infty dk \int_0^\infty dk_0\{G(k, ik_0)K[\hat{p}, \omega_1(\hat{p}); k, ik_0]T^+(k, k_0) + G^*(k, ik_0)K^*[\hat{p}, \omega_1(\hat{p}); k, ik_0]T^-(k, k_0)\} + \int_0^{k_{\max}} dk\, G_\omega(k)\{K[\hat{p}, \omega_1(\hat{p}); k, \omega_1(k)]t(k) - (k/\hat{p})K[\hat{p}, \omega_1(\hat{p}); \hat{p}, \hat{p}_0]t(\hat{p})\} + \hat{p}^{-1}K[\hat{p}, \omega_1(\hat{p}); \hat{p}, \hat{p}_0]t(\hat{p}) \cdot I(s) . \tag{7}$$

In order to get rid of the principal value singularity of the Green's function, a subtraction has been used; the subtraction integral $I(s)$ is given by

$$P\int_0^{k_{\max}} dk\cdot kG_{\omega}(k) = (2s^{1/2})^{-1}\log\left(\frac{M^2-1}{s-2s^{1/2}+1-M^2}\right).$$

For energies below the first inelastic threshold $s^{1/2} < M+2$, the kernel $K[p, \omega_1(p); k, \omega_1(k)]$ is real; this implies that t is real, and that T^+ and T^- are each other's complex conjugates.

For the inelastic region $s^{1/2} > M+2$, this kernel will be complex in general, and Eqs. (4) and (5) become independent. t, and therefore the phase shift, will also become complex now.

The real part of the phase shift δ_R and the inelasticity η, defined by $S = \eta\cdot e^{2i\delta_R}$, are then given by $\delta_R = \frac{1}{2}\arg z$, and $\eta = |z|$, with $z = (1+ift)/(1-ift)$.

Three-particle unitarity can of course also be satisfied here by replacing the nucleon propagator $S(p) = -i(i\gamma p + M)^{-1}$ in the Green's function by the renormalized "bubble" propagator in lowest order,

$$\bar{S}(p) = S(p)[1 + A(p^2) + B(p^2)(i\gamma p + M)]^{-1}, \tag{8}$$

which is shown diagrammatically in Fig. 3. The functions $A(p^2)$ and $B(p^2)$ can be expressed in the form of a dispersion integral, and are given in Appendix B.

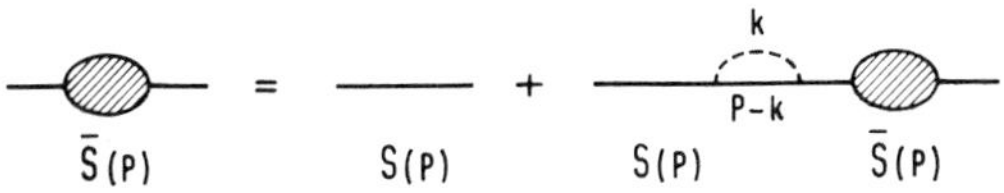

Fig. 3. The Dyson equation for the bubble propagator in diagrammatic form.

For the solution of the set of equations (4)–(7), we applied the method of Padé approximants. In the ladder approximation, the amplitude $\hat{t}(g^2)$ is approximated by the ratio of two polynomials $P_M(g^2)$ and $Q_N(g^2)$ of order M and N in g^2, respectively, and usually denoted by $[N, M]$, whose coefficients are determined from those of the formal power series expansion of $\hat{t}(g^2)$ at $g^2 = 0$ by the requirement that the expansion of $[N, M]$ at $g^2 = 0$ reproduces the first $N + M + 1$ coefficients of $\hat{t}$.

The coefficients of the series expansion of $\hat{t}(g^2)$ have been obtained by iterating the equations on an IBM 360–50, using an integration with eight Gaussian points in each variable. The accuracy of the coefficients, obtained by increasing the number of integration points, has been estimated to be $\leq 1\%$. Above inelastic threshold, the coefficients can be complex, because of the possibility $|z| < 1$, where z is the argument of the Q_l functions in $K[p, \omega_1(p); k, \omega_1(k)]$. Since this can occur only in a narrow strip of the p, k plane, one can take the slowly varying part outside the integral, and integrate the remainder analytically, in order to obtain numerically a stable result. It turns out that this contribution is negligible, however. As an independent check, the box diagram contribution was also calculated without Wick rotation; the cuts now remain on the real axis, and the Feynman integral is evaluated by contour integration.

With respect to the solution of the equations satisfying three-particle unitarity, this has been done by replacing the function $S(p)$ in the Green's functions by the calculated value of $\bar{S}(p)$ at the given coupling constant $g^2 = 14.6$. Padé technique is then used on the modified coefficients. Strictly speaking, we are of course not applying the Padé method on the power-series expansion in g^2 However, the result turns out also to be convergent.

IV. RESULTS AND DISCUSSION

With the equations obtained previously, we can analyze in principle any partial wave; however, without additional refinements, we only expect a reasonable description of the P_{33} partial wave, where N exchange is known to be the dominant force. Therefore, we mainly concentrate on this partial wave.

The cutoff Λ, which is the only free parameter here, is fixed by the mass of the N^* resonance. Choosing $M = 6.72$ and $M^* = 8.85$, we obtain $\Lambda = 15.7$ for the ladder graphs. The resulting phase shift is shown

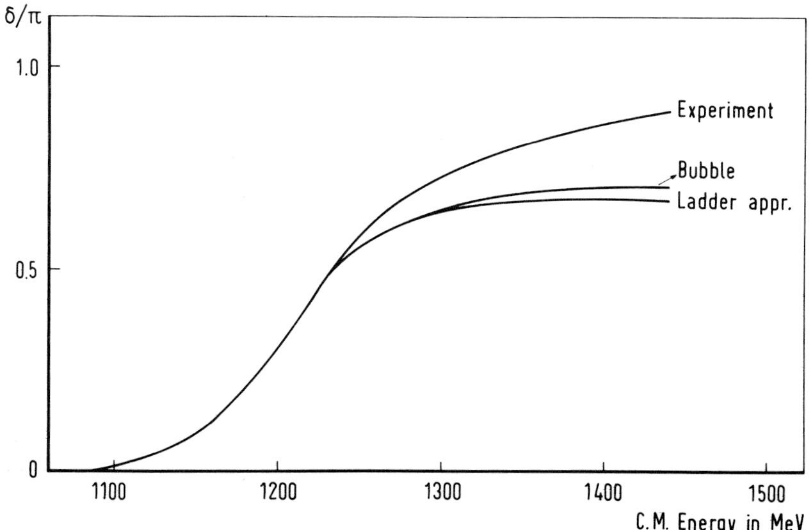

Fig. 4. The P_{33} phase shifts as a function of center-of-mass energy.

in Fig. 4. The agreement with the experimental data is satisfactory up to an energy of about 1240 MeV in the center-of-mass system. In particular, the a_{33} scattering length obtained from a Chew–Low effective range plot comes out fairly well: $a_{33} = 0.206 \pm 0.005$, to be compared with the experimental value $a_{33} = 0.215 \pm 0.005$ (Hamilton and Woolcock, 1963).

In case we also include the self-energy graphs, the forces become effectively stronger. As a result, the cutoff Λ should be adjusted to 10.5 in order to get the position of the N^* at the right position. The phase shift is shown in Fig. 4. The main difference with the ladder case is that the phase shift reaches a higher maximum. It should be noted that in the elastic region the $[N, N]$ approximant satisfies two-particle unitarity. However, in the inelastic region, the Padé approximants violate three-particle unitarity, although the exact solution of course satisfies it. In spite of this, the rate of convergence in both regions is the same.

With respect to the inelasticity, we find that it is very small, as well in the ladder approximation as after inclusion of the bubble.

The inclusion of the bubble in the original nucleon propagator introduces some features, which must be considered in more detail. First of all, the asymptotic behavior is different from the scalar case, since it can be shown (Woolcock, 1967, 1968) that $A(p^2)$ diverges logarithmically for $|p^2| \to \infty$, and that $B(p^2) \sim (\log p^2)/p^2$; therefore, the kernel of the BS equation acquires an extra converging factor $(\log p^2)^{-1}$, which makes the interaction somewhat less singular.

In order to be consequent, one should also use the same cutoff procedure for the nucleon propagator in the integrand of the Dyson equation (Fig. 3), because this state of the intermediate nucleon is the same as the exchanged nucleon. This cutoff term, being already a small correction on a correction of the Green's function nucleon propagator, has a negligible influence on the phase shift, and is therefore omitted.

Of course, the possibility of ghost solutions must be considered, too. This occurs if the equation

$$[(g^2)^{-1} + A(p^2) + MB(p^2)]^2 + B(p^2)^2 \cdot p^2 = 0 \tag{9}$$

has for some value of p^2 a solution with $0 < g^2 < \infty$. A and B are the functions given in Appendix B, but without the factor g^2.

It is clear that complex values of p^2 lead to complex g^2, and are therefore ruled out. A and B are real analytic functions of p^2, and can give rise to a root of Eq. (9) only if $p^2 > -(M+1)^2$. In addition, Eq. (9) admits a solution only if $p^2 < 0$. We find indeed a solution in the foregoing region. However, the minimum value of g^2 for which there is a root is about 250.

As has been described already in Ref. 4, the [1, 1] approximant turns out to be a bad approximation in the P_{33} channel. A similar result has been obtained by Gammel *et al.* (Chapter 13 of this volume) in a coupled channel problem. Taking as a criterion the value of g^2 for which the first bound state occurs at elastic threshold, the [1, 1] gives a result for the coupling constant, which is off by a factor of 2 at least, compared to the exact result.

In order to find out the possible reason for this phenomenon, we have looked at some potentials in the nonrelativistic case, such as the Yukawa, exponential, Gaussian, and square-well potential. It turns out that the [1, 1] predicts the S waves rather well, but higher partial waves come out badly in general. The deviation becomes larger with increasing l, which can be ascribed to the threshold factors in the Lippmann–Schwinger equation. In some cases, such as the square-well or the Gaussian potential, this effect is less enhanced. This is related to the smooth behavior of the potential near the origin. To check whether the discrepancy in the relativistic case was also due to the centrifugal barrier, we studied the S waves and D waves and have indeed found a similar behavior. From this, we expect that the [1, 1] approximant should also be bad in the scalar case for $l \geq 1$.

APPENDIX A

In this appendix, we write out explicitly the inhomogeneous term $V(p, p_0)$ and the kernel $K(p, p_0; k, k_0)$ for the $j = l \pm \frac{1}{2}$ partial wave. Consider

$$V(p, p_0; k, k_0) = \frac{4\pi g^2}{M[(E_p + M)(E_k + M)]^{1/2}} [F_+(p, p_0; k, k_0) Q_l(z) + F_-(p, p_0; k, k_0) Q_{l\pm 1}(z)] , \tag{A1}$$

with

$$E_p = (p^2 + M^2)^{1/2} ,$$
$$z = \frac{(p_0 + k_0)^2 - p^2 - k^2 - M^2 + i\varepsilon}{2pk} ; \tag{A2}$$

$$F_\pm(p, p_0; k, k_0) = N_\pm(p, k)[\pm M(1 - 2\Theta) + E_p + E_k - p_0 - k_0] ,$$

where $N_\pm$ and Θ are given in Table I for the four possible spinor combinations.

The inhomogeneous term is then given by

$$V(p, p_0) = V(p, p_0; \hat{p}, \hat{p}_0) .$$

The kernel $K(p, p_0; k, k_0)$ is obtained by multiplying $V(p, p_0; k, k_0)$ with $C(k, k_0)$, which stands for the operator $i\gamma(\frac{1}{2}P + k) - M$, taken be-

TABLE I

Spinor combination	$N_+(p,k)$	$N_-(p,k)$	Θ	$C(k,k_0)$
$\bar{u}\cdots u$	$(E_p+M)(E_k+M)$	pk	1	$-2M-(\frac{1}{2}s^{1/2}+k_0-E_k)(E_k/M)$
$\bar{u}\cdots v$	$(E_p+M)\cdot k$	$p\cdot(E_k+M)$	0	$-(\frac{1}{2}s^{1/2}+k_0-E_k)(k/M)$
$\bar{v}\cdots u$	$p\cdot(E_k+M)$	$(E_p+M)\cdot k$	0	$-(\frac{1}{2}s^{1/2}+k_0-E_k)(k/M)$
$\bar{v}\cdots v$	pk	$(E_p+M)(E_k+M)$	-1	$-(\frac{1}{2}s^{1/2}+k_0-E_k)(E_k/M)$

tween two spinors u and v (see Table I). The precise relations are

$$\begin{aligned}
K^{uu} &= -(16\pi^2)^{-1}(V^{uu}C^{uu} - V^{uv}C^{vu}) ,\\
K^{uv} &= (16\pi^2)^{-1}(V^{uu}C^{uv} - V^{uv}C^{vv}) ,\\
K^{vu} &= (16\pi^2)^{-1}(V^{vu}C^{uu} - V^{vv}C^{vu}) ,\\
K^{vv} &= -(16\pi^2)^{-1}(V^{vu}C^{uv} - V^{vv}C^{vv}) .
\end{aligned}$$

APPENDIX B

The unrenormalized "bubble" propagator is given in lowest order by

$$\bar{S}_0(p) = \frac{-i}{i\gamma p + M}\left[1 + \frac{-i}{i\gamma p + M} \cdot 4\pi g^2 \int \frac{d^4k}{(2\pi)^4} \frac{i\gamma(p-k) + M}{(p-k)^2 + M^2 - i\varepsilon} \frac{1}{k^2 + 1 - i\varepsilon}\right]^{-1} . \tag{B1}$$

Using Cutkosky rules, the integral can be written as a dispersion integral; the result is

$$\frac{i}{32\pi^2}\int_{(M+1)^2}^{\infty} \frac{ds}{s + p^2 - i\varepsilon}\,\vartheta(s)\left[i\gamma p\left(1 + \frac{M^2-1}{s}\right) + 2M\right], \tag{B2}$$

with

$$\vartheta(s) = \left[1 - \frac{(M-1)^2}{s}\right]^{1/2}\left[1 - \frac{(M+1)^2}{s}\right]^{1/2} .$$

The finite part of this expression is obtained by expanding the integrand in powers of $(i\gamma p + M)$ and removing the first two divergent terms by renormalisation. This leads to the renormalized propagator:

$$\bar{S}(p) = \frac{-i}{i\gamma p + M}[1 + A(p^2) + B(p^2)(i\gamma p + M)]^{-1} , \tag{B3}$$

where

$$A(p^2) = \frac{g^2}{8\pi}(p^2 + M^2)\int_{(M+1)^2}^{\infty} \frac{ds}{s(s + p^2 - i\varepsilon)}\,\vartheta(s)\left(\frac{s + M^2}{(s - M^2)^2} - 1\right), \tag{B4}$$

and

$$B(p^2) = \frac{g^2}{8\pi} M \int_{(M+1)^2}^{\infty} \frac{ds}{s(s + p^2 - i\varepsilon)} \vartheta(s) \left(\frac{1}{s - M^2} + 1 \right). \tag{B5}$$

Here in the coupling constant g^2, the isospin factor of the direct Born term, which is $+3$, should also be included.

ACKNOWLEDGMENT

This investigation is part of the research program of the Stichting voor Fundamenteel Onderzoek der Materie (F. O. M.), which is financially supported by the Nederlandse Organisative voor Zuiver Wetenschappelijk Onderzoek (Z. W. O.).

REFERENCES

Hamilton, J., and Woolcock, W. S. (1963). *Rev. Mod. Phys.* **35,** 737.
Kowalski, K. L. (1965). *Phys. Rev. Letters* **15,** 798.
Levine, M. J., Wright, J. and Tjon, J. A. (1967). *Phys. Rev.* **157,** 1416.
Nieland, H. M., and Tjon, J. A. (1968). *Phys. Letters* **27B,** 309.
Noyes, H. P. (1965). *Phys. Rev. Letters* **15,** 538.
Schwartz, C., and Zemach, C. (1966). *Phys. Rev.* **141,** 1454,
Woolcock, W. S. (1967, 1968). *J. Math. Phys.* **8,** 1270; **9,** 1350.

CHAPTER 13 APPLICATION OF THE PADÉ APPROXIMANT TO A COUPLED CHANNEL PROBLEM: MESON-BARYON SCATTERING

J. L. Gammel and M. T. Menzel

University of California
Los Alamos Scientific Laboratory

J. J. Kubis

Cambridge University
England

The results of this chapter are mainly negative. The work began with the question of whether or not the coupling of the $\pi^+ - p$ channel to the $K^+ - \Sigma^+$ channel might not affect the parameters (position and width) describing the (3, 3) resonance. We therefore develop a matrix Padé approximant to the tangent matrix in order to take account of this coupling. The usefulness of the formalism is tested by applying it to a coupled channel potential model that simulates the field theory situation as closely as possible. It turns out that, because of the analytic structure of the tangent matrix elements as functions of energy, the formalism is not applicable at the (3, 3) resonance. At energies at which the formalism is applicable to the potential model, the [1, 1] matrix Padé approximant is not accurate. Ordinary [2, 2] Padé approximants to each element of the tangent matrix separately are accurate at all energies. This evidence, like the evidence of Tjon and Nieland (Chapter 12 of this volume), suggests that at least [2, 2] approximants will be required in the case of field theory.

I. INTRODUCTION

Let the strong interaction Hamiltonian be

$$\mathscr{H}_I = \lambda^{1/2} G_{ijk} \bar{\psi}_i \gamma_5 \psi_j \varphi_k ,$$

where λ is a parameter that either takes the value unity in all calculations or scales the strength of the interaction, the $\bar{\psi}$ and ψ are creation and annihilation operators for the basic octet of baryons (nucleons N, lambda Λ, sigmas Σ, and xi particles Ξ), and the φ annihilate or create one of the basic octet of mesons (pions π, kaons K, antikaons $\bar{K}$, and eta particles η).

The coupling constants G_{ijk} might be chosen in one of the following ways:

(1) When the strange particles are taken into account, isospin symmetry and strangeness conservation reduce the number of coupling constants to 12.

(2) SU_3 symmetry further reduces this number to two.

Mainly we have used the second choice.

When the strange particles are taken into account, the problem of coupled channels arises. By this problem, we mean that, for example, the reaction $\pi^+ - p \rightleftarrows K^+ - \Sigma^+$ couples the $\pi^+ - p$ and $K^+ - \Sigma^+$ channels. Above the threshold for this reaction, the scattering amplitude becomes a 2×2 matrix (Wigner and Eisenbud, 1947) that describes simultaneously $\pi^+ - p$ and $K^+ - \Sigma^+$ elastic scattering and the reaction referred to. Even below threshold, the coupling must be taken into account. How this situation is handled in the Padé method is described in Sect. II.

In order to form the [1, 1] Padé approximant, we need the expansion of the scattering amplitude through second order in the parameter λ. We have published separately (Kubis and Gammel, 1968) an account of the derivation of the expressions for the first- and second-order amplitudes that we have used in the numerical work. Sections III and IV are a straightforward tally of our numerical results on the (3, 3) phase shift for pion-nucleon scattering.

Although, as may be seen from the foregoing discussion, our purpose is to develop the Padé formalism for a coupled channel problem and to calculate the (3, 3) phase shift for pion–nucleon scattering as a specific example of the method, our goal is to fit the (3, 3) phase shift found by Roper (1964) in his phase shift analysis of the pion-nucleon scattering data. We should like to achieve this fit with a value of the pion–nucleon coupling constant in agreement with the value obtained by other methods (Kim, 1967) and a value of the F/D ratio in agreement with the value obtained by other methods. We have not reached this goal for several reasons.

The first reason is that, in view of the work of Tjon and Nieland (Chapter 12), it is apparent that the [1, 1] Padé approximant may not be sufficiently accurate. From our potential model calculations, we shall conclude that such is the case.

The second reason is that we have not included the so-called $\lambda\varphi^4$ term,

$$\mathscr{H}_I' = \lambda_{ijkl}\varphi_i\varphi_j\varphi_k\varphi_l ,$$

in our interaction Hamiltonian. In their work on $\pi - \pi$ scattering, Bessis and Pusterla (1967), and also Newton (1967), used a pure $\lambda\varphi^4$ interaction with pions only. Mignaco *et al.* (1968) calculations on pion nucleon scattering which include this term indicate that it is not important.

The third reason that we have not reached this goal is that the Padé formalism that we develop cannot be applied to pion-nucleon scattering at the energy of the (3, 3) resonance. The reason for this fact will be set out at length below (Sect. II.F); briefly stated, the reason has to do with the analytic structure of the first and second Born approximations to the $K^+ - \Sigma^+$ elastic scattering amplitude as a function of the K^+ laboratory kinetic energy (K^+ incident on Σ^+ at rest). This energy is negative (-532.6 MeV) at the energy corresponding to the (3, 3) resonance, and a branch point in the first Born approximation occurs at a higher (less negative) value of the energy (-500 MeV). There is nothing incomprehensible in these facts; indeed, the same things happen for coupled channel potential models (Sect. II.F-2) when the Q-value for the reaction is so great that the range parameter calculated from the Q-value [(range parameter)$^{-2} = 2 \times$ (reduced mass in $K^+ - \Sigma^+ \times Q/\hbar^2$)] is greater than the range of the potential.

II. THE PADÉ APPROXIMANT TO THE SCATTERING AMPLITUDE IN THE CASE OF COUPLED CHANNEL POTENTIAL MODELS

We consider the coupled channel Schrödinger equations

$$-\frac{\hbar^2}{2M_i}\left(\frac{d^2u_i}{dr^2} - \frac{l_i(l_i+1)}{r^2}u_i\right) + V_{ij}u_j = (E - Q_i)u_i . \tag{1}$$

The index j is summed over. Here, as in the following, we use the convention that indices appearing on both sides of an equation are not summed over, and indices appearing on one side only of an equation are summed over. The range of the indices i and j is the number of coupled channels. M_i is the reduced mass in channel i, l_i the angular momentum in channel i, the V_{ij}'s are potentials and arbitrary functions of r, $V_{ij} = V_{ji}$, and E is the center-of-mass energy. The Q's are Q values for various reactions; in a two-channel problem, one would take $Q_1 = 0$, $Q_2 = Q$. We shall be thinking in terms of a two-channel case ($\pi^+ -$ P and $K^+ - \Sigma^+$) in the following, but all of the results have obvious generalizations to more channels.

A. Above Threshold Case ($E > Q_1$, $E > Q_2$)

These Schrödinger equations may be converted to integral equations in the usual way:

$$u_i^{(k)} = j_i \delta_{ik} + (2M_i/\hbar^2) \int G_i V_{ij} u_j^{(k)} \, dr' \tag{2}$$

j_i is a radial Bessel function $j_n(k_i r)$, $n = l_i$, $k_i^2 = 2M_i(E - Q_i)/\hbar^2$. [$j_0(kr) = \sin kr$, and the corresponding irregular radial Bessel function $n_0 = \cos kr$.]

The superscript (k) has been introduced because the system of equations Eq. (1) has a number (equal to the number of channels) of linearly independent solutions that are regular (that is, vanish) at the origin. [A solution may be thought of as a column matrix whose components are the u_i; the $u_i^{(k)}$ then form a square matrix whose columns are labeled by (k) and whose rows are labeled by i.] These linearly independent solutions are obtained by using an inhomogeneous term in the equation for $u_k^{(k)}$ as indicated by the Kronecker delta in Eq. (2).

Depending on the choice of the Green's functions G_i in Eq. (2), different, but linearly related, solutions of Eq. (1) are obtained. One choice is

$$G_i(r, r') = (k_i)^{-1} j_i(k_i r_<) n_i(k_i r_>) , \tag{3}$$

where $r_<$ is the lesser of r and r', and $r_>$ the greater.

Thus, asymptotically (as $r \to \infty$),

$$u_i^{(k)} \sim j_i \delta_{ik} + \frac{2M_i}{k_i \hbar^2} u_i \int_0^\infty j_i V_{ij} u_j^{(k)} \, dr' . \tag{4}$$

With a suitable and obvious definition of the unsymmetrized tangent or x-matrix,

$$u_i^{(k)} \sim j_i \delta_{ik} + n_i x_{ik} . \tag{5}$$

Integral equations for the unsymmetrized x-matrix elements may be obtained as follows. For the Green's functions, introduce integral representations,

$$G_i = \frac{2P}{\pi} \int_0^\infty \frac{j_i(\kappa r) j_i(\kappa r')}{\kappa^2 - k_i^2} \, d\kappa . \tag{6}$$

P stands for principal part of the integral. Also, extend the definition of the x-matrix elements as implied by Eqs. (4) and (5) to off-energy-shell elements as follows:

$$x_{ik}(\kappa) = \frac{2M_i}{\kappa \hbar^2} \int j_i(\kappa r') V_{ij} u_j^{(k)} \, dr' . \tag{7}$$

Then, multiplying Eq. (2) by

$$\frac{2M_l}{\kappa\hbar^2} j_l(\kappa r) V_{li}(r) ,$$

integrating over r, and summing over i yields

$$x_{ik}(\kappa) = \kappa^{-1} V_{ij}(\kappa, k_j)\,\delta_{jk} + \frac{2P}{\pi}\int_0^\infty \frac{\kappa'}{\kappa} V_{ij}(\kappa, \kappa')(\kappa'^2 - k_j^2)^{-1} x_{jk}(\kappa')\,d\kappa' , \qquad (8)$$

where

$$V_{ij}(\kappa, \kappa') = \frac{2M_i}{\hbar^2}\int j_i(\kappa r)\,V_{ij}(r) j_j(\kappa' r)\,dr . \qquad (9)$$

One defines the symmetrized x-matrix elements as follows:

$$x_{ij}(\kappa) = \left(\frac{\hbar\kappa}{M_i V_j}\right)^{1/2} x_{ij}(\kappa) , \qquad (10)$$

where $V_i = \hbar k_i / M_i$ is the relative speed of the particles in channel i. Defining symmetrized potential matrix elements as follows:

$$\overline{V}_{ij}(\kappa, \kappa') = \frac{2(M_i M_j)^{1/2}}{\hbar^2(\kappa\kappa')^{1/2}}\int j_i(\kappa r)\,V_{ij}(r) j_j(k' r)\,dr , \qquad (11)$$

one obtains

$$\bar{x}_{ij}(\kappa) = \overline{V}_{ij}(\kappa, \kappa_j) + \frac{2P}{\pi}\int_0^\infty \overline{V}_{ik}(\kappa, \kappa')\,\frac{\kappa'}{\kappa'^2 - k_k^2}\,\bar{x}_{kj}(\kappa')\,d\kappa' . \qquad (12)$$

It may be verified from Eq. (1), by using the procedure for showing that the Wronskian of two linearly independent solutions is constant and noting the constant is zero if both solutions vanish at the origin, that

$$\bar{x}_{ij} = \bar{x}_{ji} . \qquad (13)$$

B. Separable Potentials

The close connection between Padé approximants and separable potentials is well known and has been emphasized by Chisholm (1963). Therefore, let the potential be separable:

$$\overline{V}_{ik} = \varphi_{ij}^T \varphi_{jk} , \qquad (14)$$

where

$$\varphi_{ij} = \varphi_{ij}(k_j) , \qquad (15)$$

although an arbitrary function of k_j should be chosen with some regard for l_j. φ^T stands for the transposed matrix.

For this separable potential, Eq. (12) can be solved exactly, and the solution is

$$x = \varphi^T (1 - \Lambda)^{-1} \varphi , \tag{16}$$

where all quantities appearing in this equation are matrices, matrix multiplication is implied, and

$$\Lambda = \frac{2P}{\pi} \int_0^\infty d\kappa\ \varphi(\kappa) G(\kappa) \varphi^T(\kappa) . \tag{17}$$

Equation (17) is also a matrix equation: the matrix $G(\kappa)$ is

$$[G(\kappa)]_{ij} = \frac{\kappa}{\kappa^2 - k_i^{\,2}} \delta_{ij} . \tag{18}$$

The argument of φ in Eq. (16) is as shown in Eq. (15), and the argument of φ^T is k_i.

It is instructive to carry out these operations for a definite example. We have considered

$$(\varphi) = \begin{pmatrix} \dfrac{\alpha k_1^{\,3/2}}{k_1^{\,2} + \mu^2} & \dfrac{\gamma k_2^{\,3/2}}{k_2^{\,2} + \mu^2} \\ \dfrac{\beta k_1^{\,3/2}}{k_1^{\,2} + \mu^2} & \dfrac{\delta k_2^{\,3/2}}{k_2^{\,2} + \mu^2} \end{pmatrix}, \tag{19}$$

which is appropriate when $l_1 = l_2 = 1$, and found

$$\begin{aligned} \Lambda = {} & \frac{1}{2} \frac{\mu}{(k_1^{\,2} + u^2)^2} (3k_1^{\,2} + \mu^2) \begin{pmatrix} \alpha^2 & \alpha\beta \\ \alpha\beta & \beta^2 \end{pmatrix} \\ & + \frac{1}{2} \frac{\mu}{(k_2^{\,2} + \mu^2)^2} (3k_2^{\,2} + \mu^2) \begin{pmatrix} \gamma^2 & \gamma\delta \\ \gamma\delta & \delta^2 \end{pmatrix} . \end{aligned} \tag{20}$$

C. The Padé Approximant

It is important to note that Eq. (16) is exactly what one obtains from the [1, 1] Padé approximant (Baker, 1965; Wall, 1948), provided all operations are carried out using matrix multiplication. Expanding in powers of the interaction, one obtains

$$\bar{x} = \varphi^T \varphi + \varphi^T \Lambda \varphi + \cdots . \tag{21}$$

In order to obtain the Padé approximant, one writes

$$\bar{x} = A(1 + B)^{-1} , \tag{22}$$

and requires that

$$A = \varphi^T \varphi , \tag{23}$$

and

$$-AB = \varphi^T \Lambda \varphi \,. \tag{24}$$

Thus,

$$B = -\varphi^{-1} \Lambda \varphi \,, \tag{25}$$

and

$$\begin{aligned} \bar{x} &= \varphi^T \varphi (1 - \varphi^{-1} \Lambda \varphi)^{-1} \\ &= \varphi^T (1 - \Lambda)^{-1} \varphi \,, \end{aligned} \tag{26}$$

which agrees with Eq. (16).

D. Properties of the Padé Approximant

(1) The threshold dependences of the terms in the Born expansion, Eq. (21), may be divided out before the Padé approximant is formed, and the result will be the same. As may be seen from Eqs. (11) and (14), it is convenient to write

$$\varphi = \varphi_R k \,, \tag{27}$$

where k is a diagonal matrix:

$$k_{ij} = k_i{}^p \delta_{ij} \,, \qquad p = l_i + \tfrac{1}{2} \,. \tag{28}$$

Here and hereafter, the subscript R indicates that a quantity has been reduced by dividing out its leading energy dependence at threshold.

We give a formal proof of this property in discussing property (2).

(2) The basis in which the tangent matrix is calculated does not matter. For the problem we are considering, one could use as basis states $\pi^+ - p$ and $K^+ - \Sigma^+$, but for complete SU_3 symmetry, in which the masses of all mesons are taken equal and the masses of all baryons are taken equal (we hasten to add that our calculations are not limited to this equal mass case), the coupling is eliminated if one takes

$$\begin{aligned} u_1 &= \sqrt{2}^{-1}[(\pi^+ - p) + (K^+ - \Sigma^+)] \,, \\ u_2 &= \sqrt{2}^{-1}[(\pi^+ - p) - (K^+ - \Sigma^+)] \,. \end{aligned} \tag{29}$$

One of these functions belongs to a 10-dimensional representation of SU_3, and the other to a 27-dimensional representation. Since the coupling is eliminated, one forms ordinary Padé approximants to $\tan \delta(10)$ and $\tan \delta(27)$. The question that arises is this: do these ordinary Padé approximants give the same result as the result obtained by calculating in the $(\pi^+ - p)$, $(K^+ - \Sigma^+)$ basis and forming matrix Padé approximants?

To see that the answer is yes, consider a change of basis:

$$u' = Uu \,.$$

The S matrix relates initial and final states in the U basis as follows:

$$u_f = Su_i \,. \tag{30}$$

Multiply from the left by U, and on the right put $u_i = U^{-1}Uu_i$, so that

$$u_f' = USU^{-1}u_i' \,,$$

that is,

$$S' = USU^{-1} \,, \tag{31}$$

as is well known.

Let the Born series for S' be

$$S' = 1 + S_1'V + S_2'V^2 + \cdots \,,$$

and the Born series for S be

$$S = 1 + S_1V + S_2V^2 + \cdots \,.$$

Then

$$S_1' = US_1U^{-1} \,,$$
$$S_2' = US_2U^{-1} \,.$$

The Padé approximant to S' is

$$\begin{aligned} S' &= S_1'(S_1' - S_2')^{-1}S_1' \,, \\ &= US_1U^{-1}(US_1U^{-1} - US_2U^{-1})^{-1}US_1U^{-1} \,, \\ &= US_1(S_1 - S_2)^{-1}S_1U^{-1} \,. \end{aligned}$$

Q.E.D.

We claim that the proof that the threshold dependence can be factored out is entirely similar.

Properties (1) and (2) are quite desirable. Later we shall consider forming an ordinary Padé approximant to each element of the tangent matrix separately. In that case, the result depends on the basis in which one calculates, which is not desirable.

E. Below Threshold Case ($E > Q_1$, $E < Q_2$)

We now restrict ourselves to two channels to avoid awkward writing.

1. *Standard Scattering Theory*

In the below threshold case, there is only one solution of physical interest, namely,

$$\begin{aligned} u_1^{(1)} &\sim j_1 + y_{11}u_1 \,, \\ u_2^{(1)} &\sim y_{21}h_2^+(i\alpha_2 r) \,, \end{aligned} \tag{32}$$

where $\alpha_2^2 = -k_2^2 = 2M_2(Q - E_2)/\hbar^2$, and $h_2^+ = j_2 - in_2$ approaches zero as r approaches infinity. One can persist in defining a square tangent or

y matrix by defining another solution of no immediate physical interest, namely,

$$\begin{aligned} u_1^{(2)} &\sim y_{12}n_1 , \\ u_2^{(2)} &\sim j_2(i\alpha_2 r) + y_{22}h_2^+(i\alpha_2 r) . \end{aligned} \tag{33}$$

The solution of Eq. (32) is obtained from the integral equations

$$\begin{aligned} u_1 &= j_1 + \frac{2M_1}{\hbar^2} \int G_1(V_{11}u_1 + V_{12}u_2) , \\ u_2 &= \frac{2M_2}{\hbar^2} \int G_2(V_{21}u_1 + V_{22}u_2) , \end{aligned} \tag{34}$$

where G_1 is, as before, given by Eq. (6), and

$$G_2 = \frac{2}{\pi} \int_0^\infty \frac{j_2(\kappa r) j_2(\kappa r')}{\kappa^2 + \alpha_2^2} \, d\kappa . \tag{35}$$

2. *The Analytic Tangent Matrix*

However, instead of proceeding in this way, we introduce what we call the analytic tangent matrix by using Eq. (2) just as we did before, except that the argument of j_2 contains $i\alpha_2$ as a factor, and, instead of the Green's function given by Eq. (35), we calculate a Green's function that is the analytic continuation of the principal part we had to calculate before; that is, we average the integral shown in Eq. (35) with its value calculated with the contour shown in Fig. 1. Thus we still find two solutions of Eq. (1), namely,

$$\begin{aligned} u_1^{(1)} &\sim j_1 + n_1 x_{11} & \qquad u_1^{(2)} &\sim n_1 x_{12} , \\ u_2^{(1)} &\sim n_2 x_{21} , & u_2^{(2)} &\sim j_2 + n_2 x_{22} . \end{aligned} \tag{36}$$

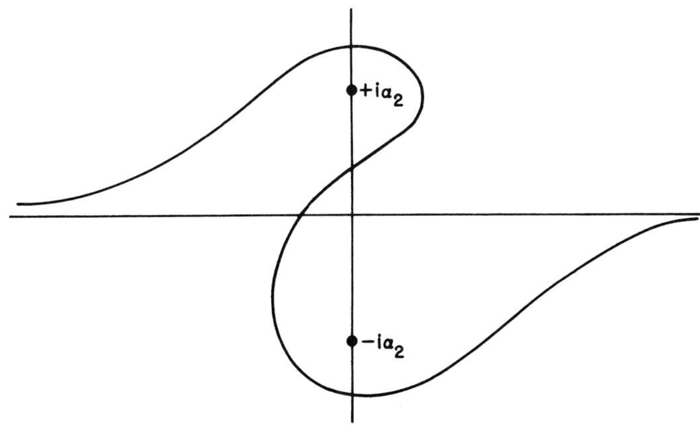

Fig. 1. A contour used in the calculation of a Green's function below threshold.

These tangent matrix elements are the analytic continuation of our previous tangent matrix elements below threshold, and so we have given them the same name (x).

3. *Connection between y and x*

To form y, one linearly combines the two solutions shown in Eq. (36) to get solutions of the form of Eqs. (32) and (33); one finds ultimately, in terms of reduced and symmetrized quantities, that

$$(\bar{y}_R)_{ij} = (\bar{x}_R)_{ij} - \frac{(-1)^p \alpha_2^{2p+1} (\bar{x}_R)_{i2} (\bar{x}_R)_{2j}}{1 + (-1)^p \alpha_2^{2p+1} (\bar{x}_R)_{22}}, \tag{37}$$

where $p = l_2$. $\bar{y}_r$ is not real above threshold, and $\bar{y}_R$ is not analytic due to the odd α_2 that appears in Eq. (37). Perhaps we should not say that $\bar{y}_R$ is not analytic; more precisely, we mean that as a function of α_2^2 it has a branch point at $\alpha_2^2 = 0$ and so cannot be numerically extrapolated from $\alpha_2^2 < 0$ (above threshold) to $\alpha_2^2 > 0$ (below threshold). $\bar{x}_R$ is real above and below threshold and can be extrapolated.

4. *The Example of Eq.* (19)

We have verified Eq. (37) for the example set forth in Eq. (19). Calculating $\bar{y}_R$, we find that it is given by an equation entirely analogous to Eq. (16) with

$$\begin{aligned}(\Lambda) = {} & \frac{1}{2} \frac{\mu}{(k_1^2 + \mu^2)^2} (3k_1^2 + \mu^2) \begin{pmatrix} \alpha^2 & \alpha\beta \\ \alpha\beta & \beta^2 \end{pmatrix} \\ & + \frac{1}{2} \frac{\mu^3 - 3\alpha_2^2\mu + 2\alpha_2^3}{(\mu^2 - \alpha_2^2)^2} \begin{pmatrix} \gamma^2 & \gamma\delta \\ \gamma\delta & \delta^2 \end{pmatrix}\end{aligned} \tag{38}$$

Notice that, because of the term $2\alpha_2^3$ in the numerator of the second term on the right-hand side of Eq. (38), this (Λ) is *not* the analytic continuation of the (Λ) shown in Eq. (20) below threshold $(k_2^2 = -\alpha_2^2)$. We then calculated $\bar{x}_R$ and found it to be given by Eq. (16) with (Λ) as shown in Eq. (38) *without* the term $2\alpha_2^3$ referred to; thus, as we expected, this (Λ) is indeed the analytic continuation of the (Λ) of Eq. (20) below threshold. Finally, with the $\bar{y}_R$ and $\bar{x}_R$ so calculated, we verified Eq. (37). The algebra is very pretty, but we shall avoid presenting the details.

5. *Consequences of Eq.* (37) *for the Born Expansion*

Since one is computing the perturbation series in field theory, it may be just as well to spell out the consequences of Eq. (37) for the Born expansion. Denoting the first and second Born terms by upper indices (1) and (2), respectively, for $l_2 = 1$,

$$(\bar{y}_R^{(1)})_{11} = (\bar{x}_R^{(1)})_{11}\,, \qquad\qquad (39)$$
$$(\bar{y}_R^{(2)})_{11} = (\bar{x}_R^{(2)})_{11} + \alpha_2{}^3(\bar{x}_R^{(1)})_{12}(\bar{x}_R^{(2)})_{21}\,.$$

Equations (39) are not the full consequences of Eq. (37); we shall not need the others. It is easy to verify these equations for the example of Eq. (19) using the result Eq. (38). Again, we avoid presenting the algebra.

F. What One Calculates in Quantum Field Theory

We assert that above threshold one naturally computes the Born expansion of the $\bar{x}_R$ matrix, and below threshold one naturally computes the Born expansion of $(\bar{y}_R)_{11}$. By the word "naturally," we do not imply that it is impossible to proceed in some other way; we mean only that it is probable that one proceeds in the way described.

We assert that the above-threshold Born elements can be numerically extrapolated to the below-threshold region, and that Eq. (39) should be satisfied.

The extrapolation may fail, but this failure will show up in two ways. First, Eq. (39) will not be satisfied. Second, when one extrapolates by fitting data with the ratio of two polynomials, or by a continued fraction technique, poles may be encountered, and one of course suspects the extrapolation in the neighborhood of these poles. However, one does believe that these poles are indications of actual poles, or possibly branch points and their associated cuts. Therefore, we turn to a discussion of the analytic structure of the first and second Born terms in the expansion of the $\bar{x}_R$ matrix in powers of the strength of the interaction.

1. *Analytic Structure for Separable Potentials*

From Eq. (19), it is already evident that poles occur, for that example, in the first Born approximation when $\alpha_2{}^2 = \mu^2$. From Eq. (20), it is evident that poles of higher order occur in the second Born approximation. Yet, for this example, neither the extrapolations nor the Padé approximant will fail.

2. *Analytic Structure for Local Potentials*

The situation is worse when branch points occur, as happens for local potentials. Consider a Yukawa potential:

$$V(r) = V_0 \exp(-\mu r)/\mu r\,.$$

Then the first Born approximation is (for $l = 0$; presumably the analytic structure for other l's is similar)

$$(\tan \delta)^{(1)} \sim (\bar{y}_R^{(1)})_{11} \sim V_0 \log\left(1 + \frac{4k^2}{\mu^2}\right).$$

The symbol "$\sim$" here means "proportional to," and k^2 is related to the energy. There is a branch point in $(\bar{y}_R)_{11}$ at $k^2 = -\mu^2/4$. Consider, further, a coupled channel problem,

$$k_1^2 = 2M_1E/\hbar^2\,,$$
$$k_2^2 = 2M_2(E - Q)/\hbar^2\,.$$

Of course we have to compute for $E = 0$ (threshold in channel 1). For $E = 0$, $k_2^2 = -2M_2Q/\hbar^2$, and if this value of k_2^2 is less than $-\mu^2/4$, where μ is the range parameter for V_{22}, then $(\bar{x}_R^{(1)})_{22}$ has a branch point in the physical E region for channel 1.

Form Eqs. (12) and (11), it may be seen that the first Born approximation is

$$(\bar{x}^{(1)})_{ij} = \frac{2}{\hbar(v_iv_j)^{1/2}} \int j_i(k_ir)\, V_{ij}(r) j_j(k_jr)\, dr\,.$$

For $(\bar{x}^{(1)})_{22}$, $k_2 = i\alpha_2$, and the j's increase exponentially at infinity like $\exp(\alpha_2 r)$. V decreases like $\exp(-\mu r)$, but, if $\alpha_2 > \mu/2$, the integral ceases to exist. Viewed this way, it is remarkable that one has *only* a branch point.

However, contemplation of the Schrödinger equation,

$$(d^2u/dr^2) - \alpha^2 u + g^2 e^{-\mu r} u = 0\,,$$

shows why it is possible analytically to continue $(\bar{x}^{(1)})_{22}$ when $\alpha > \mu/2$. Solutions of the form

$$u_1 \sim e^{\alpha r} + a_1 e^{(\alpha-\mu)r} + a_2 e^{(\alpha-2\mu)r} + \cdots\,,$$
$$u_2 \sim e^{-\alpha r} + b_1 e^{-(\alpha+\mu)r} + b_2 e^{-(\alpha+2\mu)r} + \cdots$$

exist, and the a's and b's are easily found to be

$$a_n = -\frac{g^2 a_{n-1}}{(\alpha - n\mu)^2 - \alpha^2}\,, \qquad b_n = -\frac{g^2 b_{n-1}}{(\alpha + n\mu)^2 - \alpha^2}\,.$$

The regular solution of the Schrödinger equation (the one that vanishes at the origin) is

$$u = \tfrac{1}{2}u_1 - \tfrac{1}{2}[u_1(0)/u_2(0)]u_2 \sim \sinh \alpha r + x \cosh \alpha r\,,$$

where

$$x = \frac{u_2(0) - u_1(0)}{u_2(0)}\,.$$

If $\alpha > \mu/2$, the term proportional to a_1 in the preceding expression for u_1 destroys the meaning of the symbol "$\sim$," which here should mean "the leading terms in u as $r \to \infty$." However, one may *continue* to define x via the equation immediately preceding even if $\alpha > \mu/2$.

For this example of the exponential potential, the first Born approximation has a pole at $\alpha = \mu/2$ (not a branch point), and successive Born approximations bring in additional poles at $\alpha = \mu, \frac{3}{2}\mu, 2\mu, \ldots$. x itself has this same set of poles, as is evident from the foregoing equations for a_n and b_n. Very likely, a coupled channel potential model with exponential potentials is such that neither the extrapolations nor the Padé approximant fails for $\alpha_2^2 = 2M_2Q/\hbar^2 > \mu^2/4$. We should confess that we are entering an area of analysis on which we had rather not tread.

The area of analysis revealed by coupled channel Schrödinger equations with Yukawa potentials is even more fearsome. As already seen, branch points do occur in the first Born approximation. Ordinary extrapolation procedures *must* fail in the region of branch points. Imagining for the moment that more powerful analytic continuation methods will yield the terms in the Born series on any desired Riemann sheet, this question still arises: what about the Padé approximant: will it work?

In contemplating the possibilities, we find it difficult to believe that $\bar{x}_R$ itself has branch points. Surely Eq. (37) remains true, and $(\bar{y}_R)_{11}$ exists and is real. If there are branch points in $\bar{x}_R$, do all branches lead to the same real $(\bar{y}_R)_{11}$ when substituted in Eq. (37)?

If, however, we accept the possibility that $\bar{x}_R$ may have branch points, we find three cases:

(1) There are branch points in $\bar{x}_R$ at the positions indicated by the Born expansion.

(2) There are such branch points, but their position is not the same as indicated by the Born expansion. The position depends on the strength of the interaction.

(3) There are no such branch points.

Our doubts about the validity of the Padé approximant center on case (2), and perhaps a simple example will illumine our (equally simple) thinking. Consider

$$f(g^2, E) = g^2 \ln(1 + E + g^2) .$$

The power series expansion of f is

$$f = g^2 \ln(1 + E) + \frac{g^4}{1 + E} + \cdots .$$

The first and second Born terms have singular points at $E = -1$; f itself

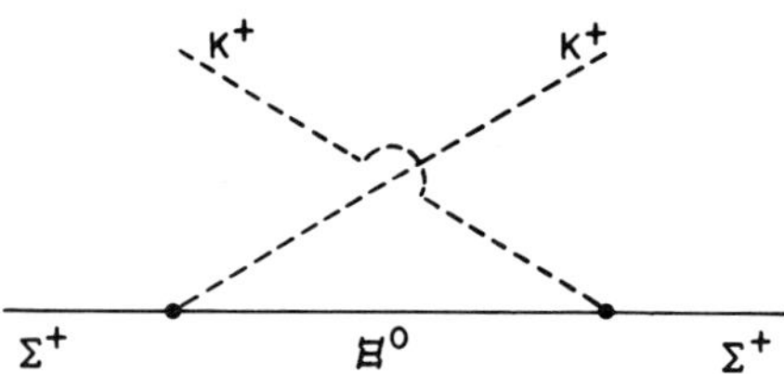

Fig. 2. A graph that occurs in the calculation.

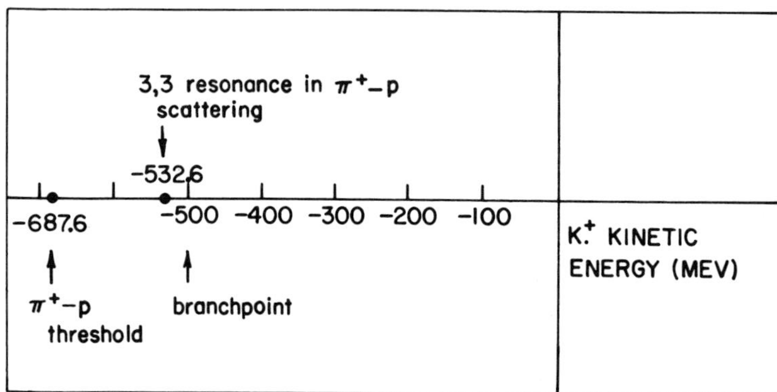

Fig. 3. Analytic structure of the contribution of the graph shown in Fig. 2 to $(\bar{x}_R{}^{(1)})_{22}$.

has no such singular point, but f does have a "moving" singular point at $E = -1 - g^2$. The Padé approximant is not a good approximation for, say, $g^2 = 1$, $E = -\frac{3}{2}$, even though this E is greater than the energy corresponding to the "moving" singular point ($E = -2$).

3. *Analytic Structure in the Case of Quantum Field Theory*

Analytic calculation of the graph shown in Fig. 2 gives, for $l = 0$ (presumably the analytic structure for other l's is similar),

$$(\bar{x}_R^{(1)})_{22} \sim \ln \frac{(M_\Xi{}^2 - M_\Sigma{}^2 - M_K{}^2) + 2p^2 + 2(p^2 + M_\Sigma{}^2)^{1/2}(p^2 + M_K{}^2)^{1/2}}{(M_\Xi{}^2 - M_\Sigma{}^2 - M_K{}^2) - 2p^2 + 2(p^2 + M_\Sigma{}^2)^{1/2}(p^2 + M_K{}^2)^{1/2}},$$

so that $(\bar{x}_R^{(1)})_{22}$ has, as a function of the K^+ laboratory kinetic energy, the analytic structure shown in Fig. 3. Because of the presence of the branch point, the extrapolations undoubtedly fail when the K^+ energy is too negative [say below about -300 MeV as indicated by the failure of Eq. (39); see Sect. IV].

G. Conditions for Resonance Below Threshold

We note that there are two ways in which Eq. (37) and the Padé approximant may lead to a resonance $[(\bar{y}_R)_{11} = \infty]$.

First, since the Padé approximant gives

$$\bar{x}_R = \bar{x}_R^{(1)}[\bar{x}_R^{(1)} - \bar{x}_R^{(2)}]^{-1}\bar{x}_R^{(1)} ,$$

all, or some, elements of $\bar{x}_R$ may become infinite if the determinant of $\bar{x}_R^{(1)} - \bar{x}_R^{(2)}$ vanishes. However, closer inspection of Eq. (37) shows that, unless $(\bar{x}_R)_{22}$ remains finite as the determinant approaches zero, Eq. (37) becomes

$$(\bar{y}_R)_{11} = \det(\bar{x}_R)/(\bar{x}_R)_{22} ,$$

and closer inspection of the Padé approximant shows this $(\bar{y}_R)_{11}$ is actually finite. We found that this situation always occurred.

The second way in which resonance may result is for the denominator in Eq. (37) to vanish. It is interesting to note that this condition is the condition for a bound state in the $K^+ - \Sigma^+$ channel in the absence of coupling. The condition is likely to be satisfied in the vicinity of a pole in $(\bar{x}_R)_{22}$, that is, in the vicinity of a zero of $\det(\bar{x}_R^{(1)} - \bar{x}_R^{(2)})$. For the (3, 3) resonance, this situation prevailed.

III. FORMAL PRELIMINARIES TO THE RESULTS

A. The Graphs

There are five types of graphs involved in the (3, 3) amplitude: crossed Born, crossed vertex, crossed self energy, direct box, and crossed box (Fig. 4). Graphs are further specified by giving the baryons as one follows the baryon lines, the initial meson, the intermediate meson, and the final meson. Since we ignore the electromagnetic mass splittings, all graphs that correspond to different possible arrangements of the electric charge in the interior of the graph are equivalent.

There are 91 nonequivalent graphs. Since we ultimately took the Λ and Σ masses to be equal, this number was further reduced to 41.

B. Masses Used in the Calculation

In our calculations, we have used the following masses:

$$M_N = 938.256 \text{ MeV} , \qquad M_\pi = 139.58 \text{ MeV} ,$$
$$M_\Sigma = M_\Lambda = 1170.9025 \text{ MeV} , \qquad M_K = 493.78 \text{ MeV} ,$$
$$M_\Xi = 1314.3 \text{ MeV} , \qquad M_\eta = 548.7 \text{ MeV} ,$$

We have used $M_\Sigma = M_\Lambda$ to avoid an anomalous threshold in the dispersion theory approach to calculating box graphs such as the one shown in Fig. 5. These anomalous thresholds occur when $M_1 + M_2 > M_3 + M_4$.

With $M_\Sigma = M_\Lambda$, only two crossed box graphs, shown in Fig. 5, have an anomalous threshold. We consider $M_\Sigma = M_\Lambda$ as reasonable approximations, especially if one is mainly interested in pion-nucleon scattering.

The crossed box graphs shown in Fig. 5 present no difficulty to calculation by ordinary Feynman procedures. Hale (1967) calculated them by Feynman procedures. We also calculated them by taking $M_3 = M_4 = a$

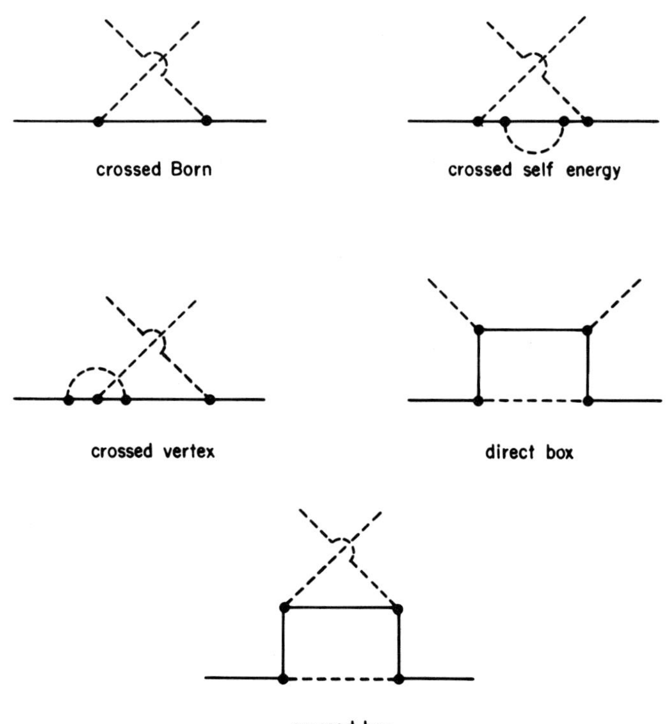

Fig. 4. Types of graphs which occur in the calculations.

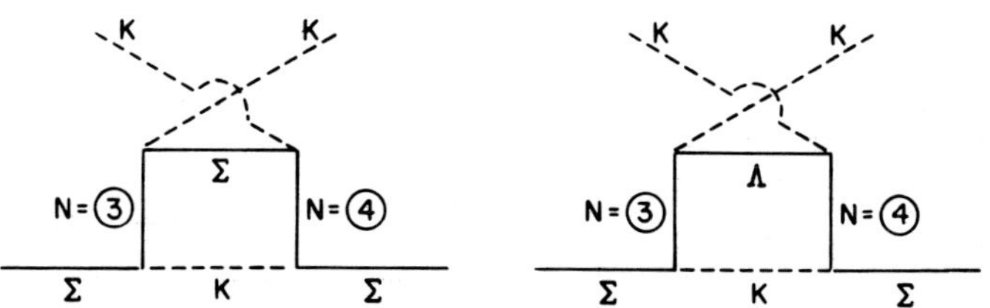

Fig. 5. Graphs with an anomalous threshold.

TABLE I[a]

NO.	EBAR	E	(K PIE−N) SQ	(K K−SIG) SQ
1	5.0000E+01	−6.4720E+02	1.3108E−02	−2.7324E−01
2	1.0000E+02	−6.0713E+02	2.8099E−02	−2.6657E−01
3	1.5000E+02	−5.6707E+02	4.4605E−02	−2.5731E−01
4	2.0000E+02	−5.2700E+02	6.2349E−02	−2.4593E−01
5	2.5000E+02	−4.8694E+02	8.1117E−02	−2.3279E−01
6	3.0000E+02	−4.4687E+02	1.0074E−01	−2.1818E−01
7	3.5000E+02	−4.0681E+02	1.2109E−01	−2.0233E−01
8	4.0000E+02	−3.6674E+02	1.4206E−01	−1.8542E−01
9	4.5000E+02	−3.2667E+02	1.6356E−01	−1.6760E−01
10	5.0000E+02	−2.8661E+02	1.8552E−01	−1.4899E−01
11	5.5000E+02	−2.4654E+02	2.0787E−01	−1.2969E−01
12	6.0000E+02	−2.0648E+02	2.3059E−01	−1.0979E−01
13	6.5000E+02	−1.6641E+02	2.5361E−01	−8.9364E−02
14	7.0000E+02	−1.2635E+02	2.7690E−01	−6.8462E−02
15	7.5000E+02	−8.6281E+01	3.0044E−01	−4.7140E−02
16	8.0000E+02	−4.6216E+01	3.2420E−01	−2.5443E−02
17	8.5892E+02	1.0000E+00	3.5245E−01	5.5510E−04
18	8.8887E+02	2.5000E+01	3.6690E−01	1.3932E−02
19	9.2007E+02	5.0000E+01	3.8202E−01	2.7973E−02
20	9.5002E+02	7.4000E+01	3.9659E−01	4.1550E−02
21	9.8247E+02	1.0000E+02	4.1244E−01	5.6358E−02
22	1.0124E+03	1.2400E+02	4.2711E−01	7.0115E−02
23	1.0449E+03	1.5000E+02	4.4306E−01	8.5108E−02
24	1.0748E+03	1.7400E+02	4.5783E−01	9.9026E−02
25	1.1073E+03	2.0000E+02	4.7388E−01	1.1418E−01
26	1.1372E+03	2.2400E+02	4.8873E−01	1.2824E−01
27	1.1697E+03	2.5000E+02	5.0486E−01	1.4355E−01
28	1.1996E+03	2.7400E+02	5.1979E−01	1.5774E−01
29	1.2321E+03	3.0000E+02	5.3600E−01	1.7318E−01
30	1.2620E+03	3.2400E+02	5.5100E−01	1.8749E−01
31	1.2945E+03	3.5000E+02	5.6728E+01	2.0304E−01
32	1.3244E+03	3.7400E+02	5.8234E−01	2.1746E−01
33	1.3569E+03	4.0000E+02	5.9868E−01	2.3312E−01
34	1.3868E+03	4.2400E+02	6.1379E−01	2.4763E−01
35	1.4193E+03	4.5000E+02	6.3020E−01	2.6340E−01
36	1.4492E+03	4.7400E+02	6.4536E−01	2.7799E−01
37	1.4817E+03	5.0000E+02	6.6181E−01	2.9385E−01

[a] The second column is the π laboratory kinetic energy in πN scattering in MeV. The third column is the K laboratory kinetic energy in ΣK scattering in MeV. Either laboratory energy gives the same center-of-mass energy (the center-of-mass energy is not tabulated). The fourth column is the square of the πN center-of-mass momentum in units of the nucleon mass ($\hbar = c = 1$). The fifth column is the square of the $K\Sigma$ center-of-mass momentum in units of the nucleon mass.

TABLE

The Reduced Tangent Matrix or $\bar{x}_R$ (or $\bar{y}_R$) Elements

	F=0.40		
NO.	X111=Y111	X112	X122
1	2.412E+01	−8.611E+00	6.792E+00
2	1.947E+01	−6.925E+00	−9.535E+00
3	1.635E+01	−5.812E+00	1.280E+02
4	1.411E+01	−5.019E+00	2.482E+01
5	1.242E+01	−4.424E+00	1.618E+01
6	1.109E+01	−3.959E+00	1.257E+01
7	1.002E+01	−3.585E+00	1.047E+01
8	9.136E+00	−3.278E+00	9.049E+00
9	8.396E+00	−3.019E+00	8.009E+00
10	7.765E+00	−2.800E+00	7.206E+00
11	7.220E+00	−2.610E+00	6.562E+00
12	6.745E+00	−2.444E+00	6.032E+00
13	6.328E+00	−2.298E+00	5.586E+00
14	5.957E+00	−2.169E+00	5.205E+00
15	5.625E+00	−2.053E+00	4.874E+00
16	5.327E+00	−1.948E+00	4.585E+00
17	5.012E+00	−1.838E+00	4.286E+00
18	4.866E+00	−1.786E+00	4.149E+00
19	4.721E+00	−1.735E+00	4.015E+00
20	4.589E+00	−1.689E+00	3.895E+00
21	4.455E+00	−1.642E+00	3.772E+00
22	4.337E+00	−1.600E+00	3.666E+00
23	4.215E+00	−1.557E+00	3.557E+00
24	4.108E+00	−1.520E+00	3.462E+00
25	3.998E+00	−1.481E+00	3.365E+00
26	3.901E+00	−1.446E+00	3.280E+00
27	3.801E+00	−1.411E+00	3.193E+00
28	3.713E+00	−1.380E+00	3.116E+00
29	3.622E+00	−1.347E+00	3.037E+00
30	3.541E+00	−1.319E+00	2.967E+00
31	3.457E+00	−1.289E+00	2.895E+00
32	3.382E+00	−1.262E+00	2.832E+00
33	3.305E+00	−1.235E+00	2.766E+00
34	3.237E+00	−1.210E+00	2.707E+00
35	3.165E+00	−1.185E+00	2.647E+00
36	3.102E+00	−1.162E+00	2.593E+00
37	3.036E+00	−1.139E+00	2.538E+00

[a] The first subscript on x or y shows the order of Born approximation. The next two is printed here X_{212}.] The results are given for $F = 0.4$ only. The y elements are cal- by extrapolation.

II

in First and Second Born Approximation[a]

F=0.40			
X211	X212	X222	Y211
−8.172E+00	−5.851E+01	4.274E+01	1.022E+00
−4.057E+00	1.528E+02	7.167E+01	1.954E+00
−1.664E+00	2.424E+01	4.914E+02	2.467E+00
−1.776E−01	1.047E+01	−7.543E+01	2.752E+00
7.812E−01	5.308E+00	−3.014E+01	2.901E+00
1.411E+00	2.678E+00	−1.679E+01	2.963E+00
1.827E+00	1.132E+00	−1.050E+01	2.968E+00
2.096E+00	1.460E−01	−6.906E+00	2.936E+00
2.264E+00	−5.129E−01	−4.619E+00	2.878E+00
2.361E+00	−9.664E−01	−3.065E+00	2.803E+00
2.405E+00	−1.283E+00	−1.962E+00	2.718E+00
2.413E+00	−1.506E+00	−1.154E+00	2.627E+00
2.393E+00	−1.662E+00	−5.490E−01	2.532E+00
2.353E+00	−1.769E+00	−8.847E−02	2.436E+00
2.299E+00	−1.839E+00	2.659E−01	2.341E+00
2.234E+00	−1.882E+00	5.405E−01	2.249E+00
2.148E+00	−1.906E+00	7.870E−01	
2.102E+00	−1.910E+00	8.870E−01	
2.053E+00	−1.909E+00	9.762E−01	
2.005E+00	−1.904E+00	1.049E−00	
1.952E+00	−1.895E+00	1.117E+00	
1.903E+00	−1.883E+00	1.169E+00	
1.849E+00	−1.868E+00	1.218E+00	
1.799E+00	−1.852E+00	1.254E+00	
1.746E+00	−1.832E+00	1.287E+00	
1.697E+00	−1.813E+00	1.312E+00	
1.644E+00	−1.790E+00	1.334E+00	
1.596E+00	−1.768E+00	1.349E+00	
1.544E+00	−1.744E+00	1.362E+00	
1.497E+00	−1.720E+00	1.370E+00	
1.446E+00	−1.694E+00	1.375E+00	
1.401E+00	−1.670E+00	1.377E+00	
1.352E+00	−1.643E+00	1.377E+00	
1.307E+00	−1.618E+00	1.375E+00	
1.260E+00	−1.591E+00	1.370E+00	
1.217E+00	−1.565E+00	1.365E+00	
1.172E+00	−1.538E+00	1.357E+00	

indices show the position of the element in the matrix. [What we call $(\bar{x}_R{}^{(2)})_{12}$ in the text
culated below the $K-\Sigma$ threshold. The x values below the $K-\Sigma$ threshold are obtained

TABLE

The (3, 3) Phase Shift for

NO.	DETERMINANT	F=0.40 DENOMINATOR	(X) 11
LAMBDA=1.00			
1	−3.6503E+03	2.2619E+00	−6.0927E+01
2	−2.7411E+04	2.1248E+00	2.6509E+00
3	−7.4486E+03	1.0195E+01	1.3723E+01
4	1.1924E+03	4.7631E−01	1.5198E+01
5	4.4427E+02	7.3120E−01	1.4180E+01
6	2.4009E+02	9.3342E−01	1.3241E+01
7	1.4953E+02	1.1230E+00	1.2513E+01
8	1.0060E+02	1.3182E+00	1.1951E+01
9	7.1146E+01	1.5305E+00	1.1511E+01
10	5.2148E+01	1.7703E+00	1.1159E+01
11	3.9286E+01	2.0506E+00	1.0875E+01
12	3.0257E+01	2.3904E+00	1.0640E+01
13	2.3736E+01	2.8224E+00	1.0445E+01
14	1.8916E+01	3.4111E+00	1.0278E+01
15	1.5286E+01	4.3154E+00	1.0134E+01
16	1.2506E+01	6.1081E+00	1.0006E+01
LAMBDA=3.00			
1	−3.3768E+04	2.2229E+00	−2.0939E−01
2	−2.2353E+05	2.1284E+00	2.8036E+00
3	−3.4895E+04	6.1132E+00	3.2149E+01
4	2.3501E+03	4.7832E−02	5.7693E+01
5	6.5974E+02	1.5597E−01	6.5461E+01
6	2.8760E+02	1.4739E−01	7.0868E+01
7	1.4179E+02	2.7210E−02	7.9761E+01
8	7.0969E+01	−2.8444E−01	9.6916E+01
9	3.2857E+01	−1.0855E+00	1.3521E+02
10	1.1191E+01	−4.1205E+00	2.6770E+02
11	−1.4862E+00	3.6145E+01	−1.4045E+03
12	−8.9795E+00	6.6851E+00	−1.6620E+02
13	−1.3373E+01	4.9624E+00	−8.1492E+01
14	−1.5866E+01	4.6853E+00	−5.1046E+01
15	−1.7173E+01	5.0695E+00	−3.5574E+01
16	−1.7735E+01	6.4775E+00	−2.6323E+01

[a] The second column gives the determinant of $\bar{x}_R^{(1)} - \bar{x}_R^{(2)}$ which is needed to form value of the denominator which appears in Eq. (37). The fourth, fifth, and sixth columns equation in Sect. II. G), and the last the phase shift resulting from the Padé approximant,

III

$F = 0.4$ and $\lambda = 1.0$ and 3.0[a]

	F=0.40	
(X) 12	(X) 22	DELTA
1.7245E+00	−1.2766E+00	−3.7521E−04
3.4581E−01	−7.0529E−01	1.2556E−02
−1.1415E+01	−3.1958E+01	1.5891E−01
−2.5711E+00	6.2626E+00	2.8202E−01
−2.7722E+00	5.7623E+00	3.6674E−01
−3.0882E+00	5.5341E+00	4.5938E−01
−3.3797E+00	5.4375E+00	5.5065E−01
−3.6455E+00	5.4153E+00	6.3707E−01
−3.8935E+00	5.4427E+00	7.1668E−01
−4.1298E+00	5.5068E+00	7.8858E−01
−4.3582E+00	5.5994E+00	8.5268E−01
−4.5805E+00	5.7152E+00	9.0943E−01
−4.7977E+00	5.8497E+00	9.5955E−01
−5.0098E+00	5.9995E+00	1.0040E+00
−5.2161E+00	6.1611E+00	1.0436E+00
−5.4153E+00	6.3315E+00	1.0796E+00
1.5408E+00	−1.1340E+00	1.2372E−04
4.2504E−01	−7.1861E−01	1.3311E−02
−2.3986E+01	−1.6096E+01	4.8814E−01
−7.5762E+00	8.0049E+00	1.3907E+00
−9.5051E+00	8.2333E+00	1.3580E+00
−1.2984E+01	9.1368E+00	1.4735E+00
−1.8119E+01	1.0853E+01	1.5614E+00
−2.6577E+01	1.4059E+01	−1.5195E+00
−4.3826E+01	2.1051E+01	−1.4774E+00
−1.0098E+02	4.5044E+01	−1.4475E+00
6.0936E+02	−2.5729E+02	−1.4255E+00
8.2214E+01	−3.3400E+01	−1.4086E+00
4.5640E+01	−1.8097E+01	−1.3946E+00
3.2189E+01	−1.2611E+01	−1.3819E+00
2.5142E+01	−9.8366E+00	−1.3691E+00
2.0772E+01	−8.1874E+00	−1.3543E+00

the Padé approximant (see the first equation in Sect. II. G). The third column gives the give the values of the $\bar{x}_R$ elements calculated from the Padé approximant (see the first that is, from the $\bar{x}_R$ elements given and Eq. (37).

TABLE

x_R and y_R Matrix Elements in First and Second Born
in the Table Heading for the

MU11=7.000000E−01 MU12=9.000000E−01 MU22=1.1000000E+00
DELR=2.0223912E−02 USES 4 PTS FOR DERIVATIVES

NO.	X111	X112	X122	X211
1	2.128E+01	−9.220E+00	−7.182E+00	1.014E+01
2	1.914E+01	−7.958E+00	−2.516E+00	1.028E+01
3	1.718E+01	−6.937E+00	3.022E+00	1.028E+01
4	1.542E+01	−6.101E+00	7.534E+00	1.016E+01
5	1.388E+01	−5.408E+00	9.981E+00	9.940E+00
6	1.253E+01	−4.828E+00	1.069E+01	9.632E+00
7	1.136E+01	−4.337E+00	1.042E+01	9.261E+00
8	1·033E+01	−3.919E+00	9.731E+00	8.846E+00
9	9.437E+00	−3.560E+00	8.917E+00	8.405E+00
10	8.652E+00	−3.249E+00	8.116E+00	7.953E+00
11	7.961E+00	−2.978E+00	7.378E+00	7.502E+00
12	7.350E+00	−2.740E+00	6.720E+00	7.060E+00
13	6.808E+00	−2.531E+00	6.140E+00	6.634E+00
14	6.325E+00	−2.345E+00	5.630E+00	6.228E+00
15	5.893E+00	−2.180E+00	5.183E+00	5.844E+00
16	5.505E+00	−2.032E+00	4.788E+00	5.482E+00
17	5.097E+00	−1.877E+00	4.381E+00	5.087E+00
18	4.907E+00	−1.806E+00	4.195E+00	4.898E+00
19	4.721E+00	−1.735E+00	4.015E+00	4.710E+00
20	4.553E+00	−1.672E+00	3.853E+00	4.537E+00
21	4.381E+00	−1.607E+00	3.690E+00	4.358E+00
22	4.231E+00	−1.551E+00	3.548E+00	4.201E+00
23	4.078E+00	−1.494E+00	3.405E+00	4.038E+00
24	3.944E+00	−1.444E+00	3.281E+00	3.894E+00
25	3.806E+00	−1.392E+00	3.154E+00	3.746E+00
26	3.686E+00	−1.348E+00	3.045E+00	3.616E+00
27	3.563E+00	−1.302E+00	2.933E+00	3.481E+00
28	3.454E+00	−1.261E+00	2.835E+00	3.362E+00
29	3.343E+00	−1.220E+00	2.735E+00	3.239E+00
30	3.245E+00	−1.184E+00	2.648E+00	3.131E+00
31	3.144E+00	−1.146E+00	2.558E+00	3.019E+00
32	3.055E+00	−1.113E+00	2.480E+00	2.920E+00
33	2.963E+00	−1.079E+00	2.400E+00	2.818E+00
34	2.882E+00	−1.050E+00	2.329E+00	2.728E+00
35	2.798E+00	−1.019E+00	2.257E+00	2.635E+00
36	2.724E+00	−9.914E−01	2.193E+00	2.552E+00
37	2.647E+00	−9.632E−01	2.127E+00	2.467E+00

MAX RELATIVE ERROR IN CALCULATION OF X12 AND X21 IS 1.3972591E−06

[a] The potential is chosen so that Tables II and IV resemble each other as closely as

IV

Approximation Calculated from the Potential Shown
Energies Siven in Table I[a]

V11=−2.472416E+01 V12=8.378579E+00 V22=−1.530028E+01
WITH DEL LAMBDA=1.0000000E−02

X212	X222	Y211	Y311	Y411
5.497E+01	−1.272E+01	1.314E+01	1.433E+01	8.501E+00
2.952E+01	−1.210E+01	1.337E+01	1.539E+01	1.557E+01
1.681E+01	−1.113E+01	1.319E+01	1.600E+01	1.896E+01
9.564E+00	−9.836E+00	1.272E+01	1.620E+01	2.207E+01
5.138E+00	−8.275E+00	1.209E+01	1.605E+01	2.328E+01
2.332E+00	−6.572E+00	1.137E+01	1.566E+01	2.385E+01
5.189E−01	−4.869E+00	1.062E+01	1.510E+01	2.368E+01
−6.609E−01	−3.291E+00	9.886E+00	1.445E+01	2.318E+01
−1.424E+00	−1.920E+00	9.178E+00	1.375E+01	2.253E+01
−1.910E+00	−7.917E−01	8.512E+00	1.304E+01	2.177E+01
−2.206E+00	9.718E−02	7.893E+00	1.234E+01	2.097E+01
−2.374E+00	7.717E−01	7.323E+00	1.166E+01	2.007E+01
−2.454E+00	1.267E+00	6.801E+00	1.102E+01	1.918E+01
−2.475E+00	1.618E+00	6.325E+00	1.042E+01	1.830E+01
−2.454E+00	1.857E+00	5.892E+00	9.858E+00	1.746E+01
−2.406E+00	2.011E+00	5.499E+00	9.340E+00	1.668E+01
−2.328E+00	2.113E+00			
−2.282E+00	2.140E+00			
−2.232E+00	2.153E+00			
−2.182E+00	2.155E+00			
−2.126E+00	2.147E+00			
−2.074E+00	2.133E+00			
−2.018E+00	2.111E+00			
−1.966E+00	2.085E+00			
−1.911E+00	2.054E+00			
−1.861E+00	2.022E+00			
−1.807E+00	1.986E+00			
−1.759E+00	1.950E+00			
−1.708E+00	1.910E+00			
−1.662E+00	1.873E+00			
−1.613E+00	1.832E+00			
−1.569E+00	1.793E+00			
−1.524E+00	1.752E+00			
−1.483E+00	1.714E+00			
−1.440E+00	1.674E+00			
−1.401E+00	1.637E+00			
−1.361E+00	1.597E+00			

possible.

variable mass $\geq M_\Sigma$ and extrapolating to $M_3 = M_4 = M_N$. The result of the extrapolation agreed with Hale' results.

IV. RESULTS FOR THE (3, 3) STATE ($\pi^+ - p$ AND $K^+ - \Sigma^+$ ELASTIC SCATTERINGS AND REACTION)

The center-of-mass momenta are given by

$$k^2_{\pi-\text{nuc}} = \frac{(\bar{E} + M_\pi)^2 - M_\pi^2}{2M_N\bar{E} + (M_N + M_\pi)^2},$$

$$k^2_{\kappa-\Sigma} = \frac{M_\Sigma^2}{M_N^2}\frac{(E + M_\kappa)^2 - M_\kappa^2}{2M_\Sigma E + (M_\Sigma + M_\kappa)^2},$$

$$\bar{E} = \frac{M_\Sigma}{M_N}E + \frac{(M_\Sigma + M_\kappa)^2 - (M_N + M_\pi)^2}{2M_N},$$

where $\bar{E}$ and E are the laboratory kinetic energies of the π^+ and K^+, respectively (π^+ incident on p at rest, or K^+ incident on Σ^+ at rest; another way of saying the same thing is that $\bar{E}$ is the kinetic energy of the π^+ in a system in which the p is at rest, and E is the kinetic energy of the K^+ in a system in which the Σ^+ is at rest). The values of E for which we computed are shown in Table I, together with the corresponding $\bar{E}, k^2_{\pi-\text{nuc}}$, and $k^2_{\kappa-\Sigma}$.

The results are given in Table II, together with the extrapolated values. The extrapolations were done using a continued fraction method (Hildebrand, 1956).

In order to check Eq. (39), we calculated $(\bar{y}_R{}^{(1)})_{11}$ and $(\bar{y}_R{}^{(2)})_{11}$ below threshold. As an example of the check of Eq. (39), consider $F = 0.4$, $\bar{E} = 200$ MeV:

$$(\bar{y}_R{}^{(2)})_{11} - (\bar{x}_R{}^{(2)})_{11} = 2.9296,$$

$$\alpha_2{}^3(x_R{}^{(1)})^2_{12} = 3.072.$$

The check is surprisingly good in spite of the branch point. Consider $F = 0.4$, $\bar{E} = 500$ MeV.

$$(\bar{y}_R{}^{(2)})_{11} - (\bar{x}_R{}^{(2)})_{11} = 0.44290,$$

$$\alpha_2{}^3(\bar{x}_R{}^{(1)})^2_{12} = 0.45074.$$

This check works quite well, so that we conclude that we are safely away from the branch point at this energy.

Finally, we have used Eq. (37) to compute $\tan\delta(3, 3)$. In Table III, we give $\tan\delta(3, 3)$ as a function of $\bar{E}$ for $F = 0.4$ and $\lambda = 1.0$ and 3.0. We also give the factors that went into this computation, namely, the

TABLE V

Exact $\bar{x}_R$ Matrix Elements Compared with the $\bar{x}_R$ Matrix Elements Computed from the Padé Approximant for two Values of the Parameter λ Which Scales the Potential ($\lambda = 0.3$ and $\lambda = 0.6$)

	XBARS—APPROXIMATE AND EXACT FOR LAMBDA=3.0000000E−01					
NO.	X11 EXACT	X11 APPRO	X12 EXACT	X12 APPRO	X22 EXACT	X22 APPRO
1	8.236E+00	4.180E+00				
2	7.680E+00	4.215E+00				
3	7.122E+00	1.758E+00				
4	6.582E+00	7.892E+00				
5	6.074E+00	5.977E+00				
6	5.603E+00	5.296E+00				
7	5.170E+00	4.800E+00				
8	4.777E+00	4.381E+00				
9	4.419E+00	4.015E+00				
10	4.096E+00	3.692E+00				
11	3.803E+00	3.404E+00				
12	3.538E+00	3.148E+00				
13	3.299E+00	2.919E+00				
14	3.083E+00	2.713E+00				
15	2.887E+00	2.528E+00				
16	2.710E+00	2.361E+00				
17	2.523E+00	2.186E+00	−1.051E+00	−8.693E−01	1.668E+00	1.560E+00
18	2.438E+00	2.105E+00	−1.024E+00	−8.430E−01	1.617E+00	1.509E+00
19	2.353E+00	2.026E+00	−9.961E−01	−8.164E−01	1.566E+00	1.457E+00
20	2.275E+00	1.953E+00	−9.700E−01	−7.918E−01	1.519E+00	1.410E+00
21	2.194E+00	1.879E+00	−9.424E−01	−7.659E−01	1.470E+00	1.361E+00
22	2.124E+00	1.814E+00	−9.177E−01	−7.429E−01	1.426E+00	1.318E+00
23	2.051E+00	1.747E+00	−8.916E−01	−7.188E−01	1.381E+00	1.274E+00
24	1.986E+00	1.688E+00	−8.682E−01	−6.974E−01	1.342E+00	1.234E+00

TABLE V (continued)

XBARS—APPROXIMATE AND EXACT FOR LAMBDA=3.0000000E−01						
NO.	X11 EXACT	X11 APPRO	X12 EXACT	X12 APPRO	X22 EXACT	X22 APPRO
25	1.920E+00	1.627E+00	−8.437E−01	−6.751E−01	1.300E+00	1.193E+00
26	1.861E+00	1.574E+00	−8.217E−01	−6.553E−01	1.263E+00	1.157E+00
27	1.801E+00	1.519E+00	−7.987E−01	−6.346E−01	1.225E+00	1.120E+00
28	1.747E+00	1.471E+00	−7.781E−01	−6.162E−01	1.191E+00	1.087E+00
29	1.692E+00	1.421E+00	−7.565E−01	−5.970E−01	1.156E+00	1.053E+00
30	1.643E+00	1.378E+00	−7.372E−01	−5.800E−01	1.125E+00	1.023E+00
31	1.592E+00	1.332E+00	−7.169E−01	−5.623E−01	1.092E+00	9.911E−01
32	1.547E+00	1.293E+00	−6.989E−01	−5.466E−01	1.063E+00	9.633E−01
33	1.501E+00	1.251E+00	−6.800E−01	−5.302E−01	1.033E+00	9.344E−01
34	1.460E+00	1.215E+00	−6.631E−01	−5.156E−01	1.007E+00	9.088E−01
35	1.417E+00	1.178E+00	−6.455E−01	−5.005E−01	9.789E−01	8.823E−01
36	1.379E+00	1.144E+00	−6.297E−01	−4.870E−01	9.542E−01	8.587E−01
37	1.340E+00	1.110E+00	−6.132E−01	−4.730E−01	9.285E−01	8.343E−01

XBARS—APPROXIMATE AND EXACT FOR LAMBDA=6.0000000E−01						
NO.	X11 EXACT	X11 APPRO	X12 EXACT	X12 APPRO	X22 EXACT	X22 APPRO
1	7.625E+01	5.143E+00				
2	1.304E+02	5.285E+00				
3	7.316E+02	2.328E+00				
4	−1.736E+02	−4.871E+01				
5	−7.305E+01	2.438E+01				

6	−4.464E+01	1.828E+01				
7	−3.136E+01	1.628E+01				
8	−2.373E+01	1.494E+01				
9	−1.881E+01	1.380E+01				
10	−1.540E+01	1.278E+01				
11	−1.289E+01	1.186E+01				
12	−1.099E+01	1.101E+01				
13	−9.495E+00	1.024E+01				
14	−8.292E+00	9.536E+00				
15	−7.301E+00	8.894E+00				
16	−6.468E+00	8.313E+00				
17	−5.620E+00	7.716E+00	3.946E+00	−3.343E+00	2.755E−01	4.139E+00
18	−5.229E+00	7.451E+00	3.772E+00	−3.282E+00	2.631E−01	4.064E+00
19	−4.867E+00	7.184E+00	3.606E+00	−3.215E+00	2.495E−01	3.984E+00
20	−4.556E+00	6.937E+00	3.459E+00	−3.148E+00	2.360E−01	3.907E+00
21	−4.254E+00	6.679E+00	3.312E+00	−3.074E+00	2.211E−01	3.822E+00
22	−4.003E+00	6.449E+00	3.185E+00	−3.005E+00	2.074E−01	3.743E+00
23	−3.758E+00	6.210E+00	3.058E+00	−2.929E+00	1.927E−01	3.656E+00
24	−3.553E+00	5.998E+00	2.949E+00	−2.858E+00	1.793E−01	3.577E+00
25	−3.351E+00	5.777E+00	2.838E+00	−2.781E+00	1.651E−01	3.491E+00
26	−3.182E+00	5.581E+00	2.742E+00	−2.710E+00	1.523E−01	3.411E+00
27	−3.014E+00	5.378E+00	2.645E+00	−2.634E+00	1.389E−01	3.326E+00
28	−2.872E+00	5.197E+00	2.560E+00	−2.565E+00	1.268E−01	3.249E+00
29	−2.730E+00	5.010E+00	2.474E+00	−2.491E+00	1.143E−01	3.165E+00
30	−2.610E+00	4.843E+00	2.399E+00	−2.423E+00	1.031E−01	3.090E+00
31	−2.490E+00	4.671E+00	2.322E+00	−2.352E+00	9.144E−02	3.009E+00
32	−2.387E+00	4.518E+00	2.255E+00	−2.287E+00	8.109E−02	2.936E+00
33	−2.284E+00	4.359E+00	2.186E+00	−2.218E+00	7.033E−02	2.858E+00
34	−2.196E+00	4.219E+00	2.126E+00	−2.156E+00	6.081E−02	2.788E+00
35	−2.107E+00	4.073E+00	2.064E+00	−2.091E+00	5.091E−02	2.714E+00
36	−2.031E+00	3.944E+00	2.009E+00	−2.032E+00	4.216E−02	2.647E+00
37	−1.953E+00	3.810E+00	1.953E+00	−1.971E+00	3.309E−02	2.576E+00

denominator that appears in Eq. (37), the value of the determinant $|\bar{x}_R^{(1)} - \bar{x}_R^{(2)}|$, and the elements of $\bar{x}_R$ as computed from the Padé approximant.

It may be seen that no resonance occurs for $\lambda = 1.0$, whereas $\lambda = 3.0$ produces a resonance at about $\bar{E} = 350$ MeV. If one believes field theory, he will have to conclude that the [1, 1] matrix Padé approximant vastly underestimates the strength of the interaction. We turn now to a study of a potential model to determine whether or not such is the case, and to determine what order Padé approximant may be required to produce more acceptable results.

V. STUDY OF A POTENTIAL MODEL

A potential that has first and second Born approximations similar to the first and second Born approximations computed from field theory is

$$\mu_{11} = 0.7\,, \qquad V_{11} = -24.72416\,,$$
$$\mu_{12} = 0.9\,, \qquad V_{12} = 8.378580\,,$$
$$\mu_{22} = 1.1\,, \qquad V_{22} = -14.3002805\,,$$
$$\hbar = c = M_N = 1\,.$$

In these units, $2M/\hbar^2$ in the $\pi^+ - p$ channel is 0.2430087, and in the $K^+ - \Sigma^+$ channel 0.694290. The k^2 in each channel were computed from the relativistic formulas given at the beginning of Sect. IV and are tabulated in Table I.

The x- and y- matrix elements calculated from this potential are shown in Table IV and may be compared with the corresponding elements shown in Table II.

This potential does not give the (3, 3) resonance at the right energy: it is too strong; a scale factor of $\lambda = 0.6$ is required if the (3, 3) resonance is to occur at the correct energy.

In Table V, exact and approximate x-matrix elements are compared for $\lambda = 0.6$. The one negative approximate x_{11} is spurious; it is a result of nonsense due to the branch point. The results for $\lambda = 0.3$ are also shown in order to illustrate the point that, for sufficiently weak potentials, the matrix Padé approximants do provide an adequate approximation. The matrix Padé approximants produce a resonance at $\bar{E} = 350$ MeV for $\lambda = 1.05$. (The exact calculations show that $\lambda = 1.05$ produces a bound state and in addition is so strong that a resonance occurs at $\bar{E} = 250$: a *second* resonance as a function of λ). The matrix Padé approximants calculated from field theory behave in this same way: they underestimate the strength of the interaction indicated by the experimental data, requiring a scale factor $\lambda = 3.0$ to produce a resonance at $E = 350$ MeV, whereas in the potential case the corresponding scale factor is $1.05/0.6 \approx 1.8$. Tjon and Nieland (Chapter 12) observe that, in the case of the Bethe–Salpeter equation, ordinary [1, 1] Padé approximants also underestimate the strength of the interaction by similar ratios.

It *has* to be concluded that higher Padé approximants will be required. But one may reasonably ask whether or not the matrix formalism is necessary. Perhaps ordinary Padé approximants to $\bar{x}_{11}$ will suffice. From Table IV, for $\bar{E} = 200$ MeV [the energy of the (3, 3) resonance],

$$\bar{x}_{11} = 15.42\lambda + 12.72\lambda^2 + 16.20\lambda^3 + 22.07\lambda^4 + \cdots .$$

The [1, 1] Padé approximant,

$$\bar{x}_{11} = 15.42\lambda/(1 - 0.8249\lambda) ,$$

requires $\lambda = 1.212$ for resonance (the exact result is $\lambda = 0.6$, as already mentioned). The [1, 2] Padé approximant,

$$\bar{x}_{11} = 15.42\lambda + \frac{12.72\lambda^2}{1 - 1.274\lambda},$$

requires $\lambda = 0.785$. The [2, 2] Padé approximant,

$$\bar{x}_{11} = \frac{15.42\lambda - 10.80\lambda^2}{1 - 1.525\lambda + 0.208\lambda^2},$$

requires $\lambda = 0.728$. (The *second* resonance occurs for $\lambda = 6.6$, which is a very bad approximation to $\lambda = 1.05$.)

Our final conclusion, based on the results of the potential model calculation, is that [2, 2] Padé approximants will be required in the field theory case, and that matrix Padé approximants will not be necessary, since the ordinary [2, 2] approximant apparently suffices. Comparison of our field theory results with experimental data suggests the same thing.

REFERENCES

Baker, G. A. (1965). "Advances in Theoretical Physis" (K. A. Brueckner, ed.), Vol, 1, pp. 1–58. Academic Press, New York.

Bessis, D., and Pusterla, M. (1967). CERN Rept. TH. 794.

Chisholm, J. S. R. (1963). *J. Math. Phys.* **4,** 1506.

Hale, G. M. (1967). Thesis, Texas A and M Univ.

Hildebrand, F. B. (1965). "Introduction to Numerical Analysis," p. 395 *et. seq.* McGraw-Hill, New York. We thank Dr. J. Holdeman for calling this method to our attention and for the use of his computer code based on the method.

Kim, J. K. (1967). *Phys. Rev. Letters* **19,** 1079. He gives $g^2_{NN\pi}$ 14.5±0.4 and F 0.41±0.07. MacGregor, Arndt, and Wright give $g^2_{NN\pi}$ 14.72 ± 0·83 in UCRL-70075, Part VII.

Kubis, J. J., and Gammel, J. L. (1968). *Rhys. Rev.* **172,** 1664.

Newton, M. A. (1967). PhD thesis, Cambridge Univ.

Remiddi, E., Pusterla, M., and Mignaco, J. A. (1968). CERN Rep. TH. 895.

Roper, L. D. (1964). UCRL-7846.

Wall, H. S. (1948). "Continued Fractions," especially the last two chapters. Van Nostrand, Princeton, New Jersey.

Wigner, E., and Eisenbud, O. (1947). *Phys. Rev.* **72,** 29. Coupled channel calculations have been carried out by many authors. The above is a classic reference.

CHAPTER 14 THE PADÉ APPROXIMANT IN THE NUCLEON-NUCLEON SYSTEM*

W. R. Wortman

University of California
Los Alamos Scientific Laboratory

I. INTRODUCTION

Nucleon-nucleon scattering is an area of strong interactions of great value for the testing of models of strong interaction dynamics. This is due to the existence of great amounts of experimental data conveniently expressed in the form of phase shifts. However, N-N scattering is a particularly stringent area for testing, since, in contrast to π-π and π-N scattering, resonances do not dominate the amplitudes, and one's goal is not so clear. In particular, schemes whose main feature is the input of possible resonances have no place in N-N scattering, and a real understanding of the dynamics is needed.

Perhaps the greatest single contribution to the theory of N-N scattering came from Yukawa (1938) in the suggestion that particle exchange is the mechanism of the interaction. The generalization of this approach has provided the tool of the perturbation expansion of renormalizable field theory taking the interaction as

$$\mathscr{H}_I = ig(4\pi)^{1/2}\bar{\psi}\gamma_5\boldsymbol{\tau}\psi\cdot\boldsymbol{\varphi}\,. \tag{1}$$

The resulting success of the one-pion-exchange (OPE) amplitude in accounting for the higher angular momentum partial waves is now legend (Cziffra *et al.*, 1959; Breit *et*, *al.*, 1960). However, attempts to pursue

* Work performed in part under the auspices of the U. S. Atomic Energy Commission. Supported in part by the U. S. Air Force Office of Scientific Research, under Grant No. AF 918-67. Based in part on a dissertation submitted to Texas A and M University in partial fulfillment of the requirements for the Ph. D. degree.

this method by looking at additional perturbation terms have not been nearly so useful (Watson and Lepore, 1949; Taketani *et al.*, 1952).

Aside from the question of the validity of the field theory model, the use of the perturbation expansion gives rise to some highly objectionable features. In particular, it is possible that the perturbation series does not converge for any value of the coupling constant. Although this does not necessarily render the truncated series meaningless, results must be suspect. Furthermore, the truncated perturbation series is not unitary. Consequently, in order to compute phase shifts, the resulting amplitude must be altered in a somewhat arbitrary way, frequently very drastically. In addition, there is no way that the truncated series can generate a pole in the amplitude, and thus the *s*-wave *N-N* properties of the $^3S_1$ deuteron and the nearly bound $^1S_0$ state cannot possibly be accounted for.

These circumstances are of just the sort that might lead to the application of the technique of the Padé approximant to *N-N* scattering. In particular, the diagonal Padé approximants are exactly unitary, and poles appear in a very natural manner so as potentially to represent bound states and resonances.

II. THE PADÉ APPROXIMANT IN SCATTERING THEORY

Although rigorous theorems as to convergence of the Padé method are not available in this case, the examples provided by other chapters in this book, by application to the Peres model field theory, (Baker and Chisholm, 1966), and by application to the Bethe–Salpeter (B–S) equation for scalar particles (Nuttall, 1967) are encouraging.

Although the Peres model involves only two dimensions, it is most interesting that Baker and Chisholm (1966) show that, in spite of the zero radius of convergence of the perturbation expansion, the [*N*, *N*] Padé approximants converge to the correct solution. For a more realistic model, Nuttall (1967) shows that, for scalar, equal mass scattered particles in the

TABLE I

B–S Solution, Padé Approximants, and Perturbation Result for $\tan \delta_0/k|_{E=0}$ with Coupling Constant λ for $\mu/M = 0.14$

λ	B–S	[1, 1]	[1, 2]	[2, 2]	8th-order[a] perturbation
0.001	0.162	0.162	0.162	0.162	0.162
0.032	8.03	7.81	8.01	8.03	7.85
0.064	44.06	32.8	41.9	43.9	27.1
0.128	−24.6	−54.6	−30.4	−25.1	162.1

[a] The first four coefficients are 160.28, 1722.4, 20915.7, and 258919.7.

laddergraph approximation, the sequence of $[N, N]$ approximants converges to the solution of the corresponding B–S equation. In this case, however, Mattioli (1968) proves that the laddergraph series has a nonzero radius of convergence.

As an indication of the usefulness of the Padé method, Table I provides a comparison of the results for the $\tan \delta_0/k$ at $E=0$ for the actual B–S equation (Schwartz and Zemach, 1966) for the laddergraph series, the Padé approximants, and the partial perturbation series for several values of the coupling constant.

Table I indicates not only the convergence of the Padé approximants for this case but also the appropriate placement of the bound state. These results hint at the possible usefulness of this method as applied to strong interactions.

The values of the coefficients used in Table I are found by evaluation of the derivatives of the B–S solution (Schwartz, 1968). As a practical matter, this is certainly the easiest means of finding the amplitudes of the laddergraphs for scalar particles. However, this method is not applicable for nonladdergraphs and for cases such as N-N scattering for which the B–S solution is not known. As an alternative, we have investigated the method of a direct Monte Carlo integration scheme applying the $i\varepsilon$ prescription numerically for Feynman amplitudes.

As indicated by Chisholm (1952), a general Feynman integral, I, can be simplified and reduced to an integral over parameters with a zero-to-one range:

$$\begin{aligned} I &= \int d^4k_1\, d^4k_2 \cdots d^4k_r (a_1 a_2 \cdots a_l)^{-1} \\ &= \int d^4k_1 \cdots d^4k_r \int_0^1 dx_1 \int_0^{x_1} dx_2 \cdots \int_0^{x_{l-2}} dx_{l-1} \frac{(l-1)!}{Q^l} \\ &= (i\pi^2)^r (l-2r-1)! \int_0^1 dx_1 \cdots \\ &\quad \times \int_0^{x_{l-2}} dx_{l-1} ([\det A]^2 [C - B^T A^{-1} B]^{l-2r})^{-1}, \end{aligned} \tag{2}$$

where

$$\begin{aligned} Q &= a_l x_{l-1} + a_{l-1}(x_{l-2} - x_{l-1}) + \cdots a_1(1 - x_1) \\ &= \sum_{m,n} A_{mn} k_m \cdot k_n + 2 \sum_n B_n \cdot k_n + C. \end{aligned} \tag{3}$$

Here l is the number of internal lines, and r is the number of independent internal momenta. A is a matrix whose elements are independent of external momenta, B is a column vector whose elements are linear in the external momentum, and C is quadratic in the external momenta.

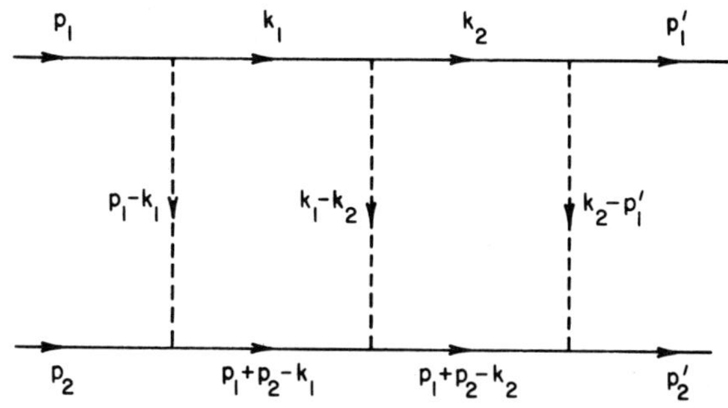

Fig. 1. Sixth-order laddergraph.

For example, the sixth-order laddergraph of Fig. 1 is evaluated from the prescription as follows:

$$\begin{aligned} F = & -(2\pi)^{-8}G^6 \int d^4k_1 \int d^4k_2 \\ & \times \{[k_1^2 + m^2][(p_1+p_2-k_1)^2+m^2][(p_1-k_1)^2+\mu^2][(k_1-k_2)^2+\mu^2]\}^{-1} \\ & \times \{[k_2^2+m^2][(p_1+p_2-k_2)^2+m^2][(p_1'-k_2)^2+\mu^2]\}^{-1} \\ = & -(2\pi)^{-8}G^6(i\pi^2)^2(7-4-1)! \int_0^1 dx_1 \cdots \\ & \times \int_0^{x_5} dx_6([\det A]^2[C - B^T A^{-1}B]^3)^{-1} . \\ = & (2\pi)^{-1}\left(\frac{4\pi}{G^2}\right)^3 \int_0^1 dx_1 \cdots \int_0^{x_5} dx_6([\det A]^2[C-B^TA^{-1}B]^3)^{-1} , \end{aligned} \tag{4}$$

where

$$\begin{aligned} A_{11} &= x_3 , \\ A_{12} &= A_{21} = x_4 - x_3 , \\ A_{22} &= 1 - x_4 , \\ B_1 &= -p_1(x_4 - x_6) - p_2(x_5 - x_6) , \\ B_2 &= -p_1(x_2 - x_3) - p_2(x_2 - x_3) - p_1'(x_1 - x_2) , \\ C &= m^2[2x_5 - x_4 + 2x_2 - x_3 - 2x_1 + 1] + \mu^2[x_3 - x_5 + x_1 - x_2] \\ & \quad + (x_5 - x_6 + x_2 - x_3)(p_1+p_2)^2 . \end{aligned} \tag{5}$$

Here $G^2/4\pi = 4\pi\lambda$ is analogous to the π-N coupling constant $g^2 \approx 14$.

Now if the energy variable s is defined $s = -(p_1+p_2)^2$, then from unitarity a branch cut in F for $s > 4m^2$ is expected, and it will be generated by the condition

$$C - B^T A^{-1} B = 0 . \tag{6}$$

(It is understood that products of B_i's are to be taken as dot products.) In order to extract the real part of F, note that the scattering amplitude is defined as

$$F(s) \equiv \lim_{\varepsilon \to 0+} F(s + i\varepsilon) \, ; \tag{7}$$

that is, the cut is approached from above. The quantity $C - B^T A^{-1} B$ may be written as

$$C - B^T A^{-1} B \equiv X = f(x_i) + sg(x_i) \, . \tag{8}$$

Therefore, taking $s \to s + i\varepsilon$ and extracting the real part,

$$\operatorname{Re} F = (2\pi)^{-1} \left(\frac{G^2}{4\pi}\right)^3 \lim_{\varepsilon \to 0} \int_0^1 dx_1 \cdots$$
$$\times \int_0^{x_5} dx_6 ([\det A]^2)^{-1} \left\{ \frac{X^3 - 3Xg^2\varepsilon^2}{[X^3 - 3g^2\varepsilon^2]^2 + [3X^2 g\varepsilon - \varepsilon^3 g^3]^2} \right\} . \tag{9}$$

This integral is well defined for $\varepsilon > 0$, and the only problem that remains is to investigate numerically the behavior as a function of ε to determine if extrapolation for $\varepsilon \to 0$ is feasible.

In the absence of detailed knowledge of the behavior of the integrand, the only practical means of evaluating such multidimensional integrals is by Monte Carlo techniques (Schreider, 1964). The rate of convergence of the Monte Carlo method is dependent upon fluctuations in the integrand. As a consequence, it is expected that the convergence of, for example, (9) will grow increasingly more difficult as $\varepsilon \to 0$.

In order to obtain improved convergence using some information about the integrand, the Monte Carlo procedure has been modified as follows. After changing to integration variables that range from 0 to 1, the integration region, a hypercube, is divided into as many equal hypercubes as is practical. Therefore, if there are n integration variables and each dimension is divided into m units, there will be m^n cubes. The Monte Carlo method is then applied to each cube, but the number of trials in each cube is chosen so as to minimize the total estimated error subject to the condition that the total number of trials is fixed.

If σ_i denotes the standard deviation in the ith cube, it may be estimated by N_i trials in that cube as

$$\sigma_i^2 = \left(\sum_{n=1}^{N_i} f_n^2 - \left(\sum_{n=1}^{N_i} f_n \right)^2 \Big/ N_i \right) \Big/ N_i = \bar{f}_n^2 - (\bar{f}_n)^2 \, . \tag{10}$$

Since for a well-behaved integrand the rate of convergence of the Monte Carlo method is $N^{1/2}$, the error estimate is

$$(\text{error})^2 = \sum_{i=1}^{m^n} \sigma_i^2 / N_i \, . \tag{11}$$

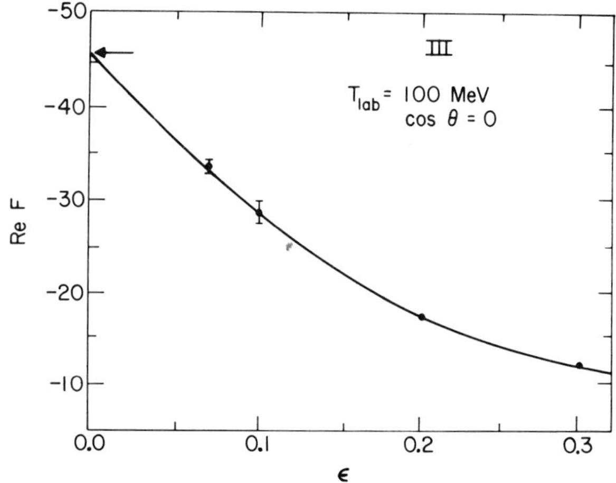

Fig. 2. Extrapolation for $\varepsilon \to 0$.

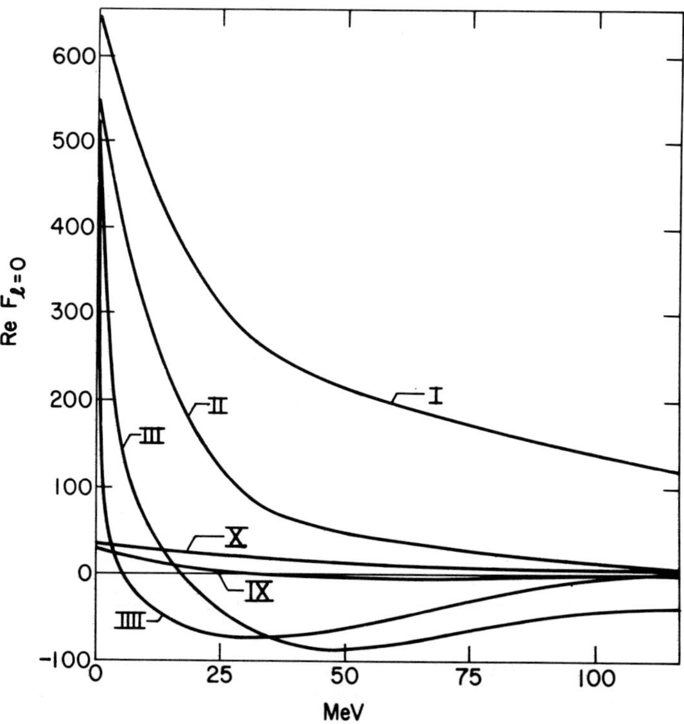

Fig. 3. Scalar diagram amplitudes.

To minimize the estimated error subject to the restriction that $N = \Sigma N_i$ is constant, it must be the case that

$$N_i \propto \sigma_i \, . \tag{12}$$

Therefore, the following procedure has been devised to evaluate the required integrals:

(a) Divide the region into hypercubes.

(b) Estimate σ_i by using an equal number of samples in each hypercube.

(c) Integrate each hypercube independently by the Monte Carlo method using $N_i \propto \sigma_i$, and add the results appropriately.

The procedure can be repeated as many times as needed to obtain the desired precision. It should be noted that estimates become increasingly unreliable as $\varepsilon \to 0$ because of development of narrow peaks in the integrand.

An example of the behavior of the scattering amplitude as a function of ε is shown in Fig. 2 for the sixth-order laddergraph. The correct result is obtained from the B–S solution and allows us to check the validity of the extrapolation procedure.

Using the same procedure, the sixth-order graphs and eighth-order laddergraphs have been evaluated. Several amplitudes are shown in Fig. 3. Note that the two diagrams that have crossed lines are generally small compared with their uncrossed counterparts.

III. THE NUCLEON–NUCLEON SYSTEM

A. Basic Physics

The fundamental physics of the N-N system for a field theory approach is conveniently expressed in terms of the S matrix. The S matrix is often decomposed in the following manner:

$$S = 1 + i(2\pi)^4 N_1 N_2 N_3 N_4 \delta(\Sigma p_i - \Sigma p_f) F \, . \tag{13}$$

Here the N_i are spinor normalization factors $N_i = [m/(2\pi)^3 E_i]^{1/2}$. The chief virtue of this decomposition is that F is a Lorentz scalar, and therefore it is a function of only invariant scalar variables. The standard choice of such variables is

$$s = -(p_1 + p_2)^2 \, , \qquad t = -(p_1 - p_1')^2 \, , \qquad u = -(p_1 - p_2')^2 \, , \tag{14}$$

where the momenta are defined by Fig. 4.

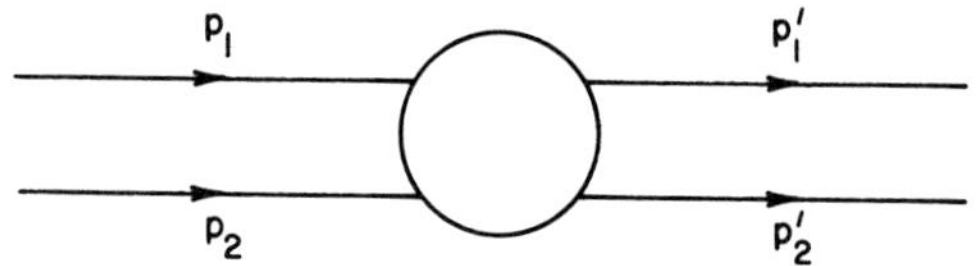

Fig. 4. Definition of external momenta.

These variables are not independent, since $s + t + u = 4m^2$. In the center-of-mass frame, the momenta are all equal, and, if θ is the scattering angle and p the three momentum,

$$\begin{aligned} s &= 4(p^2 + m^2) = 4m^2 + 2(\text{K. E.})_{\text{lab}} , \\ t &= -2p^2(1 - \cos\theta) , \\ u &= -2p^2(1 + \cos\theta) . \end{aligned} \tag{15}$$

Therefore, for physical processes, $s \geq 4m^2$, $-4p^2 \leq t \leq 0$, and $-4p^2 \leq u \leq 0$.

In addition to the usual elementary considerations for scattering, *N-N* scattering requires inclusion of the Pauli principle and an account of the spin of the nucleons. *A priori*, the number of scalar amplitudes needed will be given by the product of the number of initial spin states and the number of final spin states. Therefore, spin $\frac{1}{2}$ on spin $\frac{1}{2}$ scattering will require at most 16 amplitudes. Physical assumptions will limit the actual number.

The total angular momentum J is conserved, and, if L is the orbital angular momentum and S the spin, the possible states are shown in Table II.

TABLE II

Possible States with Total Angular Momentum J

$S = 1$	$S = 0$
$L = J - 1$ $L = J$ $L = J + 1$	$L = J$

The usual assumption of parity conservation forbids transitions such as $L = J \pm 1 \rightleftarrows L = J$, which leaves only eight possible transitions. Time reversal invariance forces the transitions $L = J + 1 \rightleftarrows L = J - 1$ to have equal amplitudes, leaving seven independent amplitudes.

In order to treat *p-p* and *n-p* scattering in a unified way, it is convenient to assume charge independence or, equivalently, isospin conservation. As a consequence of this, the Pauli principle can be expanded to include isospin coordinates. The Pauli principle so modified requires that, upon interchange of space, spin, and isospin coordinates, the scattering amplitude must change sign. If I denotes isospin (0 or 1), $L + S + I$ must be odd. Therefore, the transition $S = 1,\ L = J \rightarrow S = 0,\ L = J$ is forbidden, and spin is conserved. Consequently, only five scalar amplitudes are needed in a given isospin state.

The choice of the five scalar amplitudes (MacGregor *et al.*, 1960) is somewhat arbitrary, but, since the contributions of the Feynman diagrams are to be evaluated by dispersion relations, care must be taken that the amplitudes do not have any kinematic singularities. Nuttall (1961) has shown that, if the following decomposition of the amplitudes as proposed by Amati *et al.* (1960) is carried out, the functions $A_i(s, t, u)$ have the same analytic structure on the physical sheet as the corresponding function for the same diagram with scalar particles:

$$F = A_1P_1 + A_2P_2 + A_3P_3 + A_4P_4 + A_5P_5 ,$$

where

$$\begin{aligned} P_1 &= \bar{u}_1u_1\bar{u}_2u_2 , \\ P_2 &= \bar{u}_1 i\gamma\cdot p_2 u_1\bar{u}_2u_2 + \bar{u}_1u_1\bar{u}_2 i\gamma\cdot p_1u_2 , \\ P_3 &= \bar{u}_1 i\gamma\cdot p_2u_1\bar{u}_2 i\gamma\cdot p_1u_2 , \\ P_4 &= \bar{u}_1\gamma_\mu u_1\bar{u}_2\gamma_\mu u_2 , \\ P_5 &= \bar{u}_1\gamma_5u_1\bar{u}_2\gamma_5u_2 . \end{aligned} \tag{16}$$

The conventions used are $\hbar = c = 1$, $\bar{u}u = 1$, $\gamma_\nu\gamma_\mu + \gamma_\mu\gamma_\nu = 2\delta_{\mu\nu}$ ($\mu, \nu = 1, 2, 3, 4$), and $\gamma_5^2 = 1$. The Dirac equation is $(i\gamma\cdot p + m)u(p) = 0$.

The S matrix for a given total angular momentum J as taken between states $|J, L, S\rangle$ ordered $|J, J, 0\rangle$, $|J, J, 1\rangle$, $|J, J+1, 1\rangle$, $|J, J-1, 1\rangle$ is parametrized in terms of the nuclear bar phase shifts (Stapp *et al.*, 1958) as shown in Table III.

TABLE III

Parametrization of S Matrix

$e^{2i\delta_J}$	0	0	0
0	$e^{2i\delta_{JJ}}$	0	0
0	0	$\cos(2\varepsilon_J)e^{2i\delta_J^+}$	$i\sin(2\varepsilon_J)e^{i(\delta_J^+ + \delta_J^-)}$
0	0	$i\sin(2\varepsilon_J)e^{i(\delta_J^+ + \delta_J^-)}$	$\cos(2\epsilon_J)e^{2i\delta_J^-}$

The h amplitudes (Scotti and Wong, 1965) are more directly related to the S matrix than the A_i amplitude, and, in terms of the S matrix for a given J,

$$S = 1 + \frac{2imp}{E} \left\| \begin{matrix} h_J & 0 & 0 & 0 \\ 0 & h_{JJ} & 0 & 0 \\ 0 & 0 & h_J{}^+ & h^J \\ 0 & 0 & h^J & h_J{}^- \end{matrix} \right\| . \tag{17}$$

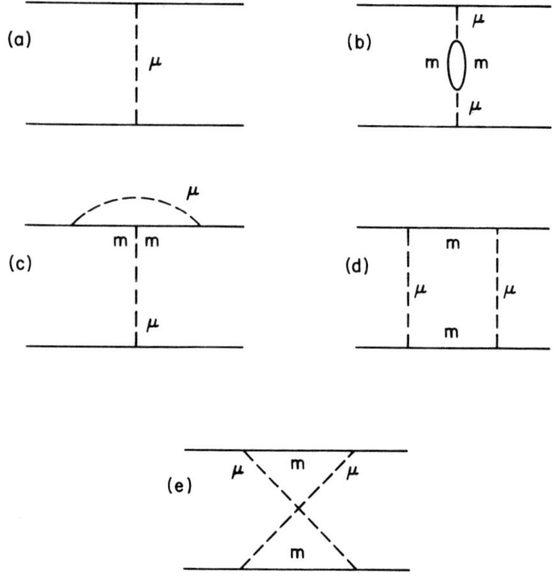

Fig. 5. Feynman diagrams through fourth order.

The Pauli principle as applied to the h amplitudes simply requires that the h's as found by examining the diagrams must be multiplied by two for allowed transitions. The relationship between the A_i amplitudes and the h amplitudes can be obtained from Amati *et al.* (1960), Goldberger *et al.* (1960), and Scotti and Wong (1965).

B. Perturbation Calculation

In order to form the [1, 1] Padé approximant, the second- and fourth-order contributions (Wortman, 1969; Haracz and Sharma, 1969) are needed, and the appropriate diagrams are shown in Fig. 5. Because of the presence of γ_5 in the interaction, there is reason to believe that even orders in powers of g^2 are more important than odd, and thus the truncation at fourth order is not completely arbitrary. In the nonrelativistic reduction, the spinors have two components, one large and one small (Schweber, 1961). The γ_5 couples the two components, and so $\bar{u}\gamma_5 u$ reduces to a product of large and small, but $\bar{u}u$ is proportional to the square of the large component. Consequently, nonrelativistically, even orders that have an even number of γ_5 factors are expected to contribute out of proportion compared to corresponding diagrams for exchange of scalar particles.

1. *One-Pion Exchange*

The OPE amplitude of Fig. 5(a) is

$$A_5^{\text{OPE}} = -4\pi g^2(\mu^2 - t)^{-1}(\boldsymbol{\tau}_1 \cdot \boldsymbol{\tau}_2) \,. \tag{18}$$

2. *Self-Energy*

The pion self-energy of Fig. 5(b) is

$$F^{\text{SE}} = i\,\frac{g^4}{\pi^2}\,\frac{\bar{u}_1\gamma_5 u_1 \bar{u}_2\gamma_5 u_2}{[\mu^2 + (p_1 - p_1')^2]^2}$$
$$\times \int d^4q\,\text{Tr}\left[\frac{\gamma_5[i\gamma\cdot(p_1 - p_1' + q) - m]\gamma_5(i\gamma\cdot q - m)}{[(p_1 - p_1' + q)^2 + m^2][q^2 + m^2]}\right](2\boldsymbol{\tau}_1\cdot\boldsymbol{\tau}_2) \,. \tag{19}$$

This requires boson mass and coupling constant renormalization and can be written as

$$A_5^{\text{SE}} = \frac{A}{[t - \mu^2]^2} + \frac{B}{t - \mu^2} + C(t) \,. \tag{20}$$

A and B are infinite constants, and, upon application of Cutkosky's (1960) method, C can be expressed as a dispersion integral in t. The result is

$$A_5^{\text{SE}} \equiv C = -2g^4 \int_{4m^2}^{\infty} \frac{dt'\,(t' - 4m^2)^{1/2} t'^{1/2}}{(t' - t)\,(t' - \mu^2)^2}\,(2\boldsymbol{\tau}_1\cdot\boldsymbol{\tau}_2) \,. \tag{21}$$

3. *Vertex*

The vertex amplitude of Fig. 5(c)

$$F^V = -i\,\frac{g^2}{\pi^2}\,\frac{\bar{u}_2\gamma_5 u_2}{\mu^2 - t}$$
$$\times \int \frac{d^4q\,\bar{u}_1\gamma_5[i\gamma\cdot(p_1' - p_1 + q) - m]\gamma_5[i\gamma\cdot q - m]\gamma_5 u_1}{[(p_1 - q)^2 + \mu^2][(p_1' - p_1 + q)^2 + m^2][q^2 + m^2]}(-\boldsymbol{\tau}_1\cdot\boldsymbol{\tau}_2) \,. \tag{22}$$

This amplitude also requires coupling constant renormalization, and it can be written as

$$A_5^V = \frac{B}{t - \mu^2} + C(t) \,. \tag{23}$$

B is an infinite constant, and, upon application of Cutkosky's method,

$$A_5^V = C = -g_4 \int_{4m^2}^{\infty} \frac{dt'[t' - 4m^2]^{1/2}}{(t' - t)\,(t' - \mu^2)t'^{1/2}}$$
$$\times \left\{1 + \frac{\mu^2}{t' - 4m^2}\ln\left(\frac{\mu^2}{\mu^2 + t' - 4m^2}\right)\right\}(\boldsymbol{\tau}_1\cdot\boldsymbol{\tau}_2) \,. \tag{24}$$

There are actually two vertex diagrams, the other of which gives a correction to the lower vertex. The two diagrams give equal contributions, and so the second is easily taken into account.

4. *Box*

The amplitude for the box diagram of Fig. 5(d) is

$$F^B = -i\frac{g^2}{\pi^2}\int d^4q\, \bar{u}_1\gamma_5[i\gamma\cdot(p_1+q) - m]\gamma_5 u_1\bar{u}_2\gamma_5$$
$$\times [i\gamma\cdot(p_2-q) - m]\gamma_5 u_2/\{[(p_1+q)^2+m^2][(p_1-p_1'+q)^2+\mu^2]$$
$$\times [(p_2-q)^2+m^2][q^2+\mu^2]\}(3-2\boldsymbol{\tau}_1\cdot\boldsymbol{\tau}_2)\,. \tag{25}$$

In order to evaluate the Mandelstam (1959) double spectral function, Cutkosky's method of finding the discontinuity across the cuts can be applied successively to the s and t channels. This gives

$$\operatorname{disc}_s(\operatorname{disc}_t F^B) = -i(2\pi i)^4\frac{g^4}{\pi^2}\int d^4q\, N(q)\,\delta_1[(p_1-p_1'+q)^2+\mu^2]$$
$$\times \delta_2(q^2+\mu^2)\,\delta_3[(p_1+q)^2+m^2]\,\delta_4[(p_2-q)^2+m^2]\,, \tag{26}$$

where $N(q)$ is the numerator in (25). The condition imposed by the four delta functions give a value of q which appears in $N(q)$. If q is written as

$$q = Ap_1 + Bp_1' + Cp_2 + DX\,, \tag{27}$$

where

$$X_\mu = \varepsilon_{\mu\nu\alpha\beta}p_{1\nu}p'_{1\alpha}p_{2\beta}\,,$$

and X is orthogonal to p_1, p_1', and p_2, then three delta functions require

$$A = -\tfrac{1}{2}\,, \qquad C = -(t-2\mu^2)/2u\,, \qquad B = -(A+C)\,. \tag{28}$$

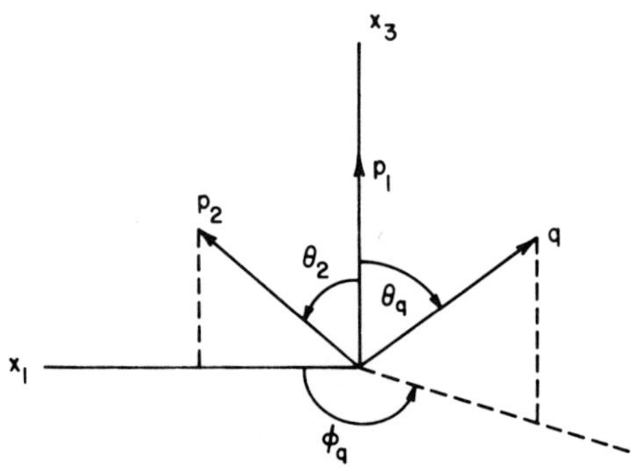

Fig. 6. Reference frame for the integration.

The condition $q^2 + \mu^2 = 0$ therefore requires

$$D^2X^2 = -\mu^2 + t/4 + (t - 2\mu^2)^2/4u \,. \tag{29}$$

That is, two values of q will simultaneously satisfy the four delta function conditions, and they differ only by the sign of the X coefficient. In order to clarify this circumstance, consider the evaluation of the integral in (26). Using the reference frame in Fig. 6, along with the conditions $\boldsymbol{p}_1 = \boldsymbol{p}_1'$ and $(p_1 - p_1')_0 = t^{1/2}$, the arguments of the delta functions are

$$\begin{aligned}
\arg_1 &= -2t^{1/2}[q_0 + t^{1/2}/2] \,, \\
\arg_2 &= \boldsymbol{q}^2 - (q_0{}^2 - \mu^2) \,, \\
\arg_3 &= 2p_1 q \left[\cos\theta_q - \frac{2p_{10}q_0 + \mu^2}{2p_1 q}\right] , \\
\arg_4 &= -2p_2 q \sin\theta_2 \sin\theta_q \left[\cos\varphi_q - \frac{2p_{20}q_0 - \mu^2 - 2p_2 q\cos\theta_2\cos\theta_q}{2p_2 q \sin\theta_2 \sin\theta_q}\right] .
\end{aligned} \tag{30}$$

Noting that $d^4q = d^3q\, dq_0 = \frac{1}{2}|\boldsymbol{q}|\, d(\boldsymbol{q}^2)\, d\Omega_q dq_0$, the first three delta functions are taken into account. The condition $\arg_4 = 0$ leads to two values of φ_q, say φ_{q_1} and φ_{q_2}, such that $\sin\varphi_{q_1} = -\sin\varphi_{q_2}$. Except for the presence of $N(q)$ in the integrand, this would lead to a factor of 2. However, the result is that $N(q)$ must be replaced by $N(q_1) + N(q_2)$. In the reference frame used, the vector X has only an x_2 component. Therefore, q_1 and q_2 differ only by the sign of the DX term, and $N(q_1) + N(q_2)$ can be replaced by $2N(q_1)$ if all terms linear in X are eliminated. That is, the presence of the numerator function in the integrand can be ignored if $Ap_1 + Bp_1' + Cp_2 + DX$ is substituted for q and all terms linear in X are eliminated.

The result is

$$\operatorname{disc}_s(\operatorname{disc}_t F^B) = -i(2\pi)^4 \frac{g^4}{\pi^2} \times \left\{\frac{-i2N(q)}{4[st^2(s-4m^2) - 4st(s-4m^2)\mu^2 - 4st\mu^4]^{1/2}}\right\}$$

where

$$N(q) = -\bar{u}_1 i\gamma\cdot DXu_1\bar{u}_2 i\gamma\cdot DXu_2 + C^2[m^2P_1 + mP_2 + P_3] \,. \tag{31}$$

Since X has only an x_2 component,

$$\begin{aligned}
\bar{u}_1 i\gamma\cdot DXu_1\bar{u}_2 i\gamma\cdot DXu_2 &= -D^2X^2\bar{u}_1\gamma_2 u_1\bar{u}_2\gamma_2 u_2 \\
&= -D^2X^2\Big[\sum_\mu \bar{u}_1\gamma_\mu u_1\bar{u}_2\gamma_\mu u_2 + \bar{u}_1 i\gamma_1 u_1\bar{u}_2 i\gamma_1 u_2 \\
&\qquad + \bar{u}_1 i\gamma_3 u_1\bar{u}_2 i\gamma_3 u_2 - \bar{u}_1\gamma_0 u_1\bar{u}_2\gamma_0 u_2\Big] \,.
\end{aligned} \tag{32}$$

Expressions for γ_1, γ_3, and γ_0 can be found by noting in this frame

$$\begin{aligned} i\gamma\cdot(p_1 - p_1') &= -\gamma_0 t^{1/2} , \\ i\gamma\cdot p_1 &= i\gamma_3 p_1 - \gamma_0 p_{10} , \\ i\gamma\cdot p_2 &= i\gamma_1 p_2 \sin\theta_2 + i\gamma_3 p_2 \cos\theta_2 - \gamma_0 p_{20} . \end{aligned} \tag{33}$$

Finally, N can be written as

$$\begin{aligned} N = &\left[m^2C^2 + m^2 \frac{u-s}{us} D^2X^2 \right] P_1 + \left[mC^2 + m \frac{t-4m^2}{us} D^2X^2 \right] P_2 \\ &+ \left[C^2 + \frac{u-s}{us} D^2X^2 \right] P_3 - [D^2X^2]\, P_4 , \end{aligned} \tag{34}$$

and the amplitude for the box diagram is

$$\begin{aligned} F^B = 2g^4 &\int_{4m^2}^{\infty} \frac{ds}{s' - s} \int_{4\mu^2 + [4\mu^4/(s'-4m^2)]}^{\infty} \frac{dt'}{t' - t} \\ &\times \frac{N(s', t')\,(3 - 2\boldsymbol{\tau}_1\cdot\boldsymbol{\tau}_2)}{(t')^{1/2}[s'(s'-4m^2)t' - 4\mu^4 s' - 4s'\mu^2(s'-4m^2)]^{1/2}} . \end{aligned} \tag{35}$$

5. *Crossed Box*

The amplitude for the crossed box diagram of Fig. 5(e) is

$$\begin{aligned} F^X = -i\frac{g^4}{\pi^2} &\int d^4q\, \bar{u}_1\gamma_5[i\gamma\cdot(p_1+q) - m]\gamma_5 u_1 \bar{u}_2\gamma_5 \\ &\times [i\gamma\cdot(p_2'+q) - m]\gamma_5 u_2/\{[(p_1+q)^2 + m^2][q^2+\mu^2] \\ &\times [(p_1 - p_1' + q)^2 + \mu^2] \\ &\times [(p_2'+q)^2 + m^2]\}(3 + 2\boldsymbol{\tau}_1\cdot\boldsymbol{\tau}_2)\} . \end{aligned} \tag{36}$$

If this amplitude is computed in just the same manner as used for the box diagram, it is found that the two are the same except for the interchange $s \rightleftarrows u$ and a sign change of the coefficients of P_2 and P_4. That is, the crossing matrix (Goldberger *et* al., 1960) is diagonal for these amplitudes. The crossed box amplitude is therefore

$$\begin{aligned} F^X = 2g^4 &\int_{4m^2}^{\infty} \frac{ds'}{s' - u} \int_{4\mu^2 + [4\mu^4/(s'-4m^2)]}^{\infty} \frac{dt'}{t' - t} \\ &\times \frac{N'(s', t')\,(3 + 2\boldsymbol{\tau}_1\cdot\boldsymbol{\tau}_2)}{(t')^{1/2}[s't'(s'-4m^2) - 4\mu^4 s' - 4s'\mu^2(s'-4m^2)]^{1/2}} \end{aligned} \tag{37}$$

where the prime on N indicates that the signs of the coefficients of P_2 and P_4 are changed relative to N.

C. Numerical Procedures

The partial wave projections of the amplitudes are needed, and this can be achieved easily because of the expression in the form of dispersion integrals. The three basic types of integrals are as follows:

$$
\begin{aligned}
I_1 &= \frac{1}{2}\int P_l(z)\,dz \int \frac{dt}{t'-t} f(t')\,, \\
I_2 &= \frac{1}{2}\int P_l(z)\,dz \int_{4m^2}^{\infty} \frac{ds'}{s'-s} \int \frac{dt'}{t'-t}\rho(s',t')\,, \\
I_3 &= \frac{1}{2}\int P_l(z)\,dz \int_{4m^2}^{\infty} \frac{ds'}{s'-u} \int \frac{dt'}{t'-t}\rho(s',t')\,.
\end{aligned}
\tag{38}
$$

The Legendre functions of the second kind, $Q_l(x)$, are defined as

$$
Q_l(x) = \frac{1}{2}\int_{-1}^{1} \frac{P_l(z)}{x-z}\,dz\,. \tag{39}
$$

Noting that $t = -2p^2(1-z)$ and $u = -2p^2(1+z)$, the required integrals become

$$
\begin{aligned}
I_1 &= (2p^2)^{-1}\int dt' f(t') Q_l(1 + t'/2p^2)\,, \\
I_2 &= (2p^2)^{-1}\int_{4m2}^{\infty} ds'/(s'-s)\int dt'\,\rho(s',t')Q_l(1+t'/2p^2)\,, \\
I_3 &= (2p^2)^{-1}\int ds' \int dt'\,\frac{\rho(s',t')[Q_l(1+t'/2p^2) + (-1)^l Q_l(1+s'/2p^2)]}{4p^2+s'+t'}\,.
\end{aligned}
\tag{40}
$$

It has been found that the required accuracy of Q_l is best obtained by using a truncated hypergeometric expansion rather than by expressing the Q_l functions in terms of the P_l functions.

The function of s defined by I_2 has a branch cut for $s > 4m^2$, that is, for the physical region. The real part of I_2 can be obtained by taking the principal part. Defining

$$
\begin{aligned}
F(s') &= \int dt' \rho(s',t') Q_l(1+t'/2p^2)\,, \\
\operatorname{Re} I_2 &= (2p^2)^{-1} P\int_{4m2}^{\infty} \frac{ds'}{s'-s} F(s') \\
&= (2p^2)^{-1}\lim_{\varepsilon\to 0}\left[\int_{4m2}^{s-\varepsilon}\frac{ds'}{s'-s}F(s') + \int_{s+\varepsilon}^{\infty}\frac{ds'}{s'-s}F(s')\right]
\end{aligned}
\tag{41}
$$

or

$$\operatorname{Re} I_2 = (2p^2)^{-1}\left[\int_s^{2s-4m^2} ds' \left[\frac{F(s') - F(2s - s')}{s' - s}\right] + \int_{2s-4m^2}^{\infty} \frac{ds'}{s' - s} F(s')\right]. \tag{42}$$

The integrals are evaluated by Gaussian quadrature with mesh size determined so as to give the desired precision. Referring to the first integral in (42) as the "improper" part and to the second as the "tail," the meshes used are shown in Table IV.

As an indication of the relative contributions of the diagrams and their dependence on energy, Tables V–VII give the values of $\operatorname{Re}(h_J)$ for the

TABLE IV

Mesh Sizes Used for the Various Graphs

Graph	Energy (MeV)	u' mesh	t' mesh	s'_{imp} mesh	s'_{tail} mesh
SE	All	—	16	—	—
Vertex	All	—	16	—	—
X-box	All	48	48	—	—
Box	0– 2	—	48	8	48
	2– 20	—	48	16	48
	20–100	—	48	32	32
	100–	—	32	48	16

TABLE V

Re h_J at 5 MeV

J	0	1	2	3	4
OPE	-6.54×10^{-1}	-5.68×10^{-1}	1.51×10^{-2}	-3.38×10^{-3}	1.15×10^{-4}
Vertex	6.70×10^{-3}	6.70×10^{-3}	-4.62×10^{-7}	4.44×10^{-10}	-5.72×10^{-14}
Self-energy	-1.39×10^{-2}	-1.39×10^{-2}	9.88×10^{-7}	-9.68×10^{-10}	1.27×10^{-13}
Box	9.47×10^{0}	1.83×10^{-1}	2.68×10^{-4}	3.70×10^{-5}	7.22×10^{-8}
X-box	1.73×10^{1}	-6.60×10^{-2}	1.11×10^{-3}	-9.56×10^{-6}	2.69×10^{-7}

TABLE VI

Re h_J at 50 MeV

J	0	1	2	3	4
OPE	-2.24×10^{0}	-1.07×10^{0}	1.20×10^{-1}	-1.29×10^{-1}	1.60×10^{-2}
Vertex	6.66×10^{-2}	6.64×10^{-2}	-4.54×10^{-5}	4.30×10^{-7}	-5.48×10^{-10}
Self-energy	-1.38×10^{-1}	-1.38×10^{-1}	9.70×10^{-5}	-9.38×10^{-7}	1.22×10^{-9}
Box	9.83×10^{0}	9.26×10^{-1}	1.05×10^{-2}	1.07×10^{-2}	1.52×10^{-4}
X-box	1.47×10^{1}	-4.60×10^{-1}	5.57×10^{-2}	-3.38×10^{-3}	6.66×10^{-4}

TABLE VII

Re h_J at 400 MeV

J	0	1	2	3	4
OPE	-3.17×10^{0}	-5.03×10^{-1}	9.98×10^{-2}	-1.89×10^{-1}	4.12×10^{-2}
Vertex	5.09×10^{-1}	4.96×10^{-1}	-2.52×10^{-3}	1.76×10^{-4}	-1.65×10^{-6}
Self-energy	-1.06×10^{0}	-1.03×10^{0}	4.18×10^{-3}	-3.81×10^{-4}	3.65×10^{-6}
Box	9.00×10^{0}	1.12×10^{0}	5.73×10^{-2}	1.71×10^{-1}	6.70×10^{-3}
X-box	5.82×10^{0}	-1.09×10^{0}	3.90×10^{-1}	-7.01×10^{-2}	3.95×10^{-2}

diagrams including isospin factors, the Pauli principle, and using $g^2 = 14$. The vertex amplitude includes the corrections to both vertices. The units are such that the nucleon mass is unity.

IV. THE PADÉ APPROXIMANT FOR *N-N* SCATTERING

The [1, 1] approximant for *N-N* scattering (Wortman, 1968; Bessis *et al.*, 1969) will be of the usual form

$$[1, 1] = [1 + g^2A]\cdot[1 + g^2B]^{-1}, \tag{43}$$

where A and B are 4×4 matrices to be determined for a given total angular momentum state. Noting the form of the S matrix in (17), it is clear that the singlet and uncoupled triplet amplitudes can be treated separately, reducing the problem to two 1×1 cases and to a 2×2 case for the coupled triplet amplitude.

Using the h functions defined by (17), the [1, 1] approximant is

$$h_p = g^2h_2\cdot(h_2 - g^2h_4)^{-1}\cdot h_2, \tag{44}$$

and the h's appearing are to be matrices for the coupled triplet state.

Prior to examination of the results, some features are evident on the basis of the behavior of OPE. In particular, the threshold behavior of the OPE amplitudes (Bessis *et al.*, 1969; Grashin, 1959) is not always the usual p^{2L+1} dependence. A check reveals that

$$\delta(^1S_0) \sim p^3, \qquad \delta(h_J^-) \sim p^{2J+1}. \tag{45}$$

Therefore, the $^1S_0$, $^3S_1$, $^3P_2$, $^3D_3$, $^3F_4$, $^3G_5$, ... OPE amplitudes have two additional powers of the momentum at threshold. This means that for these phase shifts OPE will not be dominant at low energies.

For the perturbation expansion, addition of the fourth order returns the usual threshold behavior. However, such is not necessarily the case for the [1, 1] Padé approximant (Bessis *et al.*, 1969). It is easy to check that $\delta(^1S_0[1, 1]) \sim p^5$!. This indicates that the [1, 1] result for $^1S_0$ will probably not be valid. For the coupled triplet case, the amplitude is a 2×2 matrix, and consequently the [1, 1] approximant is also a 2×2 matrix. It is most remarkable that in this case the proper threshold dependence is restored by the Padé prescription.

To assess the value of the [1, 1] Padé approximant, it is felt that the results should be compared with perturbation theory through fourth order, since both methods have the same input. It is hoped that the Padé method will improve on perturbation theory. In order to make this comparison, the perturbation expansion must be unitarized in some manner. We have chosen "geometric unitarization (Moravcsik, 1964)," so that phase shifts are defined by setting the real part of the scattering amplitude equal to the real part of the appropriate expression in Table III. The result is

$$\begin{aligned} \operatorname{Re} h_J &= \frac{E}{2mp} \sin(2\delta_J) , \\ \operatorname{Re} h_{JJ} &= \frac{E}{2mp} \sin(2\delta_{JJ}) , \\ \operatorname{Re} h_J^- &= \frac{E}{2mp} \cos(2\varepsilon_J)\sin(2\delta_J^-) , \\ \operatorname{Re} h_J^+ &= \frac{E}{2mp} \cos(2\varepsilon_J)\sin(2\delta_J^+) , \\ \operatorname{Re} h^J &= \frac{E}{2mp} \sin(2\varepsilon_J)\cos(\delta_J^+ + \delta_J^-) , \end{aligned} \tag{46}$$

where $E = \frac{1}{2}(s)^{1/2}$. One consequence of this choice is that, when $|(mp/E) \operatorname{Re} h| > \frac{1}{2}$, δ is not defined. This we consider a reasonable approach, since any unitarization scheme for $|(mp/E) \operatorname{Re} h| > \frac{1}{2}$ must drastically change the amplitude, and the results would be a reflection of the somewhat arbitrary unitarization method rather than of the perturbation calculation. The pertubation results can be meaningful only if amplitude does not approach the unitary bound.

In Fig. 7, the resulting phase shifts for OPE, fourth-order perturbation theory (Wortman, 1969) and the [1, 1] approximant as compared to the Livermore (MacGregor *et al.*, 1968) phase shifts are shown. Any comparison of the various data is difficult, since the theory is fairly crude. Furthermore, it is difficult to select a few phase shifts on which to base judgment, since, except for *s* wave, *N-N* scattering is not dominated by such features as resonances as is, for example, π-*N* scattering. Consequently, a rather subjective judgment must be made based on an overview of most of the phase shifts.

As a preface to a discussion of the results, a few observations are in order. The usefulness of fourth-order perturbation theory must be limited for at least two reasons. First, it might be expected that a perturbation

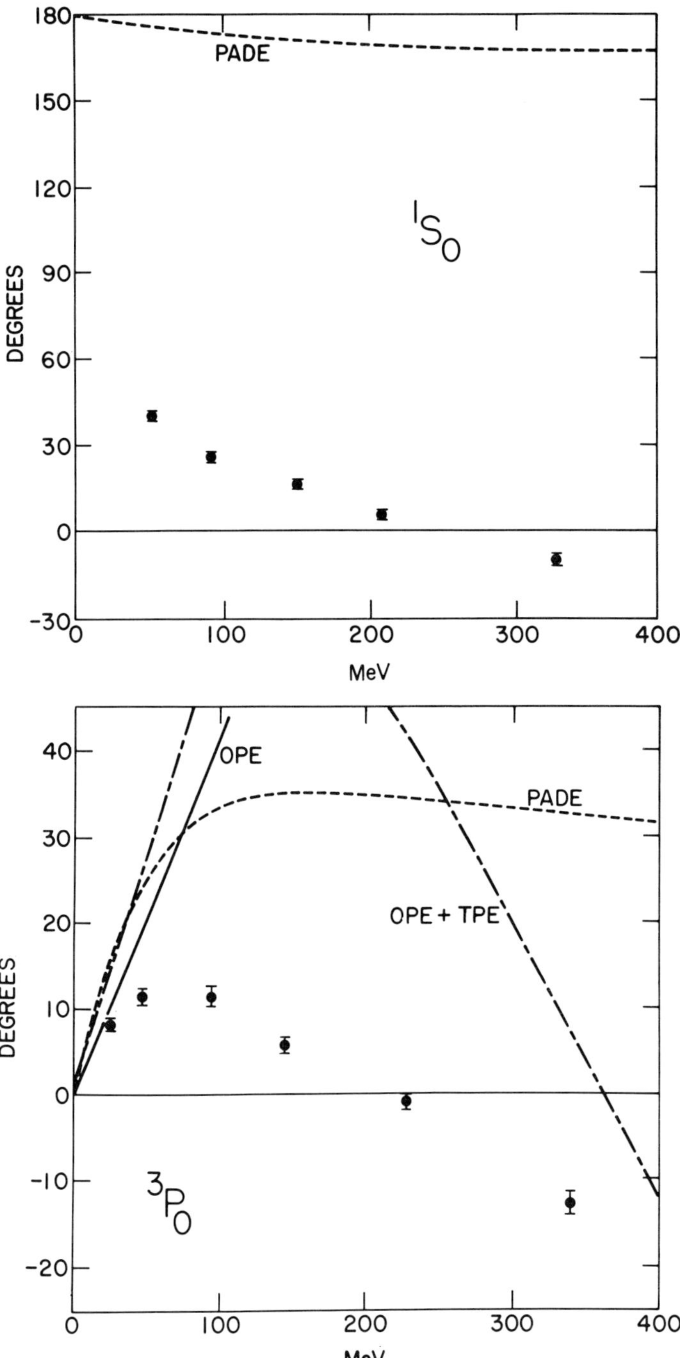

Fig. 7. Phase shifts from one-pion exchange (OPE), fourth-order perturbation theory (OPE + TPE), the [1, 1] Padé approximant (PADE), and the Livermore energy-independent analysis.

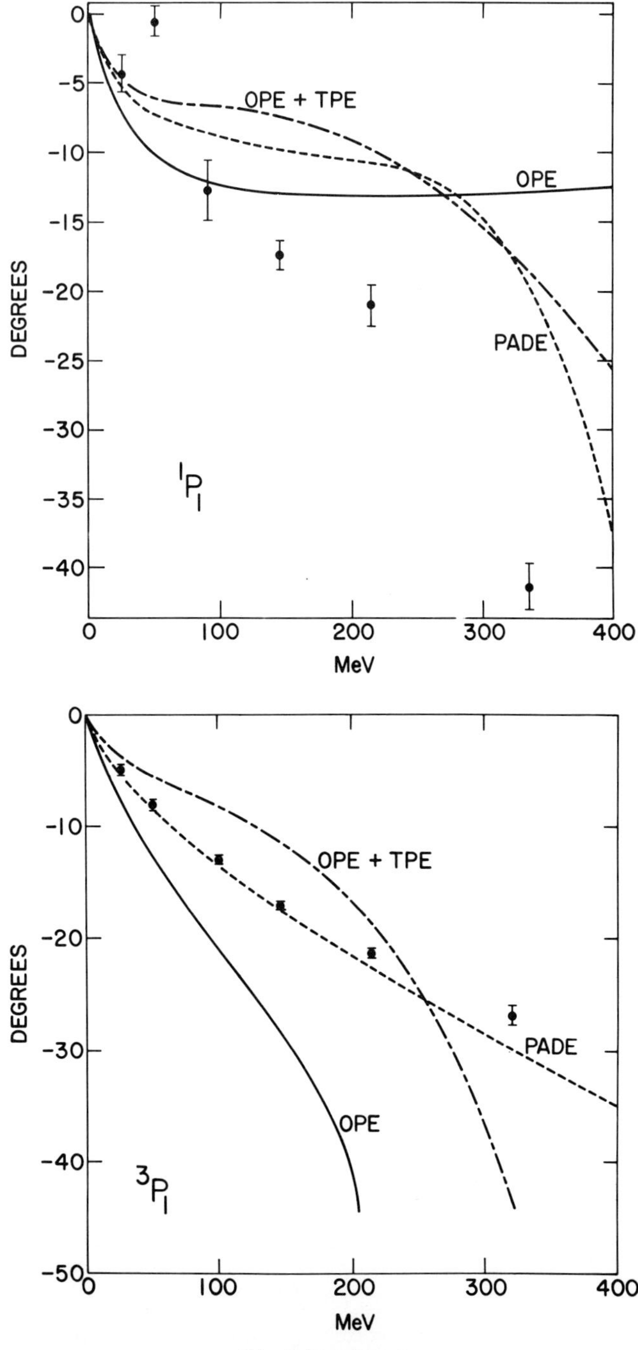

Fig. 7 (*continued*)

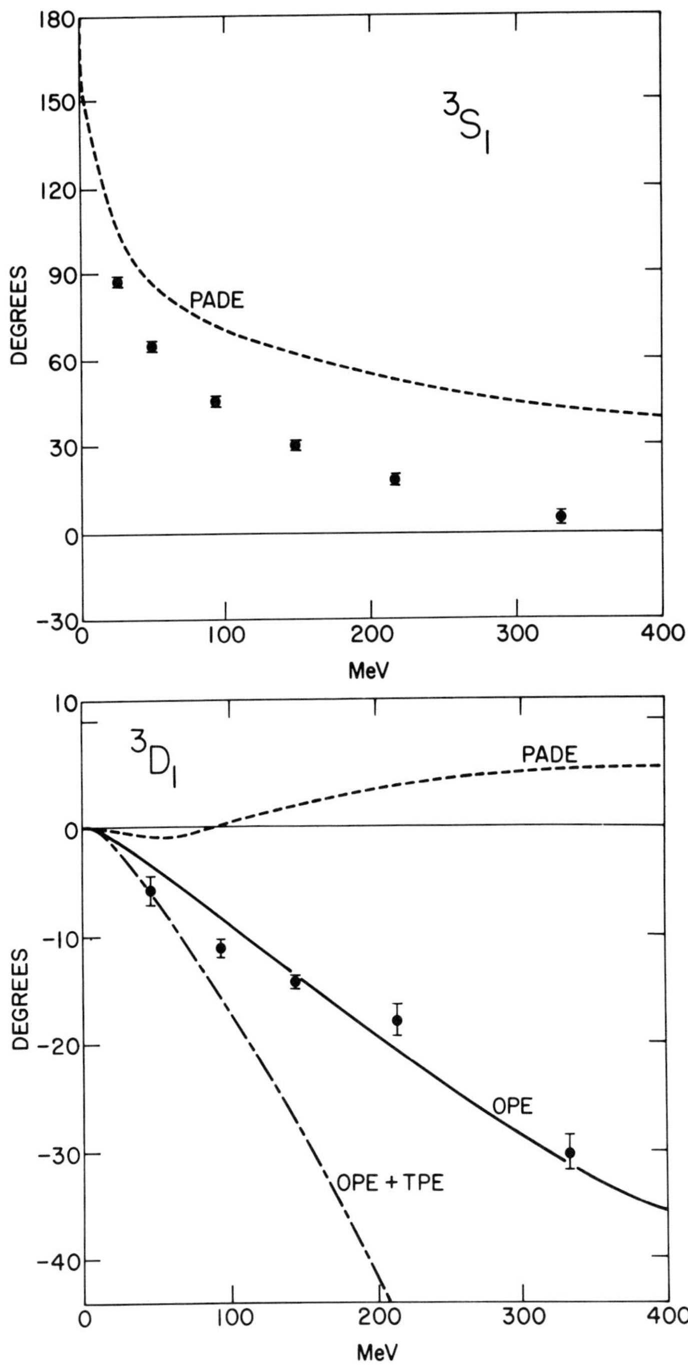

Fig. 7 *(continued)*

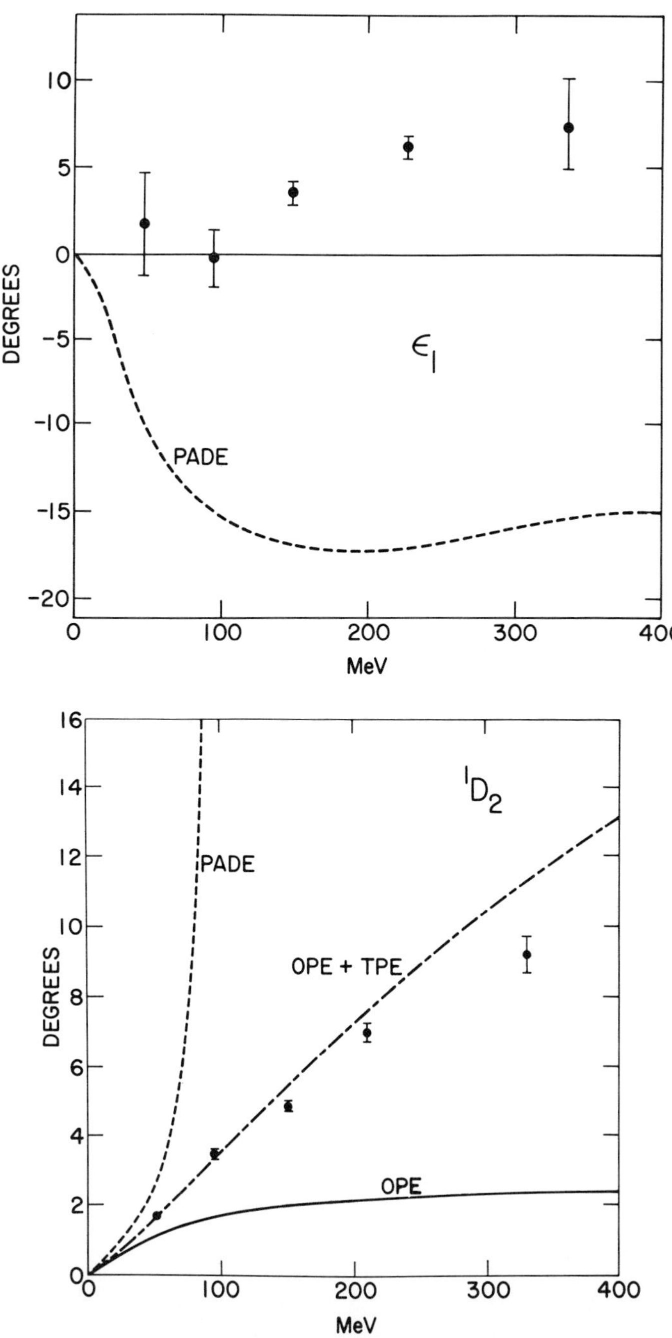

Fig. 7 *(continued)*

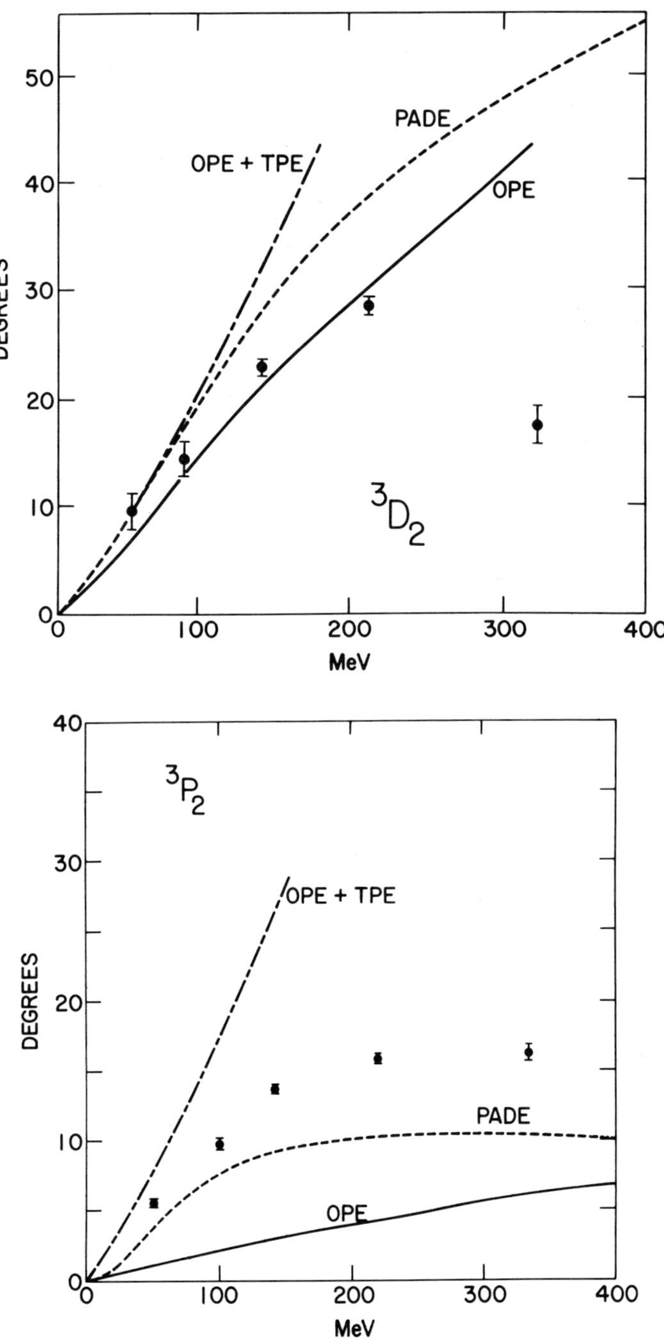

Fig. 7 *(continued)*

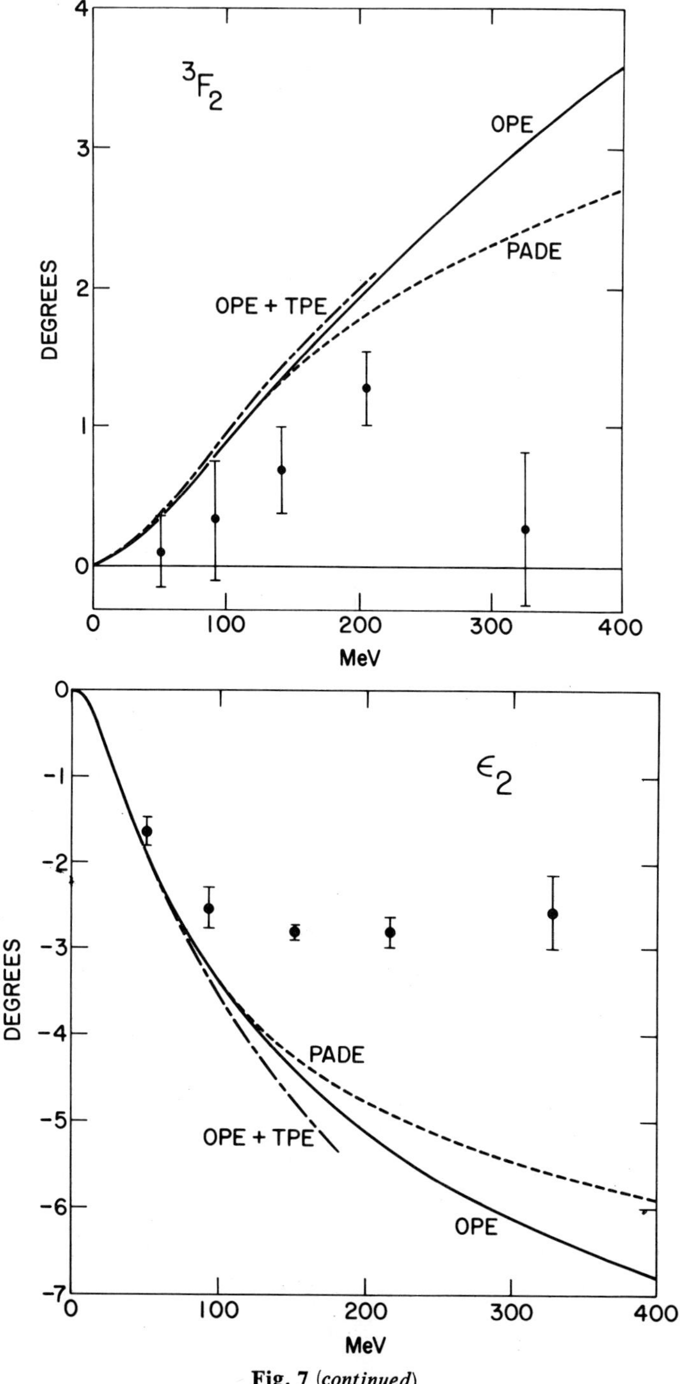

Fig. 7 (*continued*)

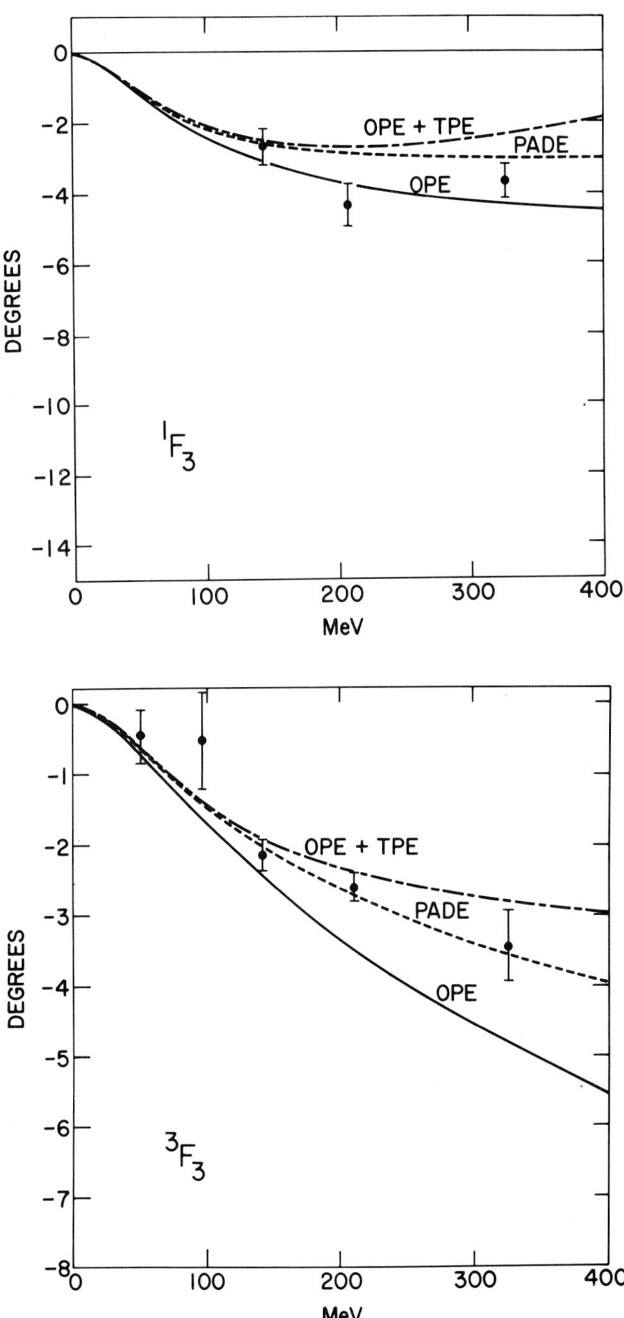

Fig. 7 *(continued)*

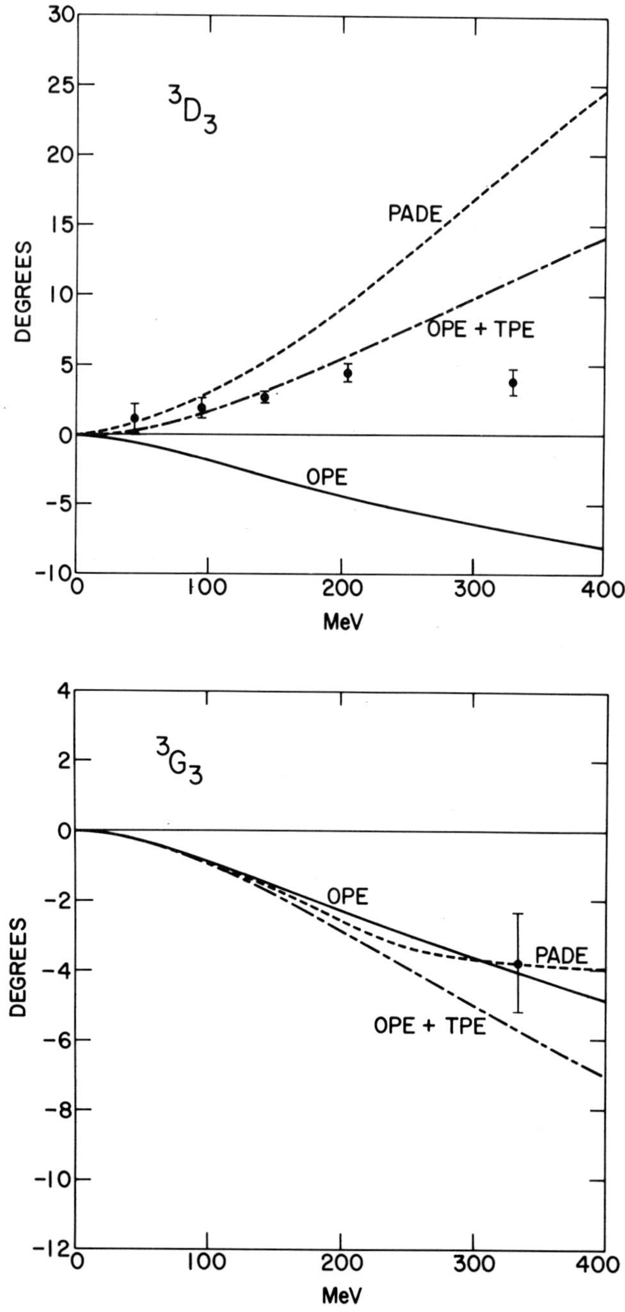

Fig. 7 *(continued)*

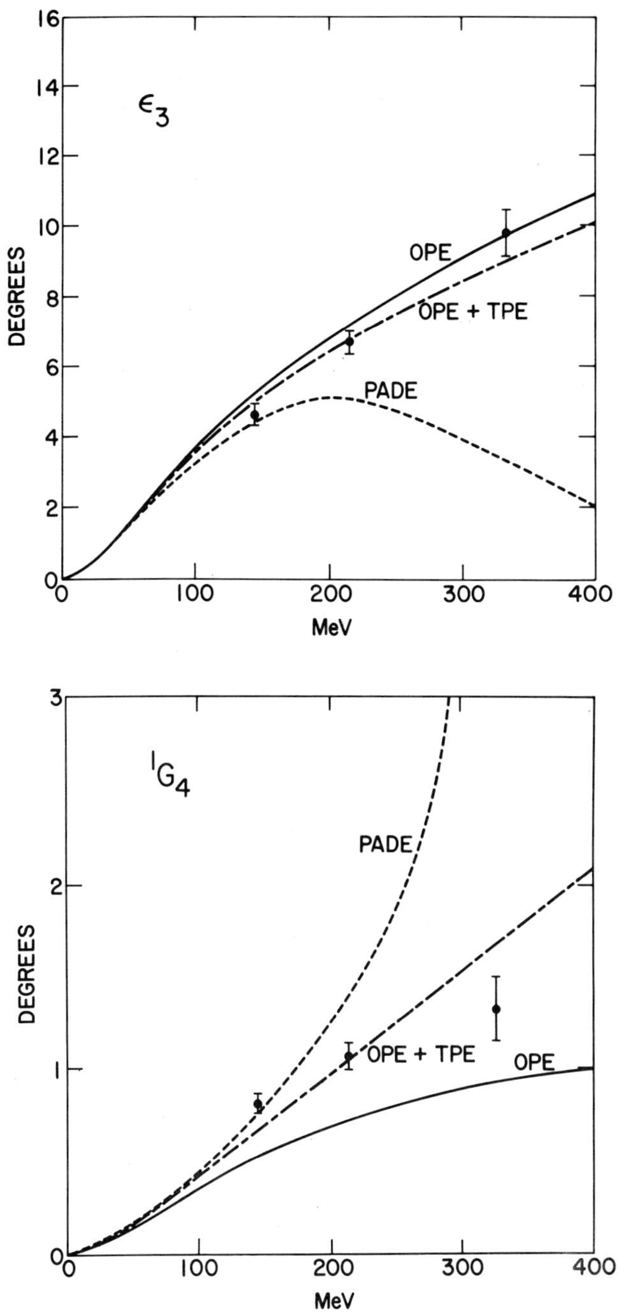

Fig. 7 (*continued*)

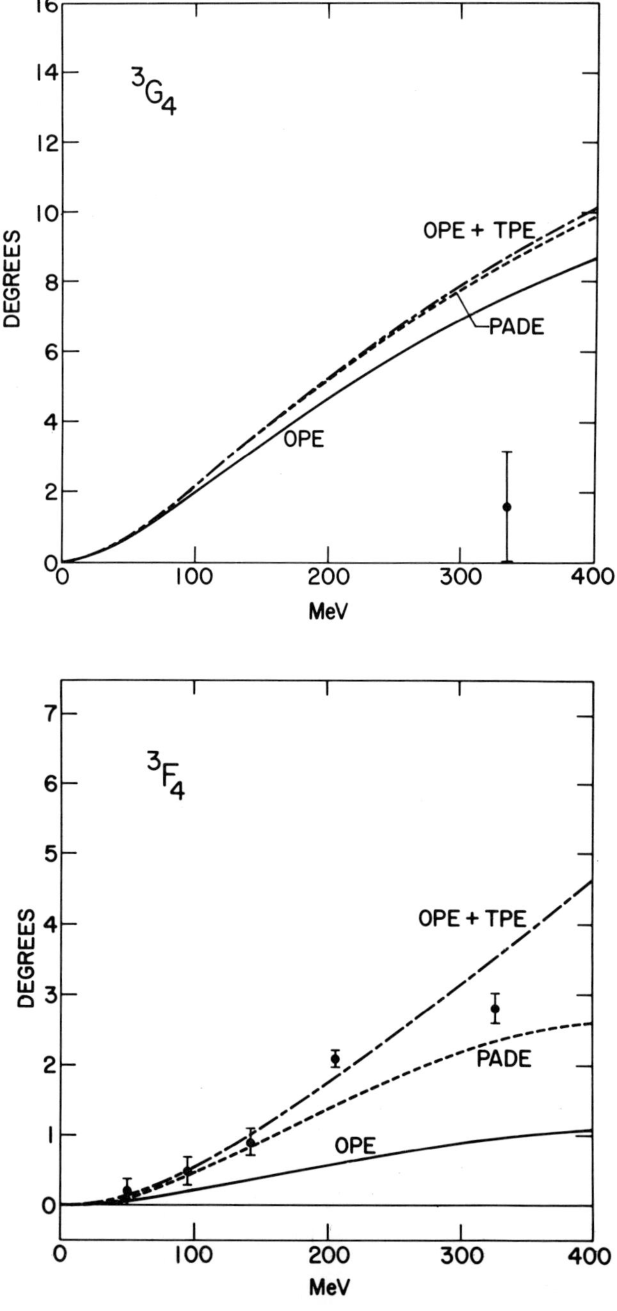

Fig. 7 (*continued*)

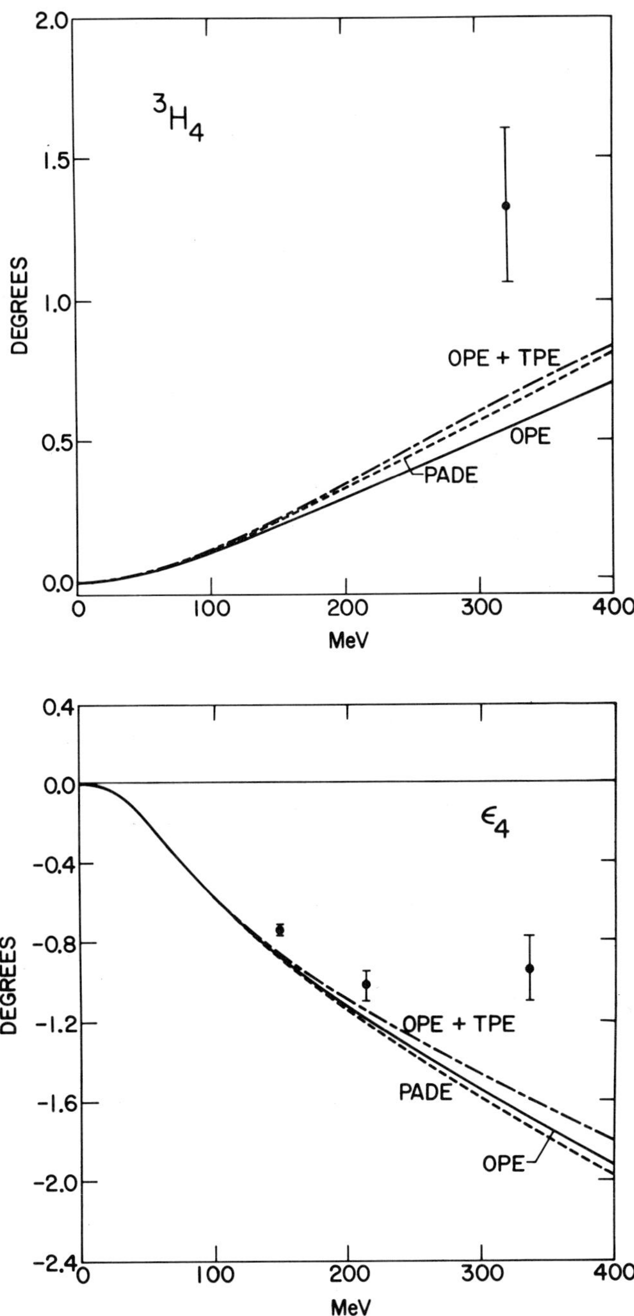

Fig. 7 (*continued*)

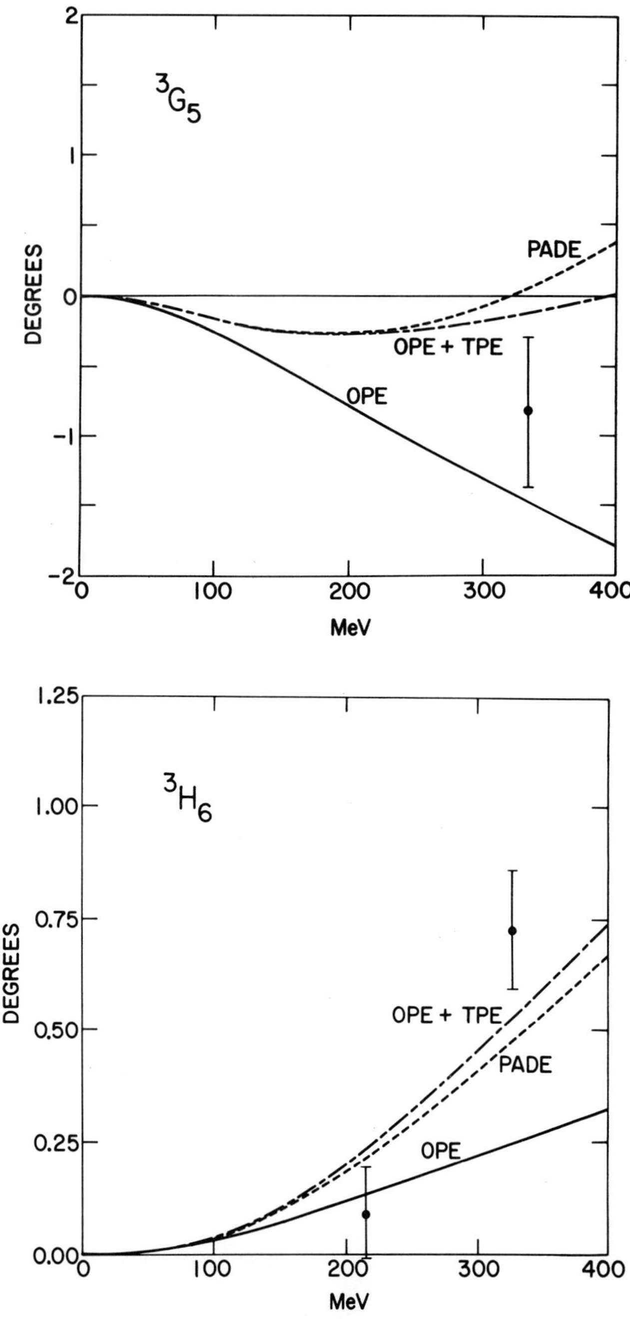

Fig. 7 *(continued)*

expansion would be of value only so long as higher-order terms decrease and all terms are generally small, that is, $1 \gg g^2h_2 \gg g^4h_4$ for this case. These conditions certainly are not always met. In such instances, useful results are not expected. However, in those cases in which the phase shifts are small and the OPE amplitude provides a fair approximation, the addition of the two-pion-exchange (TPE) amplitude should improve on the OPE result. If not, then the field theory model being used must be examined. A second factor limiting the usefulness of this approach is the exclusion of heavy mesons, which are known to play a role in *N-N* scattering. Still, it is probably the case that TPE partly takes into account the properties of some heavy meson exchanges. From this discussion, however, it would seem unlikely that the simple addition of TPE could be a solution to the *N-N* problem. Modest improvements are more to be expected.

Of course, the Padé method must not fail when the perturbation approach succeeds. Indeed, this cannot occur, since it is apparent from (44) that, if $h_2 \gg g^2h_4$, then the expression reduces to

$$h_p \approx g^2h_2 + g^4h_4\,, \tag{47}$$

which is exactly the perturbation result. The Padé results will be of most interest outside the assumed region of validity of the perturbation expansion. However, one should be cautious of assigning great significance to features of the Padé results which may be due only to enforcement of unitarity.

V. RESULTS

In order to determine whether or not the addition of TPE contributions leads to a better description of the scattering matrix, some criteria for improvement must be devised. For the purposes of this investigation, it seems reasonable to look for TPE effects in those phase shifts for which the OPE effects cannot be considered dominant and for which the experimental phase shifts are sufficiently well determined that it is clear just which direction from the OPE result would constitute the correct direction.

On the basis of such criteria, it would appear that the phase shifts best suited for consideration of TPE effects are the ${}^3P_0$, ${}^3P_1$, ${}^3P_2$, ${}^1D_2$, ${}^3D_3$, ${}^3F_2$, ${}^3F_3$, ${}^3F_4$, ${}^1G_4$, ${}^3G_5$, ${}^3H_4$, ${}^3H_6$, and ε_2 phases, as shown in Fig. 7. Of these, the ${}^3F_2$, ${}^3P_2$, ε_2, and ${}^3P_0$ TPE phase shifts show no particular improvement over the OPE result and are perhaps worse. However, it should be noted that the ${}^3F_2$, ${}^3P_2$, and ε_2 phases are coupled, whereas the ${}^3P_2$ contribution is rather large, as is that for the ${}^3P_0$ case. The ${}^3P_1$, ${}^3D_3$, ${}^3H_4$, and ${}^3H_6$ phase shifts show TPE contributions to provide corrections to the OPE result in the direction of the experimental phase, although not of exactly the correct magnitude. The $^1D_2$, $^3F_3$, $^3F_4$, $^1G_4$, and $^3G_5$ phase shifts including the TPE effects seem to be in good agreement with the experimental results.

Based on these observations, it would appear that, except for those cases where the phase shifts are large or coupled to large phases, the TPE effect provides the proper sort of correction to the OPE predictions. That is, the TPE effects are evident in the phase shifts except for those cases in which the unitary bound is approached.

Perhaps the best way to judge the success of the [1, 1] Padé approximant is to compare the TPE effects found by geometric unitarization with those results given by the [1, 1] Padé approximant to the S matrix. Again with reference to Fig. 7, it will be noted that the phase shifts given by the [1, 1] Padé approximants are (a) for the case of the $^3P_1$, ε_2, $^3F_2$, $^3P_2$, and $^3P_0$ phases, better than those given by geometric unitarization; (b) for the cases of the $^3G_5$, $^3F_4$, $^3H_4$, and $^3H_6$ phases, about the same as those given by geometric unitarization; and (c) for the case of the $^3D_3$, $^1D_2$, and $^1G_4$ phases, notably worse than those given by geometric unitarization. Note that the [1, 1] approximants for $^1D_2$ and $^1G_4$ show a resonant behavior. The $^1D_2$ state is resonant at about 90 MeV and the $^1G_4$ state is resonant at about 390 MeV. Bessis *et al.* (1969) point out that these resonances are Regge recurrences of the $^1S_0$ state.

To summarize the higher angular momentum results, it appears that the [1, 1] Padé results are not notably better than the perturbation results except in those cases where the unitary bound is approached. The latter case does not necessarily arise from the virtue of the Padé method but may reflect the restriction of unitarity which has not been placed on the perturbation expansion.

Since one virtue of the Padé approximant is the existence of poles, it is very much of interest to investigate their appearance in this model. Indeed, the $^3S_1$ state is bound as indicated by the phase shift starting from π. Noting (44) and that the bound state corresponds to a pole, the binding energy can be determined by the condition

$$\det(h_2 - g^2 h_4) = 0\,, \tag{48}$$

where h_2 and h_4 are evaluated below threshold at the binding energy. This condition is satisfied at 4.8 MeV. That is, the $^3S_1$ state is found to be bound with a binding energy of 4.8 MeV. The experimental situation is that the $^3S_1$ state of the *n-p* system is bound to form the deuteron at 2.2 MeV. However, the identification with the deuteron is not as satisfactory as might be hoped. The $^3S_1$ state is coupled to the $^3D_1$ state, and therefore the phases $\delta(^3D_1)$ and especially ε_1 should also be compared to the experimental results. A check reveals that ε_1 is both too large and of the wrong sign, whereas $\delta(^3D_1)$ is quite unlike the experimental result.

The h_p matrix has a pole at 4.8 MeV, and, if residue is found and normalized, the "residue matrix" is

$$\begin{bmatrix} 1 & -3.0 \times 10^{-4} \\ -3.0 \times 10^{-4} & 9.0 \times 10^{-8} \end{bmatrix},$$

where the order is *S-D*. Wong (1959) indicates that the ratio of the *SD* to *SS* elements is minus twice the asymptotic *D*/*S* ratio, ρ. Therefore, the [1, 1] Padé predicts

$$\rho = 1.5 \times 10^{-4},$$

whereas the experimental result is about 3×10^{-2}. The bound state is almost pure *S* state. As a consequence, the identification of the $^3S_1$ bound state with the deuteron is not complete.

The anomalous p^5 threshold behavior of the [1, 1] approximant for the $^1S_0$ phase shifts indicates that any investigation of a bound system is of questionable value. For the sake of completeness, the matter has been studied, however. Subject to the condition

$$h_2 - g^2 h_4 = 0, \tag{49}$$

a pole will appear for any of the singlet or uncoupled triplet amplitudes. For the $^1S_0$ case, the condition is satisfied for a binding energy of about 5 MeV as compared to the "almost bound" $^1S_0$ state from experiment. No other bound states appear (Bessis *et al.*, 1969).

It is interesting that, since

$$h_{J\mathrm{OPE}} \sim Q_J\left(1 + \frac{2\mu^2}{s - 4m^2}\right), \tag{50}$$

the OPE amplitudes grow infinite as s approaches a value so as to give a center-of-mass energy of about − 5.1 MeV. However, the TPE amplitudes do not approach their corresponding branch cut until the energy is four times this value. As a consequence, considering (49), if h_{OPE} and h_{TPE} are of the same sign, a bound state will appear between 0 and 5.1 MeV for a sufficiently large coupling constant. In the case of the $^1S_0$ amplitudes, $h_{\mathrm{OPE}} \sim p^2$, but $h_{\mathrm{TPE}} \sim p^0$. As a consequence, since h_{OPE} and h_{TPE} are of the same sign, a bound state will occur for any value of the coupling constant.

In the case of the $^1D_2$ and $^1G_4$ states, the smallest value of g^2 which will produce binding in the gap between 0 and 5.1 MeV is greater than 14. Therefore, these states have resonances rather than being bound.

VI. SUMMARY

In conclusion, it is found that the *N-N* scattering problem is modestly clarified by addition of TPE effects either through the perturbation expansion or by the [1, 1] Padé approximant, and there are features of the phase shifts which can be identified as due to TPE effects. However, the [1, 1] results are not particularly more appealing than the perturbation results except in these cases where the amplitudes approach the unitary bound.

The bound-state problem is confused by the irregular threshold behavior of the OPE amplitudes. This feature is inherent in single pseudoscalar exchange, and the result is

$$\delta({}^1S_0 \text{ and } {}^3S_1)_{\text{OPE}} \sim p^3 .$$

This leads to

$$\delta({}^1S_0[1, 1]) \sim p^5 ,$$

although the ${}^3S_1$ is brought back to normal by the [1, 1] Padé prescription operating on the appropriate 2×2 matrix. It is most encouraging that in spite of this difficulty a ${}^3S_1$ bound state is found at 4.8 MeV. Still, the coupling to the ${}^3D_1$ state is quite different from that observed with the deuteron.

It must be concluded that the [1, 1] Padé approximant is inadequate to account satisfactorily for many aspects of the *N-N* interaction, at least for this model. Whether this is due to the inapplicability of the Padé method, to improper dynamics of the model, or to an insufficient number of terms in the perturbation expansion is not clear. Certainly the irregular threshold behavior of the OPE amplitudes serves to cloud this issue further.

The fact that TPE effects appear seems to indicate that field theory has some validity; however, the irregular OPE threshold behavior alone indicates more terms in the perturbation series may be required. The exclusion of heavy meson exchange is a weakness of this model. It is possible that this matter could be improved by using the [2, 2] approximant, which requires knowledge of the sixth- and eighth-order contributions. Inclusion of a $\lambda\varphi^4$ interaction may also help simulate heavy meson exchange. Regrettably, direct inclusion of these mesons leads to grave renormalization troubles beyond second order.

We plan to pursue this matter by evaluation of the sixth- and eighth-order contributions, including a $\lambda\varphi^4$ interaction, making use of the techniques described in Sect. II in order to form the [2, 2] Padé approximant. It is likely that the [2, 2] result can be much more useful in assessing the value of field theory and the Padé approximant as applied to strong interactions.

ACKNOWLEDGMENTS

We should like to express our gratitude to Dr. Charles Schwartz for supplying us with solutions to the Bethe-Salpeter equation and to Dr. D. Bessis, who provided some very enlightening correspondence. The author is indebted to Dr. J. L Gammel for many discussions concerning the material discussed in Sect. II, as well as for originally suggesting the entire problem. The guidance of Dr. J. J. Kubis and helpful discussions with Dr. J. Nuttall are gratefully acknowledged.

REFERENCES

Amati, D., Leader, E., and Vitale, B. (1960). *Nuovo Cimento* **17,** 69.
Baker, G. A., Jr., and Chisholm, R. (1966). *J. Math. Phys.* **7,** 1900.
Bessis, D., Graffi, S., Grecchi, V., and Turchetti, G. (1969). *Phys. Letters* **28B,** 567.
Breit, G., Hull, M. H., Jr., Lassila, K. E., and Pyatt, K. D., Jr. (1960). *Phys. Rev.* **120,** 2227.
Chisholm, J. S. R. (1952). *Proc. Cambridge Phil. Soc.* **48,** 300.
Cutkosky, R. E. (1960). *J. Math. Phys.* **1,** 429.
Cziffra, P., MacGregor, M. H., Moravcsik, M. J., and Stapp, H. P. (1959). *Phys. Rev.* **114,** 881.
Goldberger, M. L., Grisaru, M. T., MacDowell, S. W., and Wong, D. Y. (1960). *Phys. Rev.* **120,** 2250.
Haracz, R. D., and Sharma, R. D. (1969). *Phys. Rev.* **176,** 2013.
Grashin, A. F., (1959). *Zh. Eksperim. Teor. Fiz.* **36,** 1717; English transl. *Soviet Phys.—JETP* **9,** 1223.
MacGregor, M. H., Arndt, R. A., and Wright, R. M. (1968). *Phys. Rev.* **173,** 1272.
MacGregor, M. H., Moravscik, M. J., and Stapp, H. P. (1960). *Ann. Rev. Nucl. Sci.* **10,** 291.
Mandelstam, S. (1959). *Phys. Rev.* **115,** 1741.
Mattioli, G. (1968). *Nuovo Cimento* **56,** 144.
Moravcsik, M. J. (1964). *Ann. Phys.* **30,** 10.
Nuttall, J. (1961). Ph. D. thesis, Cambridge Univ. (unpublished).
Nuttall, J. (1967). *Phys. Rev.* **157,** 1312.
Schwartz, C. (1968). (private communication).
Schwartz, C., and Zemach, C. (1966). *Phys. Rev.* **141,** 1454.
Scotti, A. and Wong, D. Y. (1965). *Phys. Rev.* **138,** 145.
Schweber, S. S. (1961). "An Introduction to Relativistic Quantum Field Theory." Harper and Row, New York.
Shreider, Y. A. (1964). "Method of Statistical Testing." Elsevier, Amsterdam.
Stapp, H. P., Ypsilantis, T. J., and Metropolis, N. (1958). *Phys. Rev.* **105,** 302.
Taketani, M., Machida, S., and Ohnuma, S. (1952). *Progr. Theoret. Phys. (Kyoto)* **7,** 45.
Watson, K. M., and Lepore, J. V. (1949). *Phys. Rev.* **76,** 1157.
Wong, D. Y. (1959). *Phys. Rev. Letters* **2,** 406.
Wortman, W. R. (1968). Ph. D. dissertation, Texas A and M Univ. (unpublished).
Wortman, W. R. (1969). *Phys. Rev.* **176,** 1762.
Yukawa, H. (1938). *Proc. Phys. Math. Soc. Japan* **17,** 48.

AUTHOR INDEX

Numbers in italics indicate pages on which complete references are listed.

F

G

H

I

J

K

L

SUBJECT INDEX

U

V

Y

Z

Mathematics in Science and Engineering

A Series of Monographs and Textbooks

Edited by RICHARD BELLMAN, ***University of Southern California***

1. T. Y. Thomas. Concepts from Tensor Analysis and Differential Geometry. Second Edition. 1965

2. T. Y. Thomas. Plastic Flow and Fracture in Solids. 1961

3. R. Aris. The Optimal Design of Chemical Reactors: A Study in Dynamic Programming. 1961

4. J. LaSalle and S. Lefschetz. Stability by by Liapunov's Direct Method with Applications. 1961

5. G. Leitmann (ed.). Optimization Techniques: With Applications to Aerospace Systems. 1962

6. R. Bellman and K. L. Cooke. Differential-Difference Equations. 1963

7. F. A. Haight. Mathematical Theories of Traffic Flow. 1963

8. F. V. Atkinson. Discrete and Continuous Boundary Problems. 1964

9. A. Jeffrey and T. Taniuti. Non-Linear Wave Propagation: With Applications to Physics and Magnetohydrodynamics. 1964

10. J. T. Tou. Optimum Design of Digital Control Systems. 1963.

11. H. Flanders. Differential Forms: With Applications to the Physical Sciences. 1963

12. S. M. Roberts. Dynamic Programming in Chemical Engineering and Process Control. 1964

13. S. Lefschetz. Stability of Nonlinear Control Systems. 1965

14. D. N. Chorafas. Systems and Simulation. 1965

15. A. A. Pervozvanskii. Random Processes in Nonlinear Control Systems. 1965

16. M. C. Pease, III. Methods of Matrix Algebra. 1965

17. V. E. Benes. Mathematical Theory of Connecting Networks and Telephone Traffic. 1965

18. W. F. Ames. Nonlinear Partial Differential Equations in Engineering. 1965

19. J. Aczel. Lectures on Functional Equations and Their Applications. 1966

20. R. E. Murphy. Adaptive Processes in Economic Systems. 1965

21. S. E. Dreyfus. Dynamic Programming and the Calculus of Variations. 1965

22. A. A. Fel'dbaum. Optimal Control Systems. 1965

23. A. Halanay. Differential Equations: Stability, Oscillations, Time Lags. 1966

24. M. N. Oguztoreli. Time-Lag Control Systems. 1966

25. D. Sworder. Optimal Adaptive Control Systems. 1966

26. M. Ash. Optimal Shutdown Control of Nuclear Reactors. 1966

27. D. N. Chorafas. Control System Functions and Programming Approaches (In Two Volumes). 1966

28. N. P. Erugin. Linear Systems of Ordinary Differential Equations. 1966

29. S. Marcus. Algebraic Linguistics; Analytical Models. 1967

30. A. M. Liapunov. Stability of Motion. 1966

31. G. Leitmann (ed.). Topics in Optimization. 1967

32. M. Aoki. Optimization of Stochastic Systems. 1967

33. H. J. Kushner. Stochastic Stability and control. 1967

34. M. Urabe. Nonlinear Autonomous Oscillations. 1967

35. F. Calogero. Variable Phase Approach to Potential Scattering. 1967

36. A. Kaufmann. Graphs, Dynamic Programming, and Finite Games. 1967

37. A. Kaufmann and R. Cruon. Dynamic Programming: Sequential Scientific Management. 1967

38. J. H. Ahlberg, E. N. Nilson, and J. L. Walsh. The Theory of Splines and Their Applications. 1967

39. Y. Sawaragi, Y. Sunahara, and T. Nakamizo. Statistical Decision Theory in Adaptive Control Systems. 1967

40. R. Bellman. Introduction to the Mathematical Theory of Control Processes Volume I. 1967 (Volumes II and III in preparation)

41. E. S. Lee. Quasilinearization and Invariant Imbedding. 1968

42. W. Ames. Nonlinear Ordinary Differential Equations in Transport Processes. 1968

43. W. Miller, Jr. Lie Theory and Special Functions. 1968

44. P. B. Bailey, L. F. Shampine, and P. E. Waltman. Nonlinear Two Point Boundary Value Problems. 1968.

45. Iu. P. Petrov. Variational Methods in Optimum Control Theory. 1968

46. O. A. Ladyzhenskaya and N. N. Ural't-seva. Linear and Quasilinear Elliptic Equations. 1968

47. A. Kaufmann and R. Faure. Introduction to Operations Research. 1968

48. C. A. Swanson. Comparison and Oscillation Theory of Linear Differential Equations. 1968

49. R. Hermann. Differential Geometry and the Calculus of Variations. 1968

50. N. K. Jaiswal. Priority Queues. 1968

51. H. Nikaido. Convex Structures and Economic Theory. 1968

52. K. S. Fu. Sequential Methods in Pattern Recognition and Machine Learning. 1968

53. Y. L. Luke. The Special Functions and Their Approximations (In Two Volumes). 1969

54. R. P. Gilbert. Function Theoretic Methods in Partial Differential Equations. 1969

55. V. Lakshmikantham and S. Leela. Differential and Integral Inequalities (In Two Volumes). 1969

56. S. H. Hermes and J. P. LaSalle. Functional Analysis and Time Optimal Control. 1969.

57. M. Iri. Network Flow, Transportation, and Scheduling: Theory and Algorithms. 1969

58. A. Blaquiere, F. Gerard, and G. Leitmann. Quantitative and Qualitative Games. 1969

59. P. L. Falb and J. L. de Jong. Successive Approximation Methods in Control and Oscillation Theory. 1969

60. G. Rosen. Formulations of Classical and Quantum Dynamical Theory. 1969

61. R. Bellman. Methods of Nonlinear Analysis, Volume I. 1970

62. R. Bellman, K. L. Cooke, and J. A. Lockett. Algorithms, Graphs, and Computers. 1970

63. E. J. Beltrami. An Algorithmic Approach to Nonlinear Analysis and Optimization. 1970

64. A. H. Jazwinski. Stochastic Processes and Filtering Theory. 1970

65. P. Dyer and S. R. McReynolds. The Computation and Theory of Optimal Control, 1970

66. J. M. Mendel and K. S. Fu (eds.). Adaptive, Learning, and Pattern Recognition Systems: Theory and Applications, 1970

67. C. Derman. Finite State Markovian Decision Processes, 1970

68. M. Mesarovic, D. Macko, and Y. Takahara. Theory of Hierarchical Multilevel Systems, 1970

69. H. H. Happ. Diakoptics and Networks, 1970

70. Karl Astrom. Introduction to Stochastic Control Theory, 1970

71. G. A. Baker, Jr. and J. L. Gammel, eds. The Padé Approximant in Theoretical Physics, 1970

In preparation

C. Berge. Principles of Combinatorics

Ya. Z. Tsypkin. Adaptation and Learning in Automatic Systems

Leon Lapidus and John H. Seinfeld. Numerical Solution of Ordinary Differential Equations

Harold Greenberg. Integer Programming

E. Polak. Computational Methods in Optimization: A Unified Approach

Leon Mirsky. Transversal Theory

Thomas G. Windeknecht. A Mathematical Introduction to General Dynamical Processes

Anwrew P. Sage and James L. Melsa. System Identification